Lecture Notes in Mathematics

Edited by J.-M. Morel, F. Takens and B. Teissier

Editorial Policy

for the publication of monographs

1. Lecture Notes aim to report new developments in all areas of mathematics – quickly, informally and at a high level. Monograph manuscripts should be reasonably self-contained and rounded off. Thus they may, and often will, present not only results of the author but also related work by other people. They may be based on specialized lecture courses. Furthermore, the manuscripts should provide sufficient motivation, examples and applications. This clearly distinguishes Lecture Notes from journal articles or technical reports which normally are very concise. Articles intended for a journal but too long to be accepted by most journals, usually do not have this "lecture notes" character. For similar reasons it is unusual for doctoral theses to be accepted for the Lecture Notes series.

2. Manuscripts should be submitted (preferably in duplicate) either to one of the series editors or to Springer-Verlag, Heidelberg. In general, manuscripts will be sent out to 2 external referees for evaluation. If a decision cannot yet be reached on the basis of the first 2 reports, further referees may be contacted: the author will be informed of this. A final decision to publish can be made only on the basis of the complete manuscript, however a refereeing process leading to a preliminary decision can be based on a pre-final or incomplete manuscript. The strict minimum amount of material that will be considered should include a detailed outline describing the planned contents of each chapter, a bibliography and several sample chapters.
Authors should be aware that incomplete or insufficiently close to final manuscripts almost always result in longer refereeing times and nevertheless unclear referees' recommendations, making further refereeing of a final draft necessary.
Authors should also be aware that parallel submission of their manuscript to another publisher while under consideration for LNM will in general lead to immediate rejection.

3. Manuscripts should in general be submitted in English.
Final manuscripts should contain at least 100 pages of mathematical text and should include
- a table of contents;
- an informative introduction, with adequate motivation and perhaps some historical remarks: it should be accessible to a reader not intimately familiar with the topic treated;
- a subject index: as a rule this is genuinely helpful for the reader.

Continued on inside back-cover

Lecture Notes in Mathematics 1776

Springer
Berlin
Heidelberg
New York
Barcelona
Hong Kong
London
Milan
Paris
Tokyo

K. Behrend C. Gómez V. Tarasov G. Tian

Quantum Cohomology

Lectures given at the C.I.M.E. Summer School
held in Cetraro, Italy, June 30 - July 8, 1997

Editors: P. de Bartolomeis
B. Dubrovin
C. Reina

Fondazione
C.I.M.E.

Springer

Authors

Kai Behrend
Dept. of Mathematics
University of British Columbia
1984 Math Rd.
Vancouver, BC scV6T 1Z2
E-mail: behrend@math.ubc.ca

César Gómez
Instituto de Matemáticas y Físicas Fundamental
Consejo Superior de Investigacíon
Calle Serrano sc123
28006 Madrid, Spain
E-mail: cesar.gomez@uam.es

Vitaly Tarasov
St. Petersburg Branch of
Steklov Mathematical Institute
Fontanka 27
St. Petersburg 191011, Russia
E-mail: vt@vt.pdmi.ras.ru

Gang Tian
Dept. of Mathematics
M.I.T.
Cambridge, MA 02139, USA
E-mail: tian@math.mit.edu

Editors

Paolo de Bartolomeis
Dipartimento di Matematica Applicata
"G. Sansone"
Via di S. Marta, 3
50139 Firenze, Italy
E-mail: deba@poincare.dma.unifi.it

Boris A. Dubrovin
Cesare Reina
C.R. - SISSA
Via Beirut, 4
34100 Trieste, Italy
E-mail: dubrovin@sissa.it
E-mail: reina@sissa.it

Cataloging-in-Publication Data applied for

Die Deutsche Bibliothek - CIP-Einheitsaufnahme

Quantum cohomology : lecture given at the CIME Summer School, held in Cetraro, Italy, Juny 30 - July 8, 1997 / K. Behrend ... Ed.: P. de Bartolomeis .... - Berlin ; Heidelberg ; New York ; Barcelona ; Hong Kong ; London ; Milan ; Paris ; Tokyo : Springer, 2002
(Lecture notes in mathematics ; Vol. 1776 : Subseries: Fondazione CIME)
ISBN 3-540-43121-7

Mathematics Subject Classification (2000): 53D45, 14N35, 81T30, 83E30

ISSN 0075-8434
ISBN 3-540-43121-7 Springer-Verlag Berlin Heidelberg New York

Springer-Verlag Berlin Heidelberg New York a member of BertelsmannSpringer Science + Business Media GmbH

http://www.springer.de

Printed in Germany

Typesetting: Camera-ready TEX output by the authors

SPIN: 10856885 41/3142/LK - 543210 - Printed on acid-free paper

Table of Contents

Quantum Cohomology

Constructing symplectic invariants

Introduction

The progress of the string theory in the last decade strongly influenced the development of many branches of geometry. In particular, new directions of researches in the enumerative geometry and symplectic topology have been created as a joint venture of physicists and mathematicians. Among the most striking achievements of this period we mention the description of the intersection theory on moduli spaces of Riemann surfaces in terms of the Korteweg - de Vries integrable hierarchy of PDEs, and the proof of mirror conjecture for Calabi - Yau complete intersections.

One of the essential ingredients of these beautiful mathematical theories is a bunch of new approaches to the problem of constructing invariants of algebraic varities and of compact symplectic manifolds known under the name *quantum cohomology*. Physical ideas from topological gravity brought into the problem of invariants new structures of the theory of integrable systems of differential equations. The discovery of dualities between different physical theories suggested existence of deep and often unexpected relationships between different types of invariants.

In order to present, by both mathematicians and physicists, these new ideas to young researchers, we have decided to organize a CIME Summer School under the general title "Quantum Cohomology". The School took place at Calabrian sea resort Cetraro from June 30 to July 8, 1997. It was organized in four courses covering various aspects of these new mathematical theories. These Lecture Notes contain the extended text of the lecture courses.

In the course of Kai Behrend "Localization and Gromov - Witten Invariants" the approach to enumerative invariants of algebraic varieties based on the Bott residue formula has been developed. Behrend gave essentially self-consistent exposition of this approach for the important particular case of Gromov - Witten invariants of projective spaces.

The lecture course of urse "Fields, Strings and Branes" by César Gómez, written in collaboration with Rafael Hernández, collect some ideas of duality in string theories important for the development of quantum cohomology. The design of the presentation looks to be a physical one. Nevertheless we are confident that those mathematicians working in the area of quantum cohomology who have no prejudices against reading physical papers will be benefitted.

The lecture notes of Vitaly Tarasov "q-Hypergeometric Functions and Representation Theory" introduces the reader to another branch of the theory of integrable systems originated in the theory of form factors in massive integrable models of quantum field theory. This branch now developed into a part of representation theory of quantum affine algebras and of the corresponding vertex operators.Tarasov explains how to compute the matrix elements of the vertex operators in the terms of solutions to the quantized

Knizhnik - Zamolodchikov equation, and derives integral representations for these solutions.

The course of Gang Tian introduces the reader to the techniques of symplectic topology involved in the construction of Gromov - Witten invariants of compact symplectic manifolds. The main technical tool is the theory of virtual fundamental class on the moduli spaces of pseudoholomorphic curves. Tian applies this technique to the definition of quantum cohomology of symplectic manifolds and to constructions of certain nontrivial examples of symplectic manifolds.

We believe that the School was successful in reaching its aims, and we express our gratitude to the speakers for the high quality of their lectures and their availability for discussions during the School.

We also thank Prof. R.Conti and CIME Scientific Committee for the invitation to organize the School.

Paolo de Bartolomeis, Boris Dubrovin, Cesare Reina

Localization and Gromov-Witten Invariants

K. Behrend

University of British Columbia, Vancouver, Canada

Summary. We explain how to apply the Bott residue formula to stacks of stable maps. This leads to a formula expressing Gromov-Witten invariants of projective space in terms of integrals over stacks of stable curves.

1. Introduction

The course is divided into three lectures. Lecture I is a short introduction to stacks. We try to give a few ideas about the philosophy of stacks and we give the definition of algebraic stack of finite type over a field. Our definition does not require any knowledge of schemes.

Lecture II introduces equivariant intersection theory as constructed by Edidin and Graham [5]. The basic constructions are explained in a rather easy special case. The localization property (in the algebraic context also due to Edidin-Graham [6]) is mentioned and proved for an example. We set up a general framework for using the localization property to localize integrals to the fixed locus, or subvarieties (substacks) containing the fixed locus.

In Lecture III we apply the localization formula to the stack of stable maps to $\mathbb{P}^r$. We deduce a formula giving the Gromov-Witten invariants of $\mathbb{P}^r$ (for any genus) in terms of integrals over stacks of stable curves $\overline{M}_{g,n}$. The proof given here is essentially complete, if sometimes sketchy. At the same time these lectures were given, Graber and Pandharipande [12] independently proved the same formula. Their approach is very different from ours. We avoid entirely the consideration of equivariant obstruction theories, on which [12] relies. The idea to use localization to compute Gromov-Witten invariants is, of course, due to Kontsevich (see [13], where the genus zero case is considered).

2. Lecture I: A short introduction to stacks

What is a variety?

We will explain Grothendieck's point of view that a variety is a functor.

Let us consider for example the affine plane curve $y^2 = x^3$. According to Grothendieck, the variety $y^2 = x^3$ is nothing but the 'system' of all solutions of the equation $y^2 = x^3$ in all rings. We restrict slightly and fix a ground filed k and consider instead of all rings only k-algebras of finite type (in other words quotients of polynomial rings in finitely many variables over k). So, following Grothendieck, we associate to every finitely generated k-algebra A, all solutions of $y^2 = x^3$ in A^2:

$$\begin{aligned} h_V : (\text{f.g. } k\text{-algebras}) &\longrightarrow (\text{sets}) \\ A &\longmapsto \{(x,y) \in A^2 \mid y^2 = x^3\} \end{aligned}$$

Notice that h_V is actually a (covariant) functor: If $\phi : A \to B$ is a morphism of k-algebras and $(x,y) \in A^2$ satisfies $y^2 = x^3$, then $(\phi(x), \phi(y)) \in B^2$ satisfies $\phi(x)^2 = \phi(y)^3$. This makes precise what we mean by 'system' of solutions: We mean this functor. Grothendieck's point of view is that the variety $V \subset \mathbb{A}^2$ defined by $y^2 = x^3$ *is* this functor h_V. At least for affine varieties this is justified by the following corollary of *Yoneda's lemma.*

The (covariant) functor

$$\begin{aligned} (\text{affine } k\text{-varieties}) &\longrightarrow \text{Funct}((\text{f.g. } k\text{-algebras}), (\text{sets})) \\ V &\longmapsto h_V \end{aligned}$$

is fully faithful. Here Funct stands for the category of functors: objects are functors from (f.g. k-algebras) to (sets), morphisms are natural transformations. Because this functor is fully faithful we may think of (affine k-varieties) as a subcategory of Funct((f.g. k-algebras), (sets)) and identify the variety V with the functor h_V.

Note 2.1. Given an affine variety V there are many ways to write it as the zero locus of a finite set of polynomials in some affine n-space. So one gets many functors h_V. This is not a problem, because all these functors are canonically isomorphic to the functor given by the affine coordinate ring $k[V]$ of V:

$$h_V(A) = \text{Hom}_{k-\text{alg}}(k[V], A)$$

For example, the affine coordinate ring of the curve $y^2 = x^3$ is $k[x,y]/(y^2 - x^3)$, and for every k-algebra A we have

$$\{(x,y) \in A^2 \mid y^2 = x^3\} = \text{Hom}_{k-\text{alg}}(k[x,y]/(y^2 - x^3), A).$$

Terminology: The functor h_V is the *functor represented by* V.

Once we have embedded the category (affine k-varieties) into Funct((f.g. k-algebras), (sets)) we may enlarge the former inside the latter to get a larger category than (affine k-varieties), still consisting of 'geometric' objects.

For example, every finitely generated k-algebra A, reduced or not, gives rise to the functor

$$\begin{aligned} h_{\text{Spec}\, A} : (\text{f.g. } k\text{-algebras}) &\longrightarrow (\text{sets}) \\ R &\longmapsto \text{Hom}_{k-\text{alg}}(A, R) \end{aligned} .$$

The functor

$$\begin{aligned} h_{\mathrm{Spec}} : (\text{f.g. } k\text{-algebras}) &\longrightarrow \mathrm{Funct}((\text{f.g. } k\text{-algebras}), (\text{sets})) \\ A &\longmapsto h_{\mathrm{Spec}\, A} \end{aligned}$$

is contravariant and fully faithful. This is Yoneda's lemma for the category (f.g. k-algebras). The above corollary of Yoneda's lemma follows from this and the equivalence of categories between affine k-varieties and their coordinate rings. Yoneda's lemma is completely formal and holds for every category in place of (f.g. k-algebras). The proof is a simple exercise in category theory.

In keeping with Grothendieck's philosophy of identifying a geometric object with the functor it represents, we write

$$\mathrm{Spec}\, A : (\text{f.g. } k\text{-algebras}) \longrightarrow (\text{sets})$$

for the functor $h_{\mathrm{Spec}\, A}$, and call it the *spectrum* of A. The full subcategory of Funct((f.g. k-algebras), (sets)) consisting of functors isomorphic to functors of the form Spec A is called the *category of affine k-schemes of finite type*, denoted (aff/k).

To construct the functor h_V for a general k-variety V is a little tricky. Unless one knows scheme theory. Then it is easy, and we can do it for any k-scheme of finite type X:

$$\begin{aligned} h_X : (\text{f.g. } k\text{-algebras}) &\longrightarrow (\text{sets}) \\ A &\longmapsto \mathrm{Hom}_{\mathrm{schemes}}(\mathrm{Spec}\, A, X) \end{aligned}$$

It is then slightly less trivial than just Yoneda's lemma that one gets a (covariant) fully faithful functor

$$\begin{aligned} h : (\text{f.t. } k\text{-schemes}) &\longrightarrow \mathrm{Funct}((\text{f.g. } k\text{-algebras}), (\text{sets})) \\ X &\longmapsto h_X \quad . \end{aligned}$$

(This is, in fact, part of what is known as *descent theory*.)

The largest subcategory of Funct((f.g. k-algebras), (sets)) which still consists of 'geometric' objects is the category of finite type *algebraic spaces* over k. We will now describe this category (without using any scheme theory).

Algebraic spaces

First of all, to get a more 'geometric' picture, we prefer to think in terms of the category (aff/k) rather than the dual category (f.g. k-algebras). Thus we replace Funct((f.g. k-algebras), (sets)) by the equivalent category $\mathrm{Funct}^*((\mathrm{aff}/k), (\mathrm{sets}))$, where Funct^* refers to the category of contravariant functors. Grothendieck calls $\mathrm{Funct}^*((\mathrm{aff}/k), (\mathrm{sets}))$ the category of *presheaves* on (aff/k).

We start by considering the covariant functor

$$\begin{aligned} h : (\mathrm{aff}/k) &\longrightarrow \mathrm{Funct}^*((\mathrm{aff}/k), (\mathrm{sets})) \\ X &\longmapsto h_X \quad , \end{aligned}$$

where $h_X(Y) = \mathrm{Hom}_{k-\mathrm{schemes}}(Y, X) = \mathrm{Hom}_{k-\mathrm{alg}}(k[X], k[Y])$.

Note 2.2. The category (aff/k) containes fibered products (the dual concept in (f.g. k-algebras) is tensor product) and a final object $\operatorname{Spec} k$. The same is true for Funct*((aff/k), (sets)). Given a diagram

$$\begin{array}{ccc} & & Z \\ & & \downarrow g \\ X & \xrightarrow{f} & Y \end{array}$$

in Funct*((aff/k), (sets)) the fibered product $W = X \times_Y Z$ is given by

$$\begin{aligned} W(\operatorname{Spec} R) &= X(\operatorname{Spec} R) \times_{Y(\operatorname{Spec} R)} Z(\operatorname{Spec} R) \\ &= \{(x, z) \in X(\operatorname{Spec} R) \times Z(\operatorname{Spec} R) \mid \\ &\qquad f(\operatorname{Spec} R)(x) = g(\operatorname{Spec} R)(z) \in Y(\operatorname{Spec} R)\} \end{aligned}$$

A final object of Funct*((aff/k), (sets)) is the constant functor $\operatorname{Spec} R \mapsto \{\emptyset\}$. Here, of course, any one-element set in place of $\{\emptyset\}$ will do. Moreover, the functor h commutes with fibered products and final objects. One says that h is *left exact.*

Note 2.3. The category (aff/k) also contains direct sums (called *disjoint sums* in this context). If X and Y are affine k-schemes then their disjoint sum $Z = X \coprod Y$ has affine coordinate ring $A_Z = A_X \times A_Y$. Also, (aff/k) contains an initial object, the empty scheme, whose affine coordinate ring is the zero ring. We do not consider the corresponding notions in Funct*((aff/k), (sets)), the functor h does not commute with disjoint sums anyway.

Definition 2.1. *Let X be an object of (aff/k) and $(X_i)_{i \in I}$ a family of objects over X (which means that each X_i comes endowed with a morphism $X_i \to X$). We call $(X_i)_{i \in I}$ a* covering *of X, if I is finite and the induced morphism $\amalg_{i \in I} X_i \to X$ is faithfully flat, i.e. flat and surjective.*

Remark 2.1. This defines a *Grothendieck topology* on (aff/k).

Now that we have the notion of covering, we can define the notion of sheaf.

Definition 2.2. *A* sheaf *on (aff/k) is an object X of* Funct*((aff/k), (sets)) *(i.e. a presheaf), satisfying the two sheaf axioms: Whenever $(U_i)_{i \in I}$ is a covering of an object U of (aff/k), we have*

1. *if $x, y \in X(U)$ are elements such that $x|U_i = y|U_i$, for all $i \in I$, then $x = y$, (Here $x|U_i$ denotes the image of x under $X(U) \to X(U_i)$.)*
2. *if $x_i \in X(U_i)$, $i \in I$, are given such that $x_i|U_{ij} = x_j|U_{ij}$, for all $(i, j) \in I \times I$, ($U_{ij} = U_i \times_U U_j$) then there exists an element $x \in X(U)$ such that $x|U_i = x_i$, for all $i \in I$.*

It is a basic fact from descent theory that for every (affine) k-scheme of finite type X, the functor h_X is a sheaf on (aff/k). The notion of covering in terms of faithful flatness is the most general notion of covering that makes this statement true.

Definition 2.3. *An* algebraic space *(of finite type) over k is a sheaf X on (aff/k) such that*

1. *the diagonal $X \xrightarrow{\Delta} X \times X$ is quasi-affine,*
2. *there exists an affine scheme U and a smooth epimorphism $U \to X$.*

Let us try to explain the meaning of *quasi-affine* and *smooth epimorphism* in this context. So let $f : X \to Y$ be an injective morphism of sheaves on (aff/k) (this means that for all objects U of (aff/k) the map $f(U) : X(U) \to Y(U)$ is injective). If U is an affine scheme and $U \to Y$ is a morphism and we form the fibered product

$$\begin{array}{ccc} V & \longrightarrow & U \\ \downarrow & & \downarrow \\ X & \xrightarrow{f} & Y \end{array}$$

in Funct*((aff/k), (sets)) then V is a subsheaf of U. Thus it makes sense to say that V is or is not a finite union of affine subschemes of U. Now the injection $f : X \to Y$ is called *quasi-affine*, if for all affine schemes U and for all morphisms $U \to Y$ (so equivalently for all elements of $Y(U)$) the pullback $V \subset U$ is a finite union of affine subschemes of U.

Now let X be a sheaf on (aff/k) such that the diagonal is quasi-affine. This implies that whenever we have two affine schemes U and V over X, then the fibered product $U \times_X V$ is a finite union of affine schemes. Now, in the situation of the above definition, the morphism $U \to X$ is called a smooth epimorphism, if for every affine scheme $V \to X$ the fibered product $U \times_X V$ can be covered by finitely many affine Zariski-open subschemes W_i such that for each i the morphism $W_i \to V$ is smooth and the induced morphism $\amalg W_i \to V$ is surjective.

Of course all k-varieties and k-schemes are algebraic k-spaces.

Definition 2.4. *A* k-scheme *is an algebraic k-space X, which is locally in the Zariski-topology an affine scheme. This means that there exist affine k-schemes $U_1, \ldots, U_n$ and open immersions of algebraic spaces $U_i \to X$ such that $\coprod U_i \to X$ is surjective. (An open immersion of algebraic spaces $X \to Y$ is a morphism such that for every affine scheme $U \to Y$ the pullback $X \times_Y U \to U$ is an isomorphism onto a Zariski open subset.)*

A k-variety *is a k-scheme which is reduced and irreducible, which means that the U_i in the definition of scheme may be chosen reduced and irreducible with dense intersection.*

One can prove that an algebraic space X is locally in the étale topology an affine scheme. This means that affine schemes $U_1, \ldots, U_n$ together with étale morphisms $U_i \to X$ can be found, such that $\coprod U_i \to X$ is an étale epimorphism. (The notion of étale epimorphism is defined as the notion of smooth epimorphism, above, using fibered products.)

Using such étale (or smooth) covers, one can do a lot of geometry on algebraic spaces. A vector bundle, for example, is a family of vector bundles E_i/U_i, together with gluing data $E_i|U_{i,j} \cong E_j|U_{i,j}$.

Groupoids

Definition 2.5. *A* groupoid *is a category in which all morphisms are invertible.*

Example 2.1. 1. Let X be a set. We think of X as a groupoid by taking X as set of objects and declaring all morphisms to be identity morphisms.
2. Let G be a group. We define the groupoid BG to have a single object with automorphism group G.
3. Let X be a G-set. Then we define the groupoid X_G to have set of objects X, and for two objects $x, y \in X$ we let $\mathrm{Hom}(x, y) = \{g \in G \mid gx = y\}$. This groupoid is called the *transformation groupoid* given by the action of G on X.
4. Let $R \subset X \times X$ be an equivalence relation on the set X. Then we define an associated groupoid by taking as objects the elements of X and as morphisms the elements of R, where the element $(x, y) \in R$ is then a unique morphism from x to y.

We think of two groupoids as 'essentially the same' if they are equivalent as categories. We say that a groupoid is *rigid* if every object has trivial automorphism group, and *connected* if all objects are isomorphic. Every rigid groupoid is equal to the groupoid given by an equivalence relation. A groupoid is rigid if and only if it is equivalent to a groupoid given by a set as in Example 1, above. A groupoid is connected if and only if it is equivalent to a groupoid of type BG, for some group G. All these follow easily from the following well-known equivalence criterion.

Proposition 2.1. *Let $f : X \to Y$ be a morphism of groupoids (i.e. a functor between the underlying categories X and Y). Then f is an equivalence of categories if and only if f is fully faithful and essentially surjective.*

Remark 2.2. Groupoids form a 2-category. This means that the category of groupoids consists of

1. objects: groupoids
2. morphisms: functors between groupoids
3. 2-morphisms, or morphisms between morphisms: natural transformations between functors.

Note that this is a special type of 2-category, since all 2-morphisms are invertible. One should think of such a 2-category as a category where for any two objects X, Y the morphisms $\mathrm{Hom}(X,Y)$ form not a set but rather a groupoid.

Example 2.2. Another important example of a 2-category with invertible 2-morphisms is the (truncated) homotopy category:

1. objects: topological spaces
2. morphisms: continuous maps
3. 2-morphisms: homotopies up to reparametrization.

One may think of groupoids as generalized sets, or rather a common generalization of sets and groups. If we replace the category (sets) in the definition of algebraic space by the 2-category (groupoids), we get algebraic stacks. This is not a completely trivial generalization because of the complications arising from the fact that (groupoids) is a 2-category rather than a 1-category, like (sets).

We call a groupoid *finite*, if it has finitely many isomorphisms classes of objects and every object has a finite automorphism group. For a finite groupoid X we define its 'number of elements' by

$$\#(X) = \sum_x \frac{1}{\#\operatorname{Aut} x},$$

where the sum is taken over a set of representatives for the isomorphism classes.

Fibered products of groupoids. The fibered product is a construction that is not only basic for the theory of groupoids and stacks, but is also a good example of the philosophy of 2-categories.

Let

$$\begin{array}{ccc} & & Z \\ & & \downarrow{\scriptstyle g} \\ X & \xrightarrow{f} & Y \end{array}$$

be a diagram of groupoids and morphisms. Then the fibered product $W = X \times_Y Z$ is the groupoids defined as follows: Objects of W are triples (x, ϕ, z), where $x \in \operatorname{ob} X$, $z \in \operatorname{ob} Z$ and $\phi : f(x) \to g(z)$ is a morphism in Y. A morphism in X from (x, ϕ, z) to (x', ϕ', z') is a pair (α, β), where $\alpha : x \to x'$ and $\beta : z \to z'$ are morphisms in X and Z, respectively, such that the diagram

$$\begin{array}{ccc} f(x) & \xrightarrow{\phi} & g(z) \\ {\scriptstyle f(\alpha)}\downarrow & & \downarrow{\scriptstyle g(\beta)} \\ f(x') & \xrightarrow{\phi'} & g(z') \end{array}$$

commutes in Y.

The groupoid W comes together with two morphisms $W \to X$ and $W \to Z$ given by projecting onto the first and last components, respectively. Moreover, W comes with a 2-morphism θ

$$\begin{array}{ccc} W & \longrightarrow & Z \\ \downarrow & \overset{\theta}{\nearrow\!\!\!\nearrow} & \downarrow{\scriptstyle g} \\ X & \underset{f}{\longrightarrow} & Y \end{array} \tag{2.1}$$

making the diagram '2-commute', which just means that θ is an isomorphism from the composition $W \to X \to Y$ to the composition $W \to Z \to Y$. The 2-isomorphism θ is given by $\theta(x, \phi, z) = \phi$. It is a natural transformation by the very definition of W.

Example 2.3. If X, Y and Z are sets, then W is (canonically isomorphic to) the fibered product $\{(x, y) \in X \times Y \mid f(x) = g(y)\}$ in the category of sets.

The 2-fibered product W satisfies a universal mapping property in the 2-category of groupoids. Namely, given any groupoid V with morphisms $V \to X$ and $V \to Z$ and a 2-isomorphism from $V \to X \to Y$ to $V \to Z \to Y$ (depicted in the diagram below by the 2-arrow crossing the dotted arrow), there exists a morphism $V \to W$ and 2-isomorphisms from $V \to X$ to $V \to W \to X$ and $V \to W \to Z$ to $V \to Z$ such that the diagram

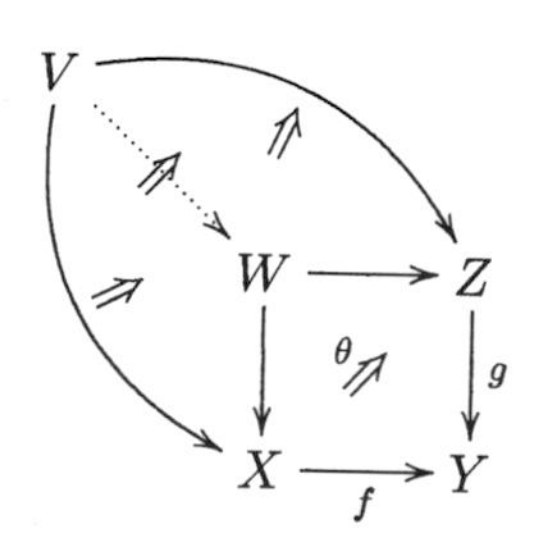

commutes, which amounts to a certain compatibility of the various 2-isomorphisms involved. (One should image this diagram as lying on the surface of a sphere.) The morphism $V \to W$ is unique up to unique isomorphism.

Whenever a diagram such as (2.1) satisfies this universal mapping property, we say that it is 2-cartesian (or just cartesian, because in a 2-category, 2-cartesian is the default value). In this case, W is equivalent to the fibered product constructed above.

If X is a G set, then we have two fundamental cartesian diagrams:

$$\begin{array}{ccc} X & \longrightarrow & pt \\ \downarrow & & \downarrow \\ X_G & \longrightarrow & BG \end{array} \tag{2.2}$$

and

$$
\begin{array}{ccc}
G \times X & \xrightarrow{\sigma} & X \\
{\scriptstyle p}\downarrow & & \downarrow \\
X & \longrightarrow & X_G
\end{array}
\qquad (2.3)
$$

Here pt denotes the groupoid with one object and one morphism (necessarily the identity morphism of the object). If we write a set, we mean the set thought of as a groupoid. By σ and p we denote the action and the projection, respectively.

Diagram (2.3) is moreover 2-cocartesian[1]. Hence X_G satisfies the universal mapping property of a quotient of X by G in the category of groupoids. Note that in the category of sets the quotient set X/G satisfies the cocartesian property, but not the cartesian property (unless the action of G on X is free, in which case the set quotient X/G is equivalent to the groupoid quotient X_G). Thus quotients taken in the category of groupoids have much better properties than quotients taken in the category of sets. For example, we have

$$\#(X_G) = \frac{\#X}{\#G}$$

if X and G are finite.

Let X be a groupoid and let X_0 be the set of objects of X and X_1 the set of all morphisms of X. Let $s : X_1 \to X_0$ be the map associating with each morphism its source object, and $t : X_1 \to X_0$ the map associating with each morphism its target object. Then the diagram

$$
\begin{array}{ccc}
X_1 & \xrightarrow{t} & X_0 \\
{\scriptstyle s}\downarrow & & \downarrow{\scriptstyle \pi} \\
X_0 & \xrightarrow{\pi} & X
\end{array}
$$

is cartesian and cocartesian, where $\pi : X_0 \to X$ is the canonical morphism. Thus a groupoid may be thought of as the quotient of its object set by the action of the morphisms.

Algebraic stacks

We will subdivide the definition of algebraic stacks into three steps.

[1] The notion of 2-cocartesian is more subtle than one might be led to believe. The correct definition is not simply the dual notion to the 2-cartesian property explained above. It involves, instead of a square, a *cube*. For our purposes it is sufficient to remark that (2.3) is cocartesian with respect to test objects which are rigid groupoids, or even just sets. For such text objects, 2-cocartesian reduces to the usual notion of cocartesian.

Prestacks. Prestacks are a generalization of presheaves (i.e. contravariant functors (aff/k) $\to$ (sets)).

Definition 2.6. *A prestack is a (lax) contravariant functor X : (aff/k) $\to$ (groupoids). This means that X is given by the data*

1. *for every affine k-scheme U a groupoid $X(U)$,*
2. *for every morphism of k-schemes $U \to V$ a morphism of groupoids $X(V) \to X(U)$,*
3. *for every composition of morphisms of k-schemes $U \to V \to W$ a natural transformation θ:*

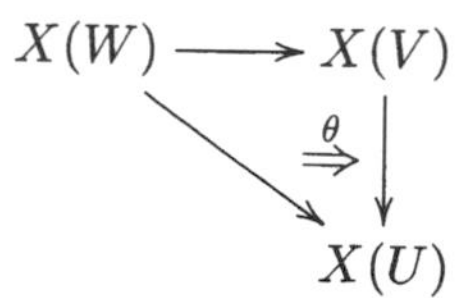

(this means that θ is a natural transformation from the functor $X(W) \to X(U)$ to the composition of the functors $X(W) \to X(V) \to X(U)$.

This data is subject to the conditions

1. *if $U \to U$ is the identity, then so is $X(U) \to X(U)$,*
2. *for each composition $U \to V \to W \to Z$ in (aff/k) a 2-cocycle condition expressing the compatibilities the various θ have to satisfy. Using the examples below as guide, this 2-cocycle condition is not difficult to write down. We leave this to the reader.*

Example 2.4. 1. Each actual functor (presheaf) (aff/k) $\to$ (sets) is a lax functor (prestack) (aff/k) $\to$ (groupoids). All θ are identities in this case.

2. The following might be thought of as a prototype stack:

$$\begin{aligned} \mathrm{Vect}_n : (\mathrm{aff}/k) &\longrightarrow \text{(groupoids)} \\ U &\longmapsto \text{(category of vector bundles of rank } n \text{ over } U \\ &\qquad\quad \text{with isomorphisms only)} \\ (U \to V) &\longmapsto \text{pullback of vector bundles} \\ (U \to V \to W) &\longmapsto \theta : \text{the canonical isomorphism of pullback} \\ &\qquad\quad \text{from } W \text{ to } U \text{ directly with pullback in two} \\ &\qquad\quad \text{steps via the intermediate } V. \end{aligned}$$

3. In this example all the θ are trivial again. Let G be an algebraic group over k and consider the functor

$$\begin{aligned} \mathrm{pre}BG : (\mathrm{aff}/k) &\longrightarrow \text{(groupoids)} \\ U &\longmapsto B(G(U)) \\ (U \to V) &\longmapsto \text{the morphism of groupoids} \\ &\qquad\quad B(G(V)) \to B(G(U)) \text{ induced by the} \\ &\qquad\quad \text{morphism of groups } G(V) \to G(U) \end{aligned}$$

Let us denote the category of contravariant lax functors from (aff/k) to (groupoids) by $\underline{\mathrm{Hom}}^*(\mathrm{aff}/k, \mathrm{groupoids})$. It is, of course, a 2-category. Its objects we have just defined. We leave it to the reader to explicate the morphisms and the 2-isomorphisms.

Given a lax functor X and an object x of the groupoid $X(U)$, where U is an affine k-scheme, we get an induced morphism

$$U \longrightarrow X$$

of lax functors (i.e., a natural transformation). We denote this morphism by the same letter:

$$x : U \to X.$$

The morphism x associates to $V \to U$ the pullback $x|V$.

A basic fact about $\underline{\mathrm{Hom}}^*(\mathrm{aff}/k, \mathrm{groupoids})$ is that it admits 2-fibered products, i.e. every diagram

$$\begin{array}{ccc} & & Z \\ & & \downarrow g \\ X & \xrightarrow{f} & Y \end{array}$$

can be completed to a cartesian diagram

$$\begin{array}{ccc} W & \longrightarrow & Z \\ \downarrow & \theta\nearrow & \downarrow g \\ X & \xrightarrow[f]{} & Y \end{array}$$

This is accomplished essentially by defining $W(U)$, for U an affine k-scheme, simply as the fibered product of $X(U)$ and $Z(U)$ over $Y(U)$.

Stacks. The notion of stacks generalizes the notion of sheaf on (aff/k).

Definition 2.7. *A prestack X : (aff/k) $\to$ (groupoids) is called a* stack *if it satisfies the following two stack axioms.*

1. *If U is an affine scheme and $x, y \in X(U)$ are objects of $X(U)$ then the presheaf*

$$\begin{array}{rcl} \mathrm{Isom}(x,y) : (\mathit{aff}/U) & \longrightarrow & (\mathit{sets}) \\ V & \longmapsto & \mathrm{Isom}(x|V, y|V) \end{array}$$

   is a sheaf on (aff/U).
2. *X satisfies the* descent property*: Given an affine scheme U, with a cover (in the sense of Definition 2.1) $(U_i)_{i\in I}$, and given objects $x_i \in X(U_i)$, for all $i \in I$ and isomorphisms $\phi_{ij} : x_i|U_{ij} \to x_j|U_{ij}$, for all $(i,j) \in I \times I$, such that the (ϕ_{ij}) satisfy the obvious cocycle condition (for each $(i,j,k) \in I \times I \times I$), then there exists an object $x \in X(U)$ and isomorphisms $\phi_i : x_i \to x|U_i$, such that for all $(i,j) \in U_{ij}$ we have $\phi_j|U_{ij} \circ \phi_{ij} = \phi_i|U_{ij}$.*

The data (x_i, ϕ_{ij}) is called a *descent datum* for X with respect to the covering (U_i); if (x, ϕ_i) exists, the descent datum is called *effective*. So the second stack axiom may be summarized by saying that every descent datum is effective.

Example 2.5. 1. Of course every sheaf is in a natural way a stack. Note how the stack axioms for presheaves reduce to the sheaf axioms.
2. The prestack Vect_n is a stack, since vector bundles satisfy the decent property.
3. The prestack preBG is not a stack. A descent datum for preBG with respect to the covering (U_i) of U is a Čech cocycle with values in G. It is effective if it is a boundary. Thus the Čech cohomology groups $H^1((U_i), G)$ contain the obstructions to preBG being a stack. Thus we let BG be the prestack whose groupoid of sections over $U \in (\mathrm{aff}/k)$ is the category of principal G-bundles over U. This is then a stack. There is a general process associating to a prestack a stack, called *passing to the associated stack* (similar to sheafification). The stack BG is the stack associated to the prestack preBG.

Algebraic stacks. This notion generalizes the notion of algebraic space.

Definition 2.8. *A stack $X : (\mathit{aff}/k) \to (\mathit{groupoids})$ is an* algebraic k-stack *if it satisfies*

1. *the diagonal $\Delta : X \to X \times X$ is representable and of finite type,*
2. *there exists an affine scheme U and a smooth epimorphism $U \to X$. Any such U is called a* presentation *of X.*

The first property is a separation property. It can be interpreted in terms of the sheaves of isomorphisms occurring in the first stack axiom. It says that all these isomorphism sheaves are algebraic spaces of finite type. (The definition of representability is as follows. The morphism $X \to Y$ of stacks is *representable* if for all affine $U \to Y$ the base change $X \times_Y U$ is an algebraic space.)

The second property says that, locally, every stack is just an affine scheme. Thus one can do 'geometry' on an algebraic stack. For example, a vector bundle E over an algebraic stack X is a vector bundle E' on such an affine presentation U, together with gluing data over $U \times_X U$ (which is an algebraic space by the first property). For another example, an algebraic stack X is *smooth of dimension n*, if there exists a smooth presentation $U \to X$, where U is smooth of dimension $n + k$ and $U \to X$ is smooth of relative dimension k. (Smoothness of representable morphisms of stacks is defined 'locally', by pulling back to affine schemes, similarly to the case of algebraic spaces, above.) Note that according to this definition, negative dimensions make sense.

Example 2.6. 1. Of course, all algebraic spaces are algebraic stacks.

2. The stack Vect_n is algebraic. The isomorphism spaces are just twists of GL_n, and therefore algebraic. For a presentation, take $\operatorname{Spec} k \to \mathrm{Vect}_n$, given by the trivial vector bundle k^n over $\operatorname{Spec} k$. This is a smooth morphism of relative dimension n^2, since for any affine scheme U with rank n vector bundle E over U, the induced morphism $U \to \mathrm{Vect}_n$ pulls back to the bundle of frames of E, which is a principal GL_n-bundle, and hence smooth of relative dimension n^2. Note that this makes $Vect_n$ a smooth stack of dimension $-n^2$.
3. Let G be an algebraic group over k. To avoid pathologies assume that G is smooth (which is always the case if $\operatorname{char} k = 0$). Then BG is an algebraic stack. The proof of algebraicity is the same as for Vect_n, after all, Vect_n is isomorphic to BGL_n. Whenever P is a G-bundle over a scheme X, then we get an induced morphism $X \to BG$, giving rise to the cartesian diagram
$$\begin{array}{ccc} P & \longrightarrow & \operatorname{Spec} k \\ \downarrow & & \downarrow \\ X & \longrightarrow & BG \end{array}$$
Therefore, $\operatorname{Spec} k \to BG$ is the universal G-bundle. Moreover, BG is smooth of dimension $-\dim G$.
4. If G is a (smooth) algebraic group acting on the algebraic space X, then we define an algebraic stack X/G as follows. For an affine scheme U, the groupoid $X/G(U)$ has as objects all pairs (P, ϕ), where $P \to U$ is a principal G-bundle and $\phi : P \to X$ is a G-equivariant morphism. One checks that X/G is an algebraic stack (for example, the canonical morphism $X \to X/G$ is a presentation) and that there are 2-cartesian diagrams
$$\begin{array}{ccc} G \times X & \longrightarrow & X \\ \downarrow & & \downarrow \\ X & \longrightarrow & X/G \end{array} \tag{2.4}$$
and
$$\begin{array}{ccc} X & \longrightarrow & \operatorname{Spec} k \\ \downarrow & & \downarrow \\ X/G & \longrightarrow & BG \end{array} \tag{2.5}$$

3. Lecture II: Equivariant intersection theory

Intersection theory

For a k-scheme X let $A_*(X) = \bigoplus_k A_k(X)$, where $A_k(X)$ is the Chow group of k-cycles up to rational equivalence tensored with $\mathbb{Q}$. Readers not familiar with Chow groups may assume that the ground field is $\mathbb{C}$ and work with $A_k(X) = H_{2k}^{BM}(X^{an})_{\mathbb{Q}}$ instead. Here X^{an} is the associated analytic space with the strong topology and BM stands for Borel-Moore homology, i.e. relative homology of a space relative to its one-point compactification. Everything works with this A_*, although the results are weaker.

Let also $A^*(X) = \bigoplus_k A^k(X)$ be the operational Chow cohomology groups of Fulton-MacPherson (see [9]), also tensored with $\mathbb{Q}$. If working with Borel-Moore homology as A_*, take $A^k(X) = H^{2k}(X^{an})_{\mathbb{Q}}$, usual (singular) cohomology with $\mathbb{Q}$-coefficients.

The most basic properties of A^* and A_* are: $A^*(X)$ is a graded $\mathbb{Q}$-algebra, for every scheme X, and $A_*(X)$ is a graded $A^*(X)$-module, the operation being cap product

$$\begin{array}{rcl} A^k(X) \times A_n(X) & \longrightarrow & A_{n-k}(X) \\ (\alpha, \gamma) & \longmapsto & \alpha \cap \gamma \ . \end{array}$$

Note that A^* and A_* exist more generally for Deligne-Mumford stacks. This was shown by A. Vistoli [16]. Deligne-Mumford stacks should be considered not too far from algebraic spaces or schemes (especially concerning their cohomological properties over $\mathbb{Q}$). Many moduli stacks (certainly all $\overline{M}_{g,n}(X,\beta)$) are of Deligne-Mumford type.

A Deligne-Mumford stack is an algebraic k-stack that is locally an affine scheme with respect to the *étale* topology. Thus a Deligne-Mumford stack X admits a presentation $p : U \to X$ (U affine) such that p is étale. This conditions implies, for example, that all automorphism groups are finite and reduced.

Equivariant theory

Let G be an algebraic group over k. To work *G-equivariantly* means to work in the category of algebraic G-spaces (i.e. algebraic k-spaces with G-action).

Now there is an equivalence of categories

$$\begin{array}{rcl} (\text{algebraic } G\text{-spaces}) & \longrightarrow & (\text{algebraic spaces } /BG) \\ X & \longmapsto & X/G \ . \end{array} \tag{3.1}$$

Here (algebraic G-spaces) is the category of algebraic k-spaces with G-action and equivariant morphisms, (algebraic spaces $/BG$) is the category of algebraic stacks over BG which are representable over BG. So an object of (algebraic spaces $/BG$) is an algebraic stack X together with a representable morphism $X \to BG$. A morphism in (algebraic spaces $/BG$) from $X \to BG$ to $Y \to BG$ is an isomorphism class of pairs (f, η), where $f : X \to Y$ is a morphism of algebraic stacks and η a 2-morphism making the diagram

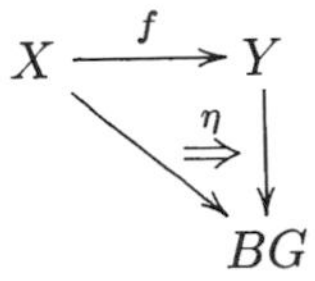

$$\begin{array}{ccc} X & \xrightarrow{f} & Y \\ & \searrow \overset{\eta}{\Rightarrow} & \downarrow \\ & & BG \end{array}$$

commute. The inverse of the functor (3.1) is defined using the construction of Diagram (2.5).

Defining equivariant Chow groups $A_G^*(X)$ and $A_*^G(X)$, for a G-space X, is equivalent to defining Chow groups $A^*(X/G)$ and $A_*(X/G)$ for stacks of the form X/G, i.e. quotient stacks.

If the quotient stack X/G is an algebraic space, then $A_*^G(X) = A_*(X/G)$ and $A_G^*(X) = A^*(X/G)$. In the general case, the construction is due to Edidin-Graham [5]. They proceed as follows. Assume that G is linear (and separable, to avoid certain pathologies in positive characteristic).

First define $A_p^G(X) = A_p(X/G)$ for p fixed. Choose a representation $G \to GL(V)$, such that there exists a G-invariant open subset U in the vector space V on which G acts freely (i.e. such that U/G is a space) and such that the complement $Z = V - U$ has codimension

$$\operatorname{codim}(Z, V) > \dim X - \dim G - p \quad .$$

The representation V of G associates to the principal G-bundle $X \to X/G$ a vector bundle over X/G. It is given by $X \times_G V = X \times V/G$, where G acts on $X \times V$ by $(x, v) \cdot g = (xg, g^{-1}v)$. It is not a space, but the open substack $X \times_G U \subset X \times_G V$ certainly is (the morphism $X \times_G U \to U/G$ is representable and U/G is already a space). Thus we have the following cartesian diagram.

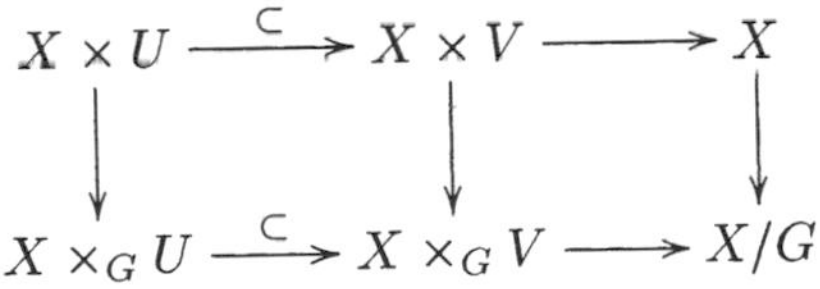

$$\begin{array}{ccccc} X \times U & \xrightarrow{\subset} & X \times V & \longrightarrow & X \\ \downarrow & & \downarrow & & \downarrow \\ X \times_G U & \xrightarrow{\subset} & X \times_G V & \longrightarrow & X/G \end{array}$$

The vertical maps are principal G-bundles, hence smooth epimorphisms. The inclusions on the left are open immersions with complement of codimension $> \dim X - \dim G - p$. The horizontal maps on the right are vector bundles of rank $\dim V$.

Having chosen V and $U \subset V$, we now define

$$A_p(X/G) = A_{p+\dim V}(X \times_G U),$$

which makes sense, because for a reasonable theory of Chow groups for quotient stacks we should have

$$A_p(X/G) = A_{p+\dim V}(X \times_G V),$$

since the Chow group of a vector bundle should be equal to the Chow group of the base, but shifted by the rank of the vector bundle, and

$$A_{p+\dim V}(X \times_G V) = A_{p+\dim V}(X \times_G U),$$

since the complement has dimension $\dim X \times_G Z < p + \dim V$, and cycles of dimension $< k$ should not affect A_k.

This definition is justified by giving rise to an adequate theory. For example, the definition is independent of the choice of V and $U \subset V$, as long as the codimension requirement is satisfied. This is proved by the 'double fibration argument', see [5].

As an example, let us work out what we get for $X/G = B\mathbb{G}_m$. Consider the action of $\mathbb{G}_m$ on $\mathbb{A}^n$, given by scalar multiplication $\mathbb{G}_m \times \mathbb{A}^n \to \mathbb{A}^n$, $(t, x) \mapsto tx$. A principal bundle quotient exists for $U = \mathbb{A}^n - \{0\}$ and $Z = \{0\}$ has codimension n. Thus this representation is good enough to calculate $A_p(B\mathbb{G}_m)$ for $n > -1 - p \Longleftrightarrow p \geq -n$. Moreover, by definition, we have for all $p \geq -n$

$$A_p(B\mathbb{G}_m) = A_{p+n}(\mathbb{P}^{n-1}).$$

In particular,

$$\begin{aligned} A_p(B\mathbb{G}_m) &= 0, \quad \text{for all } p \geq 0 \\ A_{-1}(B\mathbb{G}_m) &= A_{n-1}(\mathbb{P}^{n-1}) \\ A_{-2}(B\mathbb{G}_m) &= A_{n-2}(\mathbb{P}^{n-1}), \quad \text{etc.} \end{aligned}$$

To see how these groups fit together for various n, let $n' \geq n$ and consider a projection $\mathbb{A}^{n'} \to \mathbb{A}^n$. This induces the projection with center $\ker(\mathbb{A}^{n'} \to \mathbb{A}^n) = \mathbb{A}^{n'-n}$ from $\mathbb{P}^{n'-1}$ to $\mathbb{P}^{n-1}$.

$$\begin{array}{ccc} U & \xrightarrow{\subset} & \mathbb{P}^{n'-1} \\ \downarrow & & \\ \mathbb{P}^{n-1} & & \end{array}$$

Here the vertical map is a vector bundle of rank $n' - n$ and the horizontal map is the inclusion of the complement of the center of projection $\mathbb{P}^{n'-n-1}$. Thus we have for all $p \geq -n$

$$A_{p+n}(\mathbb{P}^{n-1}) = A_{p+n+n'-n}(U) = A_{p+n'}(U) = A_{p+n'}(\mathbb{P}^{n'-1}).$$

So we have independence of $A_p(B\mathbb{G}_m)$ on the choice of n. This is a special case of the double fibration argument.

Under the identification $A_{p+n}(\mathbb{P}^{n-1}) = A_{p+n'}(\mathbb{P}^{n'-1})$ the hyperplane $[H]$ in $\mathbb{P}^{n-1}$ corresponds to the hyperplane $[H]$ in $\mathbb{P}^{n'-1}$. The same is true for all intersections $[H]^k$. We write $h = [H]$ and thus we have for all $k \in \mathbb{Z}$

$$A_k(B\mathbb{G}_m) = \mathbb{Q}h^{-1-k},$$

where we agree that all negative powers of h are 0.

The equivariant cohomology groups $A^*_G(X) = A^*(X/G)$ are defined analogously to the usual A^*, namely by operating on $A^G_*(Y)$, for all equivariant $Y \to X$, where Y is a space (or equivalently all representable $Y \to X/G$, where Y is a stack).

In our example $B\mathbb{G}_m$ we get $A^*(B\mathbb{G}_m) = A^*_{\mathbb{G}_m}(pt) = \mathbb{Q}[c]$, where c is the Chern class of the universal line bundle and is in degree $+1$. Whenever X is a $\mathbb{G}_m$-space we get via the standard representation of $\mathbb{G}_m$ a line bundle over $X/\mathbb{G}_m$ (or equivalently an equivariant line bundle $X \times \mathbb{A}^1$ over X). The operation of $c \in A^*(B\mathbb{G}_m)$ on $A_*(X/\mathbb{G}_m)$ is through the Chern class of this line bundle. We have $c \cdot h^k = h^{k-1}$, and so we see that $A_*(B\mathbb{G}_m)$ is a free $A^*(B\mathbb{G}_m) = \mathbb{Q}[c]$-module on $h^0 \in A_{-1}(B\mathbb{G}_m)$. We may think of h^0 as the fundamental class of $B\mathbb{G}_m$ (it corresponds to $[\mathbb{P}^{n-1}]$ under any realization $A_{-1}(B\mathbb{G}_m) = A_{n-1}(\mathbb{P}^{n-1})$.)

More generally, if T is an algebraic torus with character group M, then $A^*(BT) = \mathrm{Sym}_{\mathbb{Q}} M_{\mathbb{Q}} =: R_T$, canonically. (Note how c comes from the canonical character $\mathrm{id} : \mathbb{G}_m \to \mathbb{G}_m$.) Moreover, $A_*(BT)$ is a free R_T-module of rank one on the generator $[BT]$ in degree $-\dim T$.

We shall be only interested in the case where the group $G = T$ is a torus. Then for all T-spaces X, we have that $A^*_T(X)$ is an R_T-algebra and $A^T_*(X)$ is an R_T-module. Therefore, R_T is the natural ground ring to work over. As in the usual case (the non-equivariant case, where one passes from $A^*(pt) = \mathbb{Z}$ to $\mathbb{Q}$) we want to pass from R_T to its quotient field. However, so as to not loose the grading, we only localize at the multiplicative system of homogeneous elements of positive degree, and call the resulting ring Q_T. Then we may tensor all $A^*_T(X)$ and $A^T_*(X)$ with Q_T. Still better, though, is to first pass to the completion of R_T at the augmentation ideal, $\hat{R}_T$ and then invert the homogeneous elements of positive degree to obtain $\hat{Q}_T$.

Comparing equivariant with usual intersection theory

For a G-space X, there is a canonical morphism $X \to X/G$, which is smooth of relative dimension $\dim G$. It is, in fact, a principal G-bundle. Thus flat pullback defines a homomorphism $A^G_*(X) \to A_*(X)$ of degree $\dim G$. 'Usual' pullback defines $A^*_G(X) \to A^*(X)$ preserving degrees.

Lemma 3.1. *The top-dimensional map* $A^G_{\dim X - \dim G}(X) \to A_{\dim X}(X)$ *is an isomorphism.*

Proof. By using the definitions, this reduces to proving that for a G-bundle of spaces, the top-dimensional Chow-groups agree.

This isomorphism defines the fundamental class $[X_G]$ of X/G in $A^G_{\dim X - \dim G}(X)$.

Note 3.1. If one works with cohomology one gets a Leray spectral sequence

$$H^i_G(X, H^j(G)) \Longrightarrow H^{i+j}(X, \mathbb{Q}).$$

Localization

Let X be a T-space and $Y \subset X$ a closed T-invariant subspace such that on $U = X - Y$ the torus T acts without fixed points. Then we have the proper pushforward map

$$\iota_* : A_*^T(Y) \longmapsto A_*^T(X)$$

induced by the inclusion $\iota : Y \to X$.

Proposition 3.1. *After tensoring with* Q_T

$$\iota_* : A_*^T(Y) \otimes_{R_T} Q_T \longmapsto A_*^T(X) \otimes_{R_T} Q_T$$

is an isomorphism.

Proof. Reduces the the case $Y = \emptyset$ and $X = U$, when the claim is that $A_*^T(X) \otimes_{R_T} Q_T = 0$. For details, see [6].

Rather than studying the proof of this proposition, let us study an example.

Consider the torus $T = \mathbb{G}_m{}^{n+1}$ and $M = \hat{T}$, with basis $\lambda_0, \ldots, \lambda_n$ and $A_T^*(pt) = A^*(BT) = R_T = \mathbb{Q}[\lambda_0, \ldots, \lambda_n]$. Let us denote the fundamental class of BT by $\mathfrak{t}$. Then we have $A_*^T(pt) = A_*(BT) = tR_T = t\mathbb{Q}[\lambda_0, \ldots, \lambda_n]$. Let $X = \mathbb{P}^n$ and consider the action of T on $\mathbb{P}^n$ given by

$$t \cdot \langle x_0, \ldots, x_n \rangle = \langle \lambda_0(t)x_0, \ldots, \lambda_n(t)x_n \rangle.$$

Take $Y = \{P_0, \ldots, P_n\}$, where $P_i = \langle 0, \ldots, 0, 1, 0, \ldots, 0 \rangle$, the 1 being in the ith position. Then localization (Proposition 3.1) says that

$$\iota_* : \bigoplus_{i=0}^{n} A_*^T(\{P_i\}) \otimes Q_T \longrightarrow A_*^T(\mathbb{P}^n) \otimes Q_T$$

is an isomorphism. Since everything is smooth, we may translate this into a statement about cohomology:

$$\iota_! : \bigoplus_{i=0}^{n} A_T^*(\{P_i\}) \otimes Q_T \longrightarrow A_T^*(\mathbb{P}^n) \otimes Q_T$$

is an isomorphism of degree $+n$.

To understand this isomorphism note that $\mathbb{P}^n/T \to BT$ is a $\mathbb{P}^n$-bundle, namely the projective bundle corresponding to the vector bundle E on BT given by the action of T on $\mathbb{A}^{n+1}$. Hence we have

$$\begin{aligned} A_T^*(\mathbb{P}^n) &= A^*(\mathbb{P}^n/T) \\ &= A^*(BT)[\xi]/\xi^{n+1} - c_1(E)\xi^n + \ldots + (-1)^{n+1}c_{n+1}(E) \\ &= \mathbb{Q}[\lambda_0, \ldots, \lambda_n][\xi]/\xi^{n+1} - \ldots + (-1)^{n+1}c_{n+1}(E). \end{aligned} \tag{3.2}$$

Now E is a sum of line bundles, each associated to one of the characters $\lambda_0, \dots, \lambda_n$. Hence we have $c_i(E) = \sigma_i(\lambda_0, \dots, \lambda_n)$, the symmetric function of degree i in $\lambda_0, \dots, \lambda_n$. In other words,

$$\sum_{i=0}^{n+1} (-1)^i c_i(E) \xi^{n+1-i} = \prod_{i=0}^{n} (\xi - \lambda_i),$$

so that

$$A_T^*(\mathbb{P}^n) = \mathbb{Q}[\lambda_0, \dots, \lambda_n, \xi] / \prod_{i=0}^{n} (\xi - \lambda_i).$$

Hence we have

$$\begin{aligned} A_T^*(\mathbb{P}^n) \otimes_{R_T} Q_T &= Q_T[\xi] / \textstyle\prod_{i=0}^n (\xi - \lambda_i) \\ &= \textstyle\prod_{i=0}^n Q_T[\xi]/(\xi - \lambda_i) \\ &= \textstyle\prod_{i=0}^n Q_T \\ &= \textstyle\prod_{i=0}^n A_T^*(P_i) \otimes_{R_T} Q_T, \end{aligned}$$

by the Chinese remainder theorem. This map

$$A_T^*(\mathbb{P}^n) \otimes_{R_T} Q_T \longrightarrow \prod_{i=0}^{n} A_T^*(P_i) \otimes_{R_T} Q_T$$

is of degree 0 and induced by ι^*. (Note that $\xi = c_1(\mathcal{O}(1))$ pulls back to λ_i at P_i, which is the character of the action of T on the fiber $\mathcal{O}(1)(P_i)$.) If we compose with

$$\prod_{i=0}^{n} A_T^*(P_i) \otimes Q_T \longrightarrow \prod_{i=0}^{n} A_T^*(P_i) \otimes Q_T$$

which is division by the tops Chern class of the tangent space (i.e. normal bundle) we get the inverse of the above map $\iota_!$. The tangent space $T_{\mathbb{P}^n}(P_i)$ has weights $(\lambda_j - \lambda_i)_{j \neq i}$ and so we divide by $\prod_{j \neq i} (\lambda_j - \lambda_i)$ in the ith component.

The residue formula

Let us return to the setup of Proposition 3.1. Moreover, assume that the inclusion $\iota : Y \to X$ is T-equivariantly the pullback of a regular immersion $\nu : V \to W$

$$\begin{array}{ccc} Y & \xrightarrow{\iota} & X \\ {\scriptstyle g}\downarrow & & \downarrow \\ V & \xrightarrow{\nu} & W. \end{array} \tag{3.3}$$

Then we have the self intersection formula

$$\nu^! \iota_*(\alpha) = e(g^* N_{V/W}) \alpha, \quad \text{for all } \alpha \in A_*^T(Y),$$

where e stands for the top Chern (i.e. Euler) class. So if $e(g^*N_{V/W}) \in A_T^*(Y) \otimes Q_T$ is invertible, we have

$$\alpha = \frac{\nu^! \iota_* \alpha}{e(g^*N)},$$

and we have identified the inverse of the localization isomorphism ι_*, namely

$$\frac{1}{e(g^*N)}\nu^! : A_*^T(X) \otimes Q_T \longrightarrow A_*^T(Y) \otimes Q_T.$$

That $e(g^*N)$ is invertible, is in practise easily verified, one just has to check that the weights of g^*N at the fixed points of X under T are non-zero. If X is smooth and $\iota = \nu$, then it is a theorem that these weights are always non-zero and so $e(N)$ is always invertible.

Let us from now assume that $e(g^*N)$ is, indeed, invertible in $A_*^T(Y) \otimes Q_T$. Then we have for all $\beta \in A_*^T(X)$

$$\beta = \iota_* \frac{\nu^! \beta}{e(g^*N)}.$$

If X is smooth and $\iota = \nu$, we will want to apply this to $[X_T] \in A_*^T(X)$:

$$[X_T] = \iota_* \frac{[Y_T]}{e(N_{Y/X})}.$$

So if $\alpha \in A_T^*(X)$ we have

$$\alpha[X_T] = \iota_* \frac{\iota^*(\alpha)[Y_T]}{e(N_{Y/X})}$$

in $A_*^T(X)$.

Now assume that X is moreover proper. Then $X/T \to BT$ is proper and proper pushforward gives a homomorphism $\deg^T : A_*^T(X) \otimes Q_T \to A_*^T(pt) \otimes Q_T = \mathfrak{t}\mathfrak{Q}_{\mathfrak{T}}$ and we get

$$^T\!\!\int_X \alpha := \deg^T(\alpha[X_T]) = \deg^T(\frac{\iota^*(\alpha)[Y_T]}{e(N_{Y/X})}) = {}^T\!\!\int_Y \frac{\iota^*(\alpha)}{e(N_{Y/X})},$$

an equation in $A_*^T(pt) \otimes_{R_T} Q_T = \mathfrak{t}\mathfrak{Q}_{\mathfrak{T}}$.

Now consider the cartesian diagram

$$\begin{array}{ccc} X & \longrightarrow & pt \\ \downarrow & & \downarrow \\ X/T & \longrightarrow & BT. \end{array}$$

Since flat pullback commutes with proper pushforward, we get an induced commutative diagram

$$
\begin{array}{ccc}
A_*(X) & \xrightarrow{\ \deg\ } & \mathbb{Q} \\
\uparrow & & \uparrow \theta \\
A_*^T(X) & \xrightarrow{\ \deg^T\ } & \mathfrak{t}\mathbb{Q}[\lambda_0, \ldots, \lambda_n],
\end{array}
\tag{3.4}
$$

where the homomorphism $\theta : \mathfrak{t}\mathbb{Q}[\lambda_0, \ldots, \lambda_n] \to \mathbb{Q}$ is given by sending $\mathfrak{t}$ to 1 and the λ_i to 0. Diagram (3.4) fits into the larger diagram

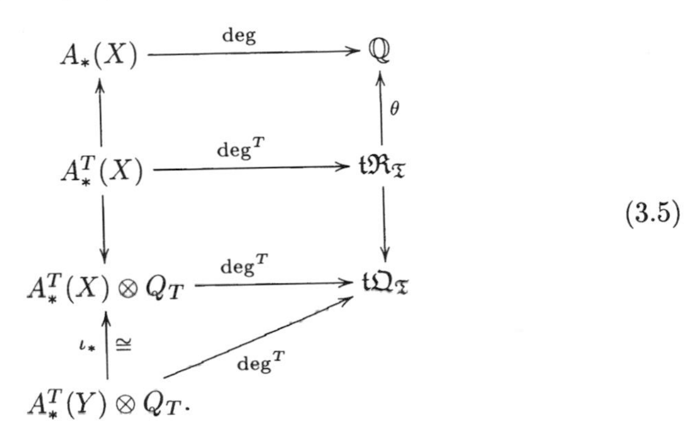

(3.5)

Corollary 3.1 (Residue Formula). *1. Assume X is smooth and $\iota = \nu$. If $a \in A^{\dim X}(X)$ comes from $\alpha \in A_T^{\dim X}(X)$, then*

$$ {}^T\!\!\int_Y \frac{\iota^*(\alpha)}{e(N_{Y/X})} \in \mathfrak{t}\mathfrak{Q}_{\mathfrak{T}} $$

is contained in the submodule $\mathfrak{t}\mathbb{Q}$ and we have

$$ \int_X a = \deg a[X] = \theta \deg^T \alpha[X_T] = \theta\, {}^T\!\!\int_Y \frac{\iota^*(\alpha)}{e(N_{Y/X})}. $$

The θ in this formula only serves to remove the factor of $\mathfrak{t}$.

2. General case. Assume $\beta \in A_*^T(X)$. Write b for the corresponding element of $A_*(X)$. Let $\alpha \in A_T^*(X)$ and write a for the corresponding element of $A^*(X)$. Then if $\deg \beta - \deg \alpha = -\dim T$, then $\deg b - \deg a = 0$ and

$$ \int_b a = \theta\, {}^T\!\!\int_\beta \alpha = \theta \deg^T \alpha \cap \beta = \theta \deg^T \alpha \cap \iota_* \frac{\nu^! \beta}{e(g^* N)} \tag{3.6} $$
$$ = \theta \deg^T \left(\iota^*(\alpha) \cap \frac{\nu^! \beta}{e(g^* N)} \right) = \theta\, {}^T\!\!\int_{\nu^! \beta} \frac{\iota^* \alpha}{e(g^* N)}. $$

Again, this is to be interpreted to mean that

$$ {}^T\!\!\int_{\nu^! \beta} \frac{\iota^* \alpha}{e(g^* N)} \in \mathfrak{t}\mathfrak{Q}_{\mathfrak{T}} $$

is contained in $\mathfrak{t}\mathbb{Q}$ and after removing $\mathfrak{t}$ we get $\int_b a$.

Proof. This is just a simple diagram chase using (3.5) and keeping track of degrees.

Remark 3.1. 1. Evaluating the rational function of degree zero $\theta\ {}^T\!\!\int_Y \frac{\iota^*(\alpha)}{e(N_{Y/X})}$ at an element $\mu \in M^\vee$ corresponds to restricting the action of T to the corresponding one-parameter subgroup. For a generic one-parameter subgroup the fixed locus of T and of the one-parameter subgroup will be the same and the denominator of ${}^T\!\!\int_Y \frac{\iota^*(\alpha)}{e(N_{Y/X})}$ will not vanish at μ. Then ${}^T\!\!\int_Y \frac{\iota^*(\alpha)}{e(N_{Y/X})}$ can be calculated by evaluating at μ. This is also how one evaluates in practise.

2. The standard way to ensure that a comes from α is to take polynomials in Chern classes of equivariant vector bundles.
3. Assume that Y is the fixed locus. Then $A^T_*(Y) = A_*(Y) \otimes_{\mathbb{Q}} A_*(BT)$ and $A^*_T(Y) \supset A^*(Y) \otimes_{\mathbb{Q}} R_T$. If $\iota^*(\alpha) \in R_T \subset A^*_T(Y)$, then
$$ {}^T\!\!\int_{\nu^!\beta} \frac{\iota^*\alpha}{e(g^*N)} = \iota^*(\alpha)\ {}^T\!\!\int_{\nu^!\beta} \frac{1}{e(g^*N)} $$
by the projection formula.
4. Also, if T acts trivially on Y, and $N_{Y/X}$ has a filtration with line bundle quotients L_i, then $e(N_{Y/X}) = \prod_i (c(L_i) + \lambda_i)$, where $c(L_i) \in A^*(Y)$ is the Chern class of L_i and $\lambda_i \in R_T$ the weight of T on L_i. This gives a very explicit form of the Bott residue formula.

Example 3.1. Let T operate on $\mathbb{P}^1$, in such a way that 0 and ∞ are the fixed points of T. Let E be an equivariant vector bundle on $\mathbb{P}^1$. Then $E(0)$ and $E(\infty)$ are representations of T. Let $\lambda_1 \ldots, \lambda_r$ be the weights of T on $E(0)$ and $\mu_1 \ldots, \mu_r$ the weights of T on $E(\infty)$. Also, let ω be the character through which T acts on $\mathbb{P}^1$, i.e. $t \cdot 1 = \omega(t)$. Assume that $H^1(\mathbb{P}^1, E) = 0$. Then we can calculate the weights of T on $H^0(\mathbb{P}^1, E)$ by equivariant Riemann-Roch: Let $\alpha_1, \ldots, \alpha_n$ be these weights. Then we have (apply Riemann-Roch to $\mathbb{P}^1/T \xrightarrow{\pi} BT$):
$$ \operatorname{ch}(H^0(\mathbb{P}^1, E)) = \deg^T(\operatorname{ch}(E)\operatorname{td}(T_\pi) \cap [\mathbb{P}^1_T]) $$
or
$$ \sum_{i=1}^n e^{\alpha_i} = \frac{\operatorname{ch}(E(0))\operatorname{td}(T_{\mathbb{P}^1}(0))}{c_1(T_{\mathbb{P}^1}(0))} + \frac{\operatorname{ch}(E(\infty))\operatorname{td}(T_{\mathbb{P}^1}(\infty))}{c_1(T_{\mathbb{P}^1}(\infty))}, $$
by localization. Now since $\operatorname{td}(x) = \frac{x}{1-e^{-x}}$ and the weight of T on $T_{\mathbb{P}^1}(0)$ is ω and on $T_{\mathbb{P}^1}(\infty)$ is $-\omega$, we get
$$ \sum_{i=1}^n e^{\alpha_i} = \frac{\operatorname{ch}(E(0))}{1-e^{-\omega}} + \frac{\operatorname{ch}(E(\infty))}{1-e^{\omega}} $$
or

$$\sum_{i=1}^{n} e^{\alpha_i} = \frac{\sum e^{\lambda_j}}{1 - e^{-\omega}} + \frac{\sum e^{\mu_j}}{1 - e^{\omega}}$$

in $\hat{Q}_T$. Note that we have uncapped with $[BT]$.

This determines the α_i uniquely. Useful to calculate the α_i in this context is the formula (which holds for all $a, b \in \mathbb{Z}$)

$$\frac{e^{a\omega}}{1 - e^{\omega}} + \frac{e^{b\omega}}{1 - e^{-\omega}} = \sum_{n=a}^{b} e^{n\omega},$$

where for $a \geq b + 1$ we set $\sum_{n=a}^{b} e^{n\omega} = -\sum_{i=b+1}^{a-1} e^{i\omega}$.

4. Lecture III: The localization formula for Gromov-Witten invariants

Using the localization formula is one of the most useful methods we have to calculate Gromov-Witten invariants, besides the WDVV-equations (i.e. the associativity of the quantum product) and its analogues for higher (but still very low) genus. The idea of applying the Bott formula in this context is due to Kontsevich [13]. It has been used by Givental [11] to verify the predictions of Mirror symmetry for complete intersections in toric varieties.

If the variety we are interested in has finitely many fixed points under a torus action, the Bott formula reduces the calculation of its Gromov-Witten invariants to a calculation on $\overline{M}_{g,n}$ and a combinatorial problem. In this lecture we will treat the case of projective space $\mathbb{P}^r$.

Let the ground field be of characteristic 0. Let $\overline{M}_{g,n}(\mathbb{P}^r, d)$ denote the stack of stable maps of degree d to $\mathbb{P}^r$, whose source is a genus g curve with n marked points. For an affine k-scheme U the groupoid

$$\overline{M}_{g,n}(\mathbb{P}^r, d)(U)$$

is the groupoid of such stable maps parameterized by U. These are diagrams

$$\begin{array}{ccc} C & \xrightarrow{f} & \mathbb{P}^r \\ {\scriptstyle\pi}\downarrow & & \\ U & & \end{array}$$

where $\pi : C \to U$ is a family of prestable curves with n sections and f is a family of maps of degree d, such that the stability condition is satisfied (see, for example, [13], [14], [10], [4], [2]). Evaluation at the n marks defines a morphism

$$\mathrm{ev} : \overline{M}_{g,n}(\mathbb{P}^r, d) \longrightarrow (\mathbb{P}^r)^n.$$

Gromov-Witten invariants are the induced linear maps

$$\begin{array}{rcl} A^*(\mathbb{P}^r)^{\otimes n} & & \longrightarrow \mathbb{Q} \\ a_1 \otimes \ldots \otimes a_n & \longmapsto & \int_{[\overline{M}_{g,n}(\mathbb{P}^r,d)]} \mathrm{ev}^*(a_1 \otimes \ldots a_n). \end{array}$$

For $g > 0$ the cycle $[\overline{M}_{g,n}(\mathbb{P}^r, d)]$ is the 'virtual fundamental class' of $\overline{M}_{g,n}(\mathbb{P}^r, d)$ (see [2], [3], [1] or [15]). This is a carefully constructed cycle giving rise to a consistent theory of Gromov-Witten invariants (i.e., a so-called cohomological field theory, [14]). The usual fundamental cycle is, in general, not even in the correct degree, as $\overline{M}_{g,n}(\mathbb{P}^r, d)$ may have higher dimension than expected, because of the presence of obstructions.

Now consider the torus $T = \mathbb{G}_m{}^{r+1}$ with character group M, whose canonical generators are denoted $\lambda_0 \ldots, \lambda_n$. Then $R_T = \mathbb{Q}[\lambda_0, \ldots, \lambda_r]$ and $Q_T \subset \mathbb{Q}(\lambda_0, \ldots, \lambda_r)$. The torus T acts on $\mathbb{P}^r$ by

$$\begin{array}{rcl} T \times \mathbb{P}^r & & \longrightarrow \mathbb{P}^r \\ (t, \langle x_0, \ldots, x_r \rangle) & \longmapsto & \langle \lambda_0(t)x_0, \ldots, \lambda_r(t)x_r \rangle. \end{array}$$

We get an induced action of T on $\overline{M}_{g,n}(\mathbb{P}^r, d)$: given $t \in T(U)$ and

$$\begin{array}{ccc} C & \xrightarrow{f} & \mathbb{P}^r \\ {\scriptstyle \pi}\downarrow & & \\ U & & \end{array}$$

in $\overline{M}_{g,n}(\mathbb{P}^r, d)(U)$ we define $t \cdot (C, f) = (C, t \circ f)$, where $(C, t \circ f)$ stands for

$$\begin{array}{ccccc} C & \xrightarrow{(\pi,f)} & U \times \mathbb{P}^r & \xrightarrow{t} & \mathbb{P}^r \\ {\scriptstyle \pi}\downarrow & & & & \\ U & & & & \end{array}$$

We leave it as an exercise, to turn this into an action of the group $T(U)$ on the groupoid $\overline{M}_{g,n}(\mathbb{P}^r, d)(U)$, i.e., actions on the morphism and object sets compatible with all the groupoid structure maps. Compatibility under change of U gives the action of the algebraic group T on the algebraic stack $\overline{M}_{g,n}(\mathbb{P}^r, d)$.

The same general arguments that allow the construction of the virtual fundamental class of $\overline{M}_{g,n}(\mathbb{P}^r, d)$ give rise to an equivariant virtual fundamental class $[\overline{M}_{g,n}(\mathbb{P}^r, d)_T] \in A_*^T(\overline{M}_{g,n}(\mathbb{P}^r, d))$, which pulls back to the usual virtual fundamental class $[\overline{M}_{g,n}(\mathbb{P}^r, d)] \in A_*(\overline{M}_{g,n}(\mathbb{P}^r, d))$. We shall apply Formula (3.6) with $\beta = [\overline{M}_{g,n}(\mathbb{P}^r, d)_T]$ and $b = [\overline{M}_{g,n}(\mathbb{P}^r, d)]$.

If $\alpha_1, \ldots, \alpha_n \in A_T^*(\mathbb{P}^r)$ and $a_1, \ldots, a_n$ are the corresponding classes in $A^*(\mathbb{P}^r)$, the induced Gromov-Witten invariants are given by

$$\int_{[\overline{M}_{g,n}(\mathbb{P}^r,d)]} \mathrm{ev}^*(a_1 \otimes \ldots \otimes a_n) = \theta \ {}^T\!\!\!\!\int_{\nu^![\overline{M}_{g,n}(\mathbb{P}^r,d)_T]} \frac{\iota^* \mathrm{ev}^*(\alpha_1 \otimes \ldots \otimes \alpha_n)}{e(g^*N)}, \tag{4.1}$$

at least if the α_i are homogeneous and $\sum_{i=1}^n \deg \alpha_i = \deg[\overline{M}_{g,n}(\mathbb{P}^r, d)]$.

To apply this formula, we need to construct a cartesian T-equivariant diagram such as (3.3):

$$\begin{array}{ccc} Y & \xrightarrow{\iota} & \overline{M}_{g,n}(\mathbb{P}^r, d) \\ {\scriptstyle g}\downarrow & & \downarrow \\ V & \xrightarrow{\nu} & W \end{array} \tag{4.2}$$

where ν is a regular closed immersion and Y contains all the fixed points of T in $\overline{M}_{g,n}(\mathbb{P}^r, d)$. Of course, we get the best results if we take Y as small as possible, namely equal to the fixed locus of T on $\overline{M}_{g,n}(\mathbb{P}^r, d)$. The point of view of more general Y is still useful, because it lets us decompose the problem into several steps. We pass successively to smaller Y until we reach the fixed locus. The regular immersions $\nu : V \to W$ will be chosen at each step in such a way that we can keep track of $\nu^![\overline{M}_{g,n}(\mathbb{P}^r, d)_T]$, i.e., we can follow what happens to the virtual fundamental class.

As we shall see, the fixed locus can be described in terms of stacks of stable curves $\overline{M}_{g,n}$. Thus Formula (4.1) reduces the computation of Gromov-Witten invariants to a computation on various $\overline{M}_{g,n}$. Since the fixed locus has many components, the combinatorics turn out to be non-trivial. Moreover, the integrals one has to evaluate on $\overline{M}_{g,n}$ are non-trivial, too. Still, this approach has been very successful in determining Gromov-Witten invariants. (See [11], [13], [12] or [7], [8] for more details.)

We shall next determine the fixed locus. The connected components of the fixed locus are indexed by marked modular graphs (τ, d, γ). Thus the right hand side of (4.1) is a sum over all marked modular graphs (τ, d, γ) involved. We can treat the fixed components given by different marked graphs separately, i.e., we determine for each (τ, d, γ) the classes $\nu^![\overline{M}_{g,n}(\mathbb{P}^r, d)_T]$ and $\frac{1}{e(g^*N)}$ restricted to the fixed locus component given by (τ, d, γ). Then we have

$$\int_{[\overline{M}_{g,n}(\mathbb{P}^r,d)]} \mathrm{ev}^*(a_1 \otimes \ldots \otimes a_n) = \theta \sum_{(\tau,d,\gamma)} {}^T\!\!\!\!\int_{\nu^![\overline{M}_{g,n}(\mathbb{P}^r,d)_T]_{(\tau,d,\gamma)}} \frac{\iota^* \mathrm{ev}^*(\alpha_1 \otimes \ldots \otimes \alpha_n)}{e(g^*N)_{(\tau,d,\gamma)}}. \tag{4.3}$$

The fixed locus

Recall that *modular graphs* are the graphs that give the degeneracy type of prestable marked curves. They consist of a set of vertices V_τ, a set of flags F_τ (which can either be tails or pair up to edges), and non-negative integer markings of the vertices, giving the vertices a genus. Tails are denoted S_τ and edges E_τ. The set of flags connected with the vertex v is denoted $F_\tau(v)$. A vertex is *stable* if its genus is at least 2, its genus is one and its valence (the number of flags it bounds) is at least 1 or its genus is 0 and its valence at least 3. The stabilization τ^s of a modular graph is obtained by contracting all edges containing unstable vertices. The set of vertices of the stabilization is equal to the set of stable vertices. For details, see [4].

Let (τ, d, γ) be a marked modular graph of the following type.

1. τ: a modular graph which is connected and whose stabilization τ^s is not empty. Moreover, the genus $g(\tau)$ is equal to g and the set of tails S_τ is $S_\tau = \{1, \ldots, n\}$.
2. $d : V_\tau \to \mathbb{Z}_{\geq 0}$ a marking of the vertices by 'degrees', such that
   a) $d(v) = 0$, for every stable vertex $v \in V_\tau^s$,
   b) $\sum_{v \in V_\tau} d(v) = d$.
   Note that we use the same letter d for the marking of the graph τ and the total degree of the graph.
3. γ consists of three maps:
   a) $\gamma : V_\tau^s \to \{P_0, \ldots, P_r\}$, where $P_i = \langle 0, \ldots, 1, \ldots, 0 \rangle$, the 1 being in the i-th position; so γ associates to every stable vertex of τ a fixed point of T on $\mathbb{P}^r$,
   b) $\gamma : V_\tau^u \to \{L_{ij} \mid 0 \leq i < j \leq r\}$, where $L_{ij} = \langle 0, \ldots, x, \ldots, y, \ldots, 0 \rangle$ and x is in the i-th, y in the j-th position; so γ associates to every unstable vertex a one-dimensional orbit closure,
   c) $\gamma : F_\tau \to \{P_0, \ldots, P_r\}$; so γ associates to every flag a fixed point.

These data are subject to the following list of compatibility requirements:

1. Every edge has an unstable vertex, i.e., no edge connects stable vertices,
2. γ is constant on edges,
3. if v is a stable vertex then $\gamma(v) = \gamma(i)$, for all $i \in F_\tau(v)$,
4. if v is an unstable vertex then
   a) $\gamma(i) \in \gamma(v)$, for all $i \in F_\tau(v)$,
   b) all $\gamma(i)$, for $i \in F_\tau(v)$ are distinct.

Fix such a marked modular graph (τ, d, γ). The following stacks will be important in what follows:

1. $\overline{M}(\tau^s) = \prod_{v \in V_\tau^s} \overline{M}_{g(v), F_\tau(v)}$,
2. $\overline{M}(\mathbb{P}^r, \tau, d)$, which is defined as the fibered product

$$\begin{array}{ccc} \overline{M}(\mathbb{P}^r, \tau, d) & \longrightarrow & \prod_{v \in V_\tau} \overline{M}_{g(v), F_\tau(v)}(\mathbb{P}^r, d(v)) \\ \downarrow & & \downarrow \\ (\mathbb{P}^r)^{E_\tau} & \xrightarrow{\Delta} & (\mathbb{P}^r \times \mathbb{P}^r)^{E_\tau}, \end{array} \tag{4.4}$$

where the vertical maps are evaluation maps,

3. $\overline{M}(\mathbb{P}^r, \tau, d; \gamma)$, which is the substack of $\overline{M}(\mathbb{P}^r, \tau, d)$ defined by requiring that $f_{\partial(i)}(x_i) = \gamma(i) \in \{P_0, \ldots, P_r\}$, for all $i \in F_\tau$. Here $\partial(i)$ is the vertex incident with i, $f_{\partial(i)}$ is the stable map indexed by this vertex and x_i is the mark of the source curve of $f_{\partial(i)}$ indexed by i. Clearly, $\overline{M}(\mathbb{P}^r, \tau, d; \gamma)$ is a closed substack of $\overline{M}(\mathbb{P}^r, \tau, d)$.

Stacks of type $\overline{M}(\mathbb{P}^r, \tau, d)$ are studied in great detail in [4]. Given a collection $(C_v, x_i, f_v)_{v \in V_\tau, i \in F_\tau}$, representing an element of $\overline{M}(\mathbb{P}^r, \tau, d)$, we can associate a stable map in $\overline{M}_{g,n}(\mathbb{P}^r, d)$ by gluing, for every edge $\{i_1, i_2\}$ of τ, the curves $C_{\partial(i_1)}$ and $C_{\partial(i_2)}$ by identifying x_{i_1} with x_{i_2}. Doing this in families defines the morphism

$$\overline{M}(\mathbb{P}^r, \tau, d) \longrightarrow \overline{M}_{g,n}(\mathbb{P}^r, d). \tag{4.5}$$

In general, a morphism such as (4.5), giving rise to a boundary component of $\overline{M}_{g,n}(\mathbb{P}^r, d)$ is only a finite morphism. But because of the special nature of (τ, d) in our context, (4.5) is actually a finite étale morphism followed by a closed immersion. More precisely:

Proposition 4.1. *Let* $\mathrm{Aut}(\tau, d)$ *be the subgroup of the automorphism group of the modular graph* τ *preserving the degrees* d. *Then* $\mathrm{Aut}(\tau, d)$ *acts on* $\overline{M}(\mathbb{P}^r, \tau, d)$ *and (4.5) induces a closed immersion*

$$\overline{M}(\mathbb{P}^r, \tau, d)/Aut(\tau, d) \longrightarrow \overline{M}_{g,n}(\mathbb{P}^r, d).$$

Proof. One has to prove that any stable map in $\overline{M}_{g,n}(\mathbb{P}^r, d)$ of degeneracy type (τ, d) or worse, can be written uniquely (up to $\mathrm{Aut}(\tau, d)$) as the result of gluing a collection $(C_v, x_i, f_v)_{v \in V_\tau, i \in F_\tau}$. This is true because every stable vertex has degree 0 and no edge connects stable vertices.

Next, we shall construct a morphism $\overline{M}(\tau^s) \to \overline{M}(\mathbb{P}^r, \tau, d)$.
Let $(C_v, x_v)_{v \in V_\tau^s}$ be a collection of stable marked curves, $x_v = (x_i)_{i \in F_\tau(v)}$, in other words, a k-valued point of $\overline{M}(\tau^s)$. Then produce a collection of stable maps as follows:

1. for $v \in V_\tau$ a stable vertex, let $f_v : C_v \to \mathbb{P}^r$ by the constant map to $\gamma(v) \in \{P_0, \ldots, P_r\}$,
2. for $v \in V_\tau$ unstable, let $C_v = \mathbb{P}^1$ and f_v be

$$\begin{aligned} f_v : \mathbb{P}^1 &\longrightarrow \mathbb{P}^1 = \gamma(v) \subset \mathbb{P}^r \\ z &\longmapsto z^{d(v)}. \end{aligned}$$

Then put marks on $C_v = \mathbb{P}^1$: for each $i \in F_\tau(v)$ let $x_i \in C_v$ be equal to $0 = \langle 1, 0\rangle$ or $\infty = \langle 0, 1\rangle$, in the unique way such that $f_v(x_i) = \gamma(i)$.

This defines $(C_v, x_v, f_v)_{v\in V_\tau}$, an element of $\overline{M}(\mathbb{P}^r, \tau, d)(k)$. Again, this can be done in families and we obtain the desired morphism $\overline{M}(\tau^s) \to \overline{M}(\mathbb{P}^r, \tau, d)$.

This morphism is also a finite étale covering followed by a closed immersion:

Proposition 4.2. *Let $\mu = \prod_{v\in V_\tau^u} \mu_{d(v)}$, where $\mu_{d(v)}$ is the cyclic group of $d(v)$-th roots of 1. Let μ act trivially on $\overline{M}(\tau^s)$. Then we have a closed immersion*

$$\overline{M}(\tau^s)/\mu \longrightarrow \overline{M}(\mathbb{P}^r, \tau, d). \tag{4.6}$$

□

We can say more, because, in fact, the group $\operatorname{Aut}(\tau, d)$ acts on the morphism (4.6). More precisely,

Proposition 4.3. *The semidirect product $G = \mu \rtimes \operatorname{Aut}(\tau, d)$ acts on $\overline{M}(\tau^s)$ and (4.6) induces a closed immersion*

$$\overline{M}(\tau^s)/G \longrightarrow \overline{M}(\mathbb{P}^r, \tau, d)/\operatorname{Aut}(\tau, d).$$

□

Putting Propositions 4.1 and 4.3 together, we obtain the composition

$$\overline{M}(\tau^s)/G \longrightarrow \overline{M}(\mathbb{P}^r, \tau, d)/\operatorname{Aut}(\tau, d) \longrightarrow \overline{M}_{g,n}(\mathbb{P}^r, d), \tag{4.7}$$

(the composite arrow being labelled $\Phi_{(\tau,d,\gamma)}$)

which is a closed immersion.

Proposition 4.4. *Consider the group $T(k)$ acting on the set of isomorphism classes of $\overline{M}_{g,n}(\mathbb{P}^r, d)(k)$. An element of this set is fixed if and only if it is in the image of $\Phi_{(\tau,d,\gamma)}(k)$, for some marked modular graph (τ, d, γ) as described above.*

□

In this sense, the image of $\coprod \Phi$ is the fixed locus of $\overline{M}_{g,n}(\mathbb{P}^r, d)$. Thus we are justified in calling the image of $\Phi_{(\tau,d,\gamma)}$ the *fixed component* indexed by (τ, d, γ). But if we endow $\overline{M}(\tau^s)/G$ with the trivial action of T, then $\Phi_{(\tau,d,\gamma)}$ is not T-equivariant. To make it so, we have to pass to a larger torus.

Consider the character group $M \subset M_\mathbb{Q} = M \otimes_\mathbb{Z} \mathbb{Q}$ and let $\widetilde{M} = M + \sum_{v\in V_\tau^u} \frac{1}{d(v)}\lambda_v \subset M_\mathbb{Q}$, where λ_v is the character of T through which T acts on $\mathbb{P}^1 = L_{ij} = \gamma(v)$. Let $\widetilde{T}$ be the torus with character group $\widetilde{M}$. We have a finite homomorphism $\widetilde{T} \to T$. We can view passing from T to $\widetilde{T}$ as a way to make the character λ_v divisible by $d(v)$.

The torus $\widetilde{T}$ acts on $\overline{M}_{g,n}(\mathbb{P}^r, d)$ through $\widetilde{T} \to T$. We can now construct a 2-isomorphism θ in the diagram

$$\begin{array}{ccc} \widetilde{T} \times \overline{M}(\tau^s) & \xrightarrow{\mathrm{id} \times \Phi} & \widetilde{T} \times \overline{M}_{g,n}(\mathbb{P}^r, d) \\ \text{proj} \downarrow & \Downarrow \theta & \downarrow \text{action} \\ \overline{M}(\tau^s) & \xrightarrow{\Phi} & \overline{M}_{g,n}(\mathbb{P}^r, d). \end{array}$$

Let us describe θ on k-valued points. We need to define a natural transformation. So for each $(t, (C_v, x_v))$ of $\widetilde{T}(k) \times \overline{M}(\tau^s)(k)$ we need to define a morphism $\theta : t \cdot \Phi(C_v, x_v) \to \Phi(C_v, x_v)$. Using notation as above, we have $\Phi((C_v, x_v)_{v \in V_\tau^s}) = (C_v, x_v, f_v)_{v \in V_\tau}$ and $t \cdot \phi(C_v, x_v) = (C_v, x_v, t \circ f_v)$. Then

1. for $v \in V_\tau^s$, we let $\theta_v : C_v \longrightarrow C_v$ be the identity,
2. for $v \in V_\tau^u$, we let $\theta_v : C_v = \mathbb{P}^1 \longrightarrow C_v = \mathbb{P}^1$ be given by

$$z \longmapsto \frac{\lambda_v}{d(v)}(t) z,$$

which fits into the commutative diagram

$$\begin{array}{ccccc} C_v & \xrightarrow{f_v = (\,\cdot\,)^{d(v)}} & \gamma(v) & \xrightarrow{\subset} & \mathbb{P}^r \\ \theta_v \downarrow & & \downarrow \lambda_v & & \downarrow t \\ C_v & \xrightarrow{f_v = (\,\cdot\,)^{d(v)}} & \gamma(v) & \xrightarrow{\subset} & \mathbb{P}^r. \end{array}$$

Thus it is better to think of the image of $\Phi_{(\tau,d,\gamma)}$ as a fixed component of $\widetilde{T}$, rather than T, acting on $\overline{M}_{g,n}(\mathbb{P}^r, d)$.

Going back to Diagram (4.2), we can now say what Y is. We shall use $Y = \coprod_{(\tau,d,\gamma)} \overline{M}(\tau^s)/G_{(\tau,d)}$. The integrals on $\overline{M}(\tau^s)/G_{(\tau,d)}$ will be evaluated on $\overline{M}(\tau^s)$. This leads to the correction factor

$$\frac{1}{\# G_{(\tau,d)}} = \frac{1}{\# \operatorname{Aut}(\tau, d)} \prod_{v \in V_\tau^u} \frac{1}{d(v)}.$$

We shall next show how to obtain regular immersions $\nu : V \to W$ as in Diagram (4.2). As mentioned above, we can treat each fixed component separately. We will proceed in several steps, corresponding to the following factorization of Φ:

$$\overline{M}(\tau^s)/G \xrightarrow{\text{III.}} \overline{M}(\mathbb{P}^r, \tau, d; \gamma)/A \xrightarrow{\text{II.}} \overline{M}(\mathbb{P}^r, \tau, d)/A \xrightarrow{\text{I.}} \overline{M}_{g,n}(\mathbb{P}^r, d),$$

where $A = \operatorname{Aut}(\tau, d)$. For each step we shall construct a suitable ν and then determine $\nu^![\overline{M}_{g,n}(\mathbb{P}^r, d)_T]$ and $\frac{1}{e(g^* N)}$.

The first step

We use the following diagram for (4.2):

$$\begin{array}{ccc} \overline{M}(\mathbb{P}^r, \tau, d)/\operatorname{Aut}(\tau, d) & \longrightarrow & \overline{M}_{g,n}(\mathbb{P}^r, d) \\ {\scriptstyle g}\downarrow & & \downarrow \\ \mathfrak{M}(\tau)/\operatorname{Aut}(\tau, \mathfrak{d}) & \xrightarrow{\ \nu\ } & \mathfrak{M}_{\mathfrak{g},\mathfrak{n}} \end{array}$$

We note that this diagram is not cartesian, but $\overline{M}(\mathbb{P}^r, \tau, d)/\operatorname{Aut}(\tau, d)$ is open and closed in the cartesian product. Since we are only interested in the (τ, d, γ)-component of the fixed locus at the moment, this is sufficient. Here $\mathfrak{M}_{\mathfrak{g},\mathfrak{n}}$ stands for the (highly non-separated) Artin stack of prestable curves of genus g with n marks. Moreover,

$$\mathfrak{M}(\tau) = \prod_{\mathfrak{v} \in \mathfrak{V}_\tau} \mathfrak{M}_{\mathfrak{g}(\mathfrak{v}), \mathfrak{F}_\tau(\mathfrak{v})}$$

and the morphism $\mathfrak{M}(\tau) \to \mathfrak{M}_{\mathfrak{g},\mathfrak{n}}$ is given by gluing according to the edges of τ. The vertical maps are given by forgetting the map, retaining the prestable curve, without stabilizing. The diagram is $\widetilde{T}$-equivariant, if we endow $\mathfrak{M}(\tau)$ and $\mathfrak{M}_{\mathfrak{g},\mathfrak{n}}$ with the trivial $\widetilde{T}$-action. We also note that ν is not a closed immersion, but certainly a regular local immersion (for this terminology see [16]), which is sufficient for our purposes.

It is a general fact about virtual fundamental classes, used in the proof of the WDVV-equation, that the Gysin pullback along ν preserves virtual fundamental classes:

$$\nu^![\overline{M}_{g,n}(\mathbb{P}^r, d)_T] = [\overline{M}(\mathbb{P}^r, \tau, d)/Aut(\tau, d)].$$

(One way to define the virtual fundamental class of $\overline{M}(\mathbb{P}^r, \tau, d)$ is to set it equal to the Gysin pullback via Δ of the product of virtual fundamental classes in Diagram 4.4.)

The normal bundle of $\mathfrak{M}(\tau)$ in $\mathfrak{M}_{\mathfrak{g},\mathfrak{n}}$ splits into a direct sum of line bundles, one summand for each edge of τ. For the edge $\{i_1, i_2\}$, the normal line bundle is

$$x_{i_1}^*(\omega^\vee) \otimes x_{i_2}^*(\omega^\vee),$$

where x_{i_1} and x_{i_2} are the sections of the universal curves corresponding to the flags i_1 and i_2 of τ and ω is the relative dualizing sheaf of the universal curve, whose dual, $\omega^\vee$ is the relative tangent bundle. We use notation c_i for the Chern class of the line bundle $x_i^*(\omega)$ on $\mathfrak{M}(\tau)$. Then

$$\frac{1}{e(g^*N)} = \prod_{\{i_1,i_2\} \in E_\tau} \frac{1}{-c_{i_1} - c_{i_2}}. \tag{4.8}$$

The second step

Instead of considering $\overline{M}(\mathbb{P}^r, \tau, d; \gamma)/\operatorname{Aut}(\tau, d) \to \overline{M}(\mathbb{P}^r, \tau, d)/\operatorname{Aut}(\tau, d)$, we shall consider

$$\overline{M}(\mathbb{P}^r, \tau, d; \gamma) \longrightarrow \overline{M}(\mathbb{P}^r, \tau, d). \tag{4.9}$$

We call an edge (flag, tail) of τ *stable*, if it meets a stable vertex. Otherwise, we call it unstable. We shall need to consider stacks of the following type:

$$\overline{M}_{0,S}(\mathbb{P}^r, d; \gamma(S)),$$

where S is a finite set (we only consider the cases that S has 1 or 2 elements) and $\gamma : S \to \{P_0, \ldots, P_n\}$ is a map. The stack $\overline{M}_{0,S}(\mathbb{P}^r, d; \gamma(S)) \subset \overline{M}_{0,S}(\mathbb{P}^r, d)$ is the closed substack of stable maps f, defined by requiring that $f(x_i) = \gamma(i)$, for all $i \in S$.

Lemma 4.1. *For $\#S \leq 2$, the stack $\overline{M}_{0,S}(\mathbb{P}^r, d; \gamma(S))$ is smooth of the expected dimension $\dim \overline{M}_{0,S}(\mathbb{P}^r, d) - r\#S$.*

Proof. This follows from $H^1(C, f^*T_{\mathbb{P}^r}(-x_1 - x_2)) = 0$, for a stable map $f : C \to \mathbb{P}^r$ in $\overline{M}_{0,S}(\mathbb{P}^r, d; \gamma(S))$.

Note that we have

$$\overline{M}(\mathbb{P}^r, \tau, d; \gamma) = \prod_{v \in V^s_\tau} \overline{M}_{g(v), F_\tau(v)} \times \prod_{v \in V^u_\tau} \overline{M}_{0, F_\tau(v)}(\mathbb{P}^r, d(v); \gamma F_\tau(v)),$$

and in particular, that $\overline{M}(\mathbb{P}^r, \tau, d; \gamma)$ is smooth of the 'expected' dimension

$$\sum_{v \in V^s_\tau} \dim \overline{M}_{g(v), F_\tau(v)} + \sum_{v \in V^u_\tau} \dim \overline{M}_{g(v), F_\tau(v)}(\mathbb{P}^r, d(v)) - r\#F^u_\tau.$$

Now the morphism (4.9) fits into the $\tilde{T}$-equivariant cartesian diagram

$$\begin{array}{ccccc}
\overline{M}(\mathbb{P}^r, \tau, d; \gamma) & \longrightarrow & \overline{M}(\mathbb{P}^r, \tau, d) & \longrightarrow & \prod_{v \in V_\tau} \overline{M}_{g(v), F_\tau(v)}(\mathbb{P}^r, d(v)) \\
\downarrow g & & \downarrow & & \downarrow e \times p \\
(\mathbb{P}^r)^{F^s_\tau} & \xrightarrow{\nu} & (\mathbb{P}^r)^{E_\tau} \times (\mathbb{P}^r)^{S_\tau} \times (\mathbb{P}^r)^{V^s_\tau} & \xrightarrow{\Delta \times \mathrm{id}} & (\mathbb{P}^r)^{F_\tau} \times (\mathbb{P}^r)^{V^s_\tau} \\
\downarrow & & & & \downarrow q \\
pt & & \xrightarrow{\nu_1} & & (\mathbb{P}^r)^{F^u_\tau} \times (\mathbb{P}^r)^{V^s_\tau}.
\end{array} \tag{4.10}$$

(with $g_1 : \overline{M}(\mathbb{P}^r, \tau, d; \gamma) \to pt$ the composite of the left vertical arrows)

The morphism $e \times p$ is the product of the evaluation morphism

$$e : \prod_{v \in V_\tau} \overline{M}_{g(v), F_\tau(v)}(\mathbb{P}^r, d(v)) \longrightarrow (\mathbb{P}^r)^{F_\tau}$$

and the projection

$$\prod_{v\in V_\tau} \overline{M}_{g(v),F_\tau(v)}(\mathbb{P}^r, d(v)) =$$
$$\prod_{v\in V_\tau^u} \overline{M}_{0,F_\tau(v)}(\mathbb{P}^r, d(v)) \times \prod_{v\in V_\tau^s} (\overline{M}_{g(v),F_\tau(v)} \times \mathbb{P}^r) \xrightarrow{p} (\mathbb{P}^r)^{V_\tau^s}.$$

The morphism $\Delta \times \mathrm{id}$ is the product of the diagonal

$$(\mathbb{P}^r)^{E_\tau} \xrightarrow{\Delta} (\mathbb{P}^r \times \mathbb{P}^r)^{E_\tau} = (\mathbb{P}^r)^{F_\tau - S_\tau}$$

and the identity on $(\mathbb{P}^r)^{S_\tau} \times (\mathbb{P}^r)^{V_\tau^s}$. The square to the upper right of (4.10) is just a base change of the defining square of $\overline{M}(\mathbb{P}^r, \tau, d)$. The morphism ν is the product of the identity

$$(\mathbb{P}^r)^{F_\tau^s} \longrightarrow (\mathbb{P}^r)^{E_\tau^s} \times (\mathbb{P}^r)^{S_\tau^s}$$

and the morphism

$$\gamma : pt \longrightarrow (\mathbb{P}^r)^{E_\tau^u} \times (\mathbb{P}^r)^{S_\tau^u} \times (\mathbb{P}^r)^{V_\tau^s},$$

induced by the marking γ on the graph (τ, d). The morphism g is given by evaluation at the points corresponding to stable flags and is, in fact, constant. The morphism q projects out the factors corresponding to stable flags. Finally, ν_1 is given, again, by γ.

The stack in the upper right corner of (4.10) is smooth, but not of the 'expected' dimension. It has a virtual fundamental class given by

$$\left(\prod_{v\in V_\tau^s} e(H(v)^\vee \boxtimes T_{\mathbb{P}^r}(\gamma(v)))\right) [\prod_{v\in V_\tau} \overline{M}_{g(v),F_\tau(v)}(\mathbb{P}^r, d(v))_T]. \tag{4.11}$$

Here $H(v)$ is the 'Hodge bundle' corresponding to the vertex v. If $\pi_v : C_v \to \overline{M}_{g(v),F_\tau(v)}$ is the universal curve, then $H(v) = \pi_{v*}(\omega_{C_v})$, where ω_{C_v} is the relative dualizing sheaf.

It is part of the general compatibilities of virtual fundamental classes that (4.11) pulled back via $(\Delta \times \mathrm{id})^!$ gives the virtual fundamental class of $\overline{M}(\mathbb{P}^r, \tau, d)$. Now because there is no excess intersection in the lower rectangle of (4.10), we get the same class in $\overline{M}(\mathbb{P}^r, \tau, d; \gamma)$ by pulling back (4.11) in two steps via $(\Delta \times \mathrm{id})^!$ and $\nu^!$ or in one step via $\nu_1^!$. Thus

$$\nu^![\overline{M}(\mathbb{P}^r, \tau, d)_T] = \nu_1^!(\text{the class (4.11)}).$$

But by Lemma 4.1, the big (total) square in (4.10) has no excess intersection either. Thus $\nu_1^!$(the class (4.11)) is equal to

$$\begin{aligned}
&\nu^![\overline{M}(\mathbb{P}^r, \tau, d)_T] \\
&= \left(\prod_{v\in V_\tau^s} e(H(v)^\vee \boxtimes T_{\mathbb{P}^r}(\gamma(v)))\right) [\overline{M}(\mathbb{P}^r, \tau, d; \gamma)_T] \\
&= \prod_{v\in V_\tau^s} \prod_{i\neq\gamma(v)} (\lambda_i - \lambda_{\gamma(v)})^{g(v)} c_t(H(v))|_{t=\frac{1}{\lambda_i - \lambda_{\gamma(v)}}} [\overline{M}(\mathbb{P}^r, \tau, d; \gamma)_T].
\end{aligned}$$

Because the morphism g in (4.10) is constant, $g^*(N)$ is constant and so $e(g^*(N))$ is just the product of the weights of T on g^*N. Thus

$$\begin{aligned}\frac{1}{e(g^*N)} & \qquad\qquad\qquad\qquad\qquad\qquad\qquad\qquad\qquad\qquad (4.12)\\ &= \prod_{j\in E^u_\tau} \frac{1}{e(T_{\mathbb{P}^r}(\gamma(j)))} \prod_{j\in S^u_\tau} \frac{1}{e(T_{\mathbb{P}^r}(\gamma(j)))} \prod_{j\in V^s_\tau} \frac{1}{e(T_{\mathbb{P}^r}(\gamma(v)))}\\ &= \prod_{j\in E^u_\tau\cup S^u_\tau} \prod_{i\neq\gamma(j)} \frac{1}{\lambda_i-\lambda_{\gamma(j)}} \prod_{v\in V^s_\tau} \prod_{i\neq\gamma(v)} \frac{1}{\lambda_i-\lambda_{\gamma(v)}}\end{aligned}$$

The third step

We shall consider the morphism

$$\overline{M}(\tau^s)/\mu \longrightarrow \overline{M}(\mathbb{P}^r,\tau,d;\gamma) = \overline{M}(\tau^s) \times \prod_{v\in V^u_\tau} \overline{M}_{0,F_\tau(v)}(\mathbb{P}^r,d(v);\gamma F_\tau(v)),$$

which we may insert into the $\widetilde{T}$-equivariant cartesian diagram of smooth stacks without excess intersection

$$\begin{array}{ccc}\overline{M}(\tau^s)/\mu & \longrightarrow & \overline{M}(\tau^s) \times \prod_{v\in V^u_\tau} \overline{M}_{0,F_\tau(v)}(\mathbb{P}^r,d(v);\gamma F_\tau(v))\\ \Big\downarrow g & & \Big\downarrow \\ \prod_{v\in V^u_\tau} B\mu_{d(v)} & \xrightarrow{\ \nu\ } & \prod_{v\in V^u_\tau} \overline{M}_{0,F_\tau(v)}(\mathbb{P}^r,d(v);\gamma F_\tau(v)).\end{array} \qquad (4.13)$$

It follows that $\nu^![\overline{M}(\mathbb{P}^r,\tau,d;\gamma)_T] = [\overline{M}(\tau^s)/\mu]$.

To calculate the normal bundle of ν, factor ν into $\#V^u_\tau$ morphisms and thus reduce to considering the morphism

$$B\mu_{d(v)} \longrightarrow \overline{M}_{0,F_\tau(v)}(\mathbb{P}^r,d(v);\gamma F_\tau(v)).$$

To fix notation, let us consider a positive integer d and

$$B\mu_d \longrightarrow \overline{M}_{0,2}(\mathbb{P}^r,d;P_0,P_1) \qquad (4.14)$$

(the case of v having valence 1 we leave to the reader). The stack

$$\overline{M}_{0,2}(\mathbb{P}^r,d;P_0,P_1) \subset \overline{M}_{0,2}(\mathbb{P}^r,d)$$

is defined by requiring the image of the first marked point to be $P_0 \in \mathbb{P}^r$ and the image of the second marked point to be $P_1 \in \mathbb{P}^r$.

The particular stable map

$$\begin{aligned} f : \mathbb{P}^1 &\longrightarrow \mathbb{P}^1 = L_{01} \subset \mathbb{P}^r \\ z &\longmapsto z^d \end{aligned} \tag{4.15}$$

(where $x_1 = 0$ and $x_2 = \infty$ are the marks on $\mathbb{P}^1$) is the unique fixed point of $\widetilde{T}$ on $\overline{M}_{0,2}(\mathbb{P}^r, d; P_0, P_1)$ and gives rise to the morphism (4.14).The normal bundle to (4.14) is the tangent space to (4.15) in $\overline{M}_{0,2}(\mathbb{P}^r, d; P_0, P_1)$ and hence equal to

$$H^0(\mathbb{P}^1, f^*T_{\mathbb{P}^r}(-0-\infty)) \;/\; H^0(\mathbb{P}^1, T_{\mathbb{P}^1}(-0-\infty)). \tag{4.16}$$

We calculate the weights of $H^0(\mathbb{P}^1, f^*T_{\mathbb{P}^r}(-0-\infty))$ and $H^0(\mathbb{P}^1, T_{\mathbb{P}^1}(-0-\infty))$ using Example 3.1.

Let (α_i) denote the weights of $\widetilde{T}$ on $H^0(\mathbb{P}^1, f^*T_{\mathbb{P}^r}(-0-\infty))$. The torus $\widetilde{T}$ acts on $\mathbb{P}^1$ via the character $\omega = \frac{\lambda_1 - \lambda_0}{d}$. We also need the weights of $f^*T_{\mathbb{P}^r}(-0-\infty)(0)$ and $f^*T_{\mathbb{P}^r}(-0-\infty)(\infty)$. To calculate these, note that $T_{\mathbb{P}^r}(P_0)$ has weights $(\lambda_i - \lambda_0)_{i\neq 0}$ and $T_{\mathbb{P}^r}(P_1)$ has weights $(\lambda_i - \lambda_1)_{i \neq 1}$. The same holds after applying f^*. Twisting by (-0) and $(-\infty)$ changes the weights by $T_{\mathbb{P}^1}(0)$ and $T_{\mathbb{P}^1}(\infty)$, respectively. But $T_{\mathbb{P}^1}(0)$ has weight $\frac{\lambda_1 - \lambda_0}{d}$ and $T_{\mathbb{P}^1}(\infty)$ has weight $\frac{\lambda_0 - \lambda_1}{d}$. Thus the weights of $f^*T_{\mathbb{P}^r}(-0-\infty)$ are $(\lambda_i - \lambda_0 - \omega)_{i \neq 0}$ at (0) and $(\lambda_i - \lambda_1 + \omega)_{i\neq 1}$ at (∞). Then by Example 3.1 we have

$$\begin{aligned} \sum e^{\alpha_i} &= \tfrac{1}{1-e^{-\omega}} \textstyle\sum_{i\neq 0} e^{\lambda_i - \lambda_0 - \omega} + \tfrac{1}{1-e^{\omega}} \textstyle\sum_{i \neq 1} e^{\lambda_i - \lambda_1 + \omega} \\ &= 1 + \textstyle\sum_{\text{all } i} e^{\lambda_i - \lambda_0} \left(\tfrac{e^{-\omega}}{1-e^{-\omega}} + \tfrac{e^{\omega + \lambda_0 - \lambda_1}}{1-e^{\omega}} \right) \\ &= 1 + \textstyle\sum_i e^{\lambda_i - \lambda_0} \left(\tfrac{e^{(1-d)\omega}}{1-e^{\omega}} + \tfrac{e^{-\omega}}{1-e^{-\omega}} \right) \\ &= 1 + \textstyle\sum_i e^{\lambda_i - \lambda_0} \sum_{n=1}^{d-1} e^{-n\omega} \\ &= 1 + \textstyle\sum_{i=0}^r \sum_{\substack{n+m=d \\ n,m\neq 0}} e^{\lambda_i - \frac{n}{d}\lambda_1 - \frac{m}{d}\lambda_0}, \end{aligned} \tag{4.17}$$

by the 'useful formula' mentioned in Exercise 3.1.

Similarly, $H^0(\mathbb{P}^1, T_{\mathbb{P}^1}(-0-\infty))$ is one-dimensional and has weight 0, so that the weights of (4.16) are

$$\left(\lambda_i - \frac{n}{d}\lambda_1 - \frac{m}{d}\lambda_0 \right)_{\substack{i=0,\ldots,r \\ n+m=d \\ n,m>0}}.$$

We deduce that for the normal bundle N of the morphism ν in (4.13) we have

$$\frac{1}{e(g^*N)} = \tag{4.18}$$

$$\prod_{\substack{v\in V_\tau^u \\ |v|=2 \\ \gamma(v)=L_{ab}}} \prod_{i=0}^{r} \prod_{\substack{n+m=d(v) \\ n,m\neq 0}} \frac{1}{\lambda_i - \frac{n}{d}\lambda_a - \frac{m}{d}\lambda_b}$$

$$\prod_{\substack{v\in V^u_\tau\\ |v|=1\\ \gamma(v)=L_{ab}}}\left(\prod_{\substack{n+m=d(v)\\ n,m\neq 0}}\frac{1}{\lambda_a-\frac{n}{d}\lambda_a-\frac{m}{d}\lambda_b}\prod_{\substack{n+m=d(v)\\ n\neq 0,1}}\frac{1}{\lambda_b-\frac{n}{d}\lambda_a-\frac{m}{d}\lambda_b}\right.$$

$$\left.\prod_{\substack{i=0\\ i\neq a,b}}^{r}\prod_{\substack{n+m=d(v)\\ n\neq 0}}\frac{1}{\lambda_i-\frac{n}{d}\lambda_a-\frac{m}{d}\lambda_b}\right).$$

Conclusion

We have now completed the computation of the right hand side of (4.3). We have

$$\begin{aligned}&\nu^![\overline{M}_{g,n}(\mathbb{P}^r,d)_T]_{(\tau,d,\gamma)}\\ &=\prod_{v\in V^s_\tau}\prod_{i\neq\gamma(v)}(\lambda_i-\lambda_{\gamma(v)})^{g(v)}\; c_t(H(v))|_{t=\frac{1}{\lambda_i-\lambda_{\gamma(v)}}}\;[\overline{M}(\tau^s)/G_{(\tau,d)}]\end{aligned}$$

and $1/e(g^*N)_{(\tau,d,\gamma)}$ is the product of the three contributions (4.8), (4.12) and (4.18). When pulling back the contribution (4.8), which is

$$\prod_{\{i_1,i_2\}\in E_\tau}\frac{1}{-c_{i_1}-c_{i_2}},$$

to $\overline{M}(\tau^s)$, we replace $-c_i$, for an unstable flag $i\in F_\tau$ by the weight of $\widetilde{T}$ on $T_{\mathbb{P}^1}(x_i)$. This weight is $\frac{\lambda_j-\lambda_i}{d}$, where $\{i,j\}$ is the edge containing i.

Thus we finally arrive at the localization formula for Gromov-Witten invariants of $\mathbb{P}^r$. Our graph formalism is well-suited for our derivation of the formula. To actually perform calculations, it is more convenient to translate our formalism into the simpler graph formalism introduced by Kontsevich [13]. But this, of course, just amounts to a reindexing of our sum.

Bibliography

1. K. Behrend. Gromov-Witten invariants in algebraic geometry. *Invent. Math.*, 127(3):601–617, 1997.
2. K. Behrend. Algebraic Gromov-Witten invariants. In M. Reid, editor, *Proceedings of the conference on Algebraic Geometry, Warwick 1996.* 1999.
3. K. Behrend and B. Fantechi. The intrinsic normal cone. *Invent. Math.*, 128(1):45–88, 1997.
4. K. Behrend and Yu. Manin. Stacks of stable maps and Gromov-Witten invariants. *Duke Math. J.*, 85(1):1–60, 1996.
5. D. Edidin and W. Graham. Equivariant intersection theory. *Invent. math.*, 131:595–634, 1998.

6. D. Edidin and W. Graham. Localization in equivariant intersection theory and the Bott residue formula. *Amer. J. Math.*, 120:619–636, 1998.
7. C. Faber and R. Pandharipande. Hodge integrals and Gromov-Witten theory. Preprint, math.AG/9810173.
8. C. Faber and R. Pandharipande. Hodge integrals, partition matrices, and the λ_g conjecture. Preprint, math.AG/9908052.
9. W. Fulton. *Intersection Theory*. Ergebnisse der Mathematik und ihrer Grenzgebiete 3. Folge Band 2. Springer-Verlag, Berlin, Heidelberg, New York, Tokyo, 1984.
10. W. Fulton and R. Pandharipande. Notes on stable maps and quantum cohomology. In *Algebraic geometry—Santa Cruz 1995*, pages 45–96. Amer. Math. Soc., Providence, RI, 1997.
11. A. Givental. Equivariant Gromov-Witten invariants. *Internat. Math. Res. Notices*, 1996(13):613–663.
12. T. Graber and R. Pandharipande. Localization of virtual classes. *Invent. math.*, 135:487–518, 1999.
13. M. Kontsevich. Enumeration of rational curves via torus actions. In *The moduli space of curves (Texel Island, 1994)*, pages 335–368. Birkhäuser Boston, Boston, MA, 1995.
14. M. Kontsevich and Yu. Manin. Gromov-Witten classes, quantum cohomology, and enumerative geometry. *Communications in Mathematical Physics*, 164:525–562, 1994.
15. J. Li and G. Tian. Virtual moduli cycles and Gromov-Witten invariants of algebraic varieties. *J. Amer. Math. Soc.*, 11(1):119–174, 1998.
16. A. Vistoli. Intersection theory on algebraic stacks and on their moduli spaces. *Invent. math.*, 97:613–670, 1989.

Fields, Strings and Branes

César Gómez[1] and Rafael Hernández[2]

[1] Instituto de Matemáticas y Física Fundamental, CSIC Serrano 123, 28006 Madrid, Spain
[2] Instituto de Física Teórica, C-XVI, Universidad Autónoma de Madrid Cantoblanco, 28049 Madrid, Spain

What is your aim in phylosophy?
To show the fly the way out of the fly-bottle.

Wittgenstein. Philosophycal Investigations, 309.

Introduction

The great challenge of high energy theoretical physics is finding a consistent theory of quantum gravity. For the time being, string theory is the best candidate at hand. Many phisicists think that the solution to quantum gravity will have little, if any, practical implications in our daily way of doing physics; others, more optimistic, or simply with a less practical approach to science, hope that the forthcoming theory of quantum gravity will provide a new way of thinking of quantum physics. At present, string theory is an easy target for the criticisms of pragmatics, as no experimental evidence is yet available; however, but it is also a rich and deep conceptual construction where new ways of solving longstanding theoretical problems in quantum field theory are starting to emerge. Until now, progress in string theory is mostly "internal", a way to evolve very similar to the one underlying evolution in pure mathematics. This is not necessarily a symptom of decadence in what is traditionally considered as an experimeantal science, but maybe the only possible way to improve physical intuition in the quantum realm.

Until very recently, most of the work in string theory was restricted to perturbation theory. Different string theories are, from this perturbative point of view, defined by two dimensional field theories, satisfying a certain set of constraints such as conformal and modular invariance. Different orders in the string perturbative expansion are obtained by working out these two dimensional conformal field theories on Riemann surfaces of different genus, and string amplitudes become good measures on the moduli space of these surfaces. This set of rules constitutes what we now call the "world-sheet" approach to string theory. From this perturbative point of view, we can think of many different string theories, as many as two dimensional conformal field theories, with an appropiate value of the central extension, which is determined by the generic constraint that amplitudes should define good measures on the moduli of Riemann surfaces. Among these conformal field theories, of special interested are the ones possessing a spacetime interpretation, which means that we can interpret them as describing the dynamics of strings moving in a definite target spacetime. Different string theories will then be defined as different types of strings moving in the same spacetime. Using this definition, we find, for instance, four different types of closed superstring theories (type IIA, type IIB, $E_8 \times E_8$ heterotic and $SO(32)$ heterotic) and one open superstring. However, this image of string theory has been enormously modified in the last few years, due to the clear emergence of duality symmetries. These symmetries, of two different species, perturbative and non perturbative, relate through equivalence relations a string theory on a particular spacetime to a string theory on some different spacetime. When this equivalence is perturbative, it can be proved in the genus expansion, which in practice means a general type of Montonen-Olive duality for the two dimensional conformal field theory. These duality symmetries are usually refered to as T-duality. A more ambitious type of duality relation between string theories is known as S-duality, where the equivalence is pretended to be non perturbative, and where a transformation from strongly to weakly coupled string theory is involved. Obviously, the first thing needed in order to address non perturbative duality symmetries is searching for a definition of string theory beyond perturbation theory, i. e., beyond the worldsheet approach; it is in this direction where the most ambitious program of research in string theory is focussing.

An important step in this direction comes of course from the discovery of D-branes. These new objects, which appear as necessary ingredients for extending T-duality to open strings, are sources for the Ramond fields in string theory, a part of the string spectrum not coupling, at the worldsheet level, to the string, and that are therefore not entering the allowed set of backgrounds used in the definition of the two dimensional conformal field theory. Thus, adding this backgrounds is already going beyond the worldsheet point of view and, therefore, constitutes an open window for the desired non perturbative definition of string theory.

Maybe the simplest way to address the problem of how a non perturbative definition of string theory will look like is wondering about the strong coupled behaviour of strings. This question becomes specially neat if the string theory chosen is the closed string of type IIA, where the string coupling constant can be related to the metric of eleven dimensional supergravity, so that the strongly coupled string theory can be understood as a new eleven dimensional theory, M-theory. When thinking about the relation between D-branes and M-theory or, more precisely, trying to understand the way D-branes dynamics should be used in order to understand the eleven dimensional dynamics describing the strong coupling regime of string theory, a good answer comes again from the misterious, for a while, relation between type IIA strings and eleven dimensional supergravity: the Kaluza-Klein modes in ten dimensions are the D-0brane sources for the Ramond $U(1)$ field. What makes this, superficially ordinary Kaluza-Klein modes, very special objects is its nature of D-branes. In fact, D-branes are sources for strings, powerful enough to provide the whole string spectrum.

A very appealing way to think of these D-0branes comes recently under the name of M(atrix) theory. The phylosophical ground for M(atrix) theory goes back to the holographic principle, based on black hole bounds on quantum information packing in space. From this point of view, the hologram of eleven dimensional M-theory is a ten dimensional theory for the peculiar set of ten dimensional degrees of freedom in terms of which we can codify all eleven dimensional physics. M(atrix) theory is the conjecture that D-0brane dynamics, which is a very special type of matrix quantum mechanics, is the correct hologram of the unknown eleven dimensional M-theory. We do not know the non perturbative region of string theory, but it seems we have already its healthy radiography.

These lectures were originally adressed to mathematics audience. The content covered along them is of course only a very small part of the huge amount of material growing around string theory on these days, and needless to say that it reflects the personal point of view of the authors. References are certainly not exhaustive, so that we apologize for this in advance.

Last, but not least, C. G. would like to thank the organizers and participants of the CIME school for suggestions and interesting questions, most of them yet unanswered in the text.

1. Chapter I

1.1 Dirac Monopole.

Maxwell's equations in the absence of matter,

$$\begin{aligned} \nabla \mathbf{E} &= 0, & \nabla \mathbf{B} &= 0, \\ \nabla \times \mathbf{B} - \frac{\partial \mathbf{E}}{\partial t} &= 0, & \nabla \times \mathbf{E} + \frac{\partial \mathbf{B}}{\partial t} &= 0, \end{aligned} \tag{1.1}$$

are invariant under the duality transformation

$$\begin{aligned} \mathbf{E} &\longrightarrow -\mathbf{B}, \\ \mathbf{B} &\longrightarrow \mathbf{E}, \end{aligned} \tag{1.2}$$

or, equivalently,

$$\begin{aligned} F^{\mu\nu} &\longrightarrow {}^*F^{\mu\nu}, \\ {}^*F^{\mu\nu} &\longrightarrow -F^{\mu\nu}, \end{aligned} \tag{1.3}$$

with ${}^*F^{\mu\nu} \equiv \tilde{F}^{\mu\nu} = \frac{1}{2}\epsilon^{\mu\nu\rho\sigma}F_{\rho\sigma}$ the Hodge dual of $F^{\mu\nu} = \partial^\mu A^\nu - \partial^\nu A^\mu$. In the pressence of both electric and magnetic matter, Maxwell's equations become

$$\begin{aligned} \partial_\nu F^{\mu\nu} &= -j^\mu \\ \partial_\nu {}^*F^{\mu\nu} &= -k^\mu, \end{aligned} \tag{1.4}$$

and (1.2) must be generalized with a transformation law for the currents,

$$\begin{aligned} F^{\mu\nu} &\rightarrow^* F^{\mu\nu} & j^\mu &\rightarrow k^\mu \\ {}^*F^{\mu\nu} &\rightarrow -F^{\mu\nu} & k^\mu &\rightarrow -j^\mu. \end{aligned} \tag{1.5}$$

As is clear from the definition of $F^{\mu\nu}$, the existence of magnetic sources (monopoles) [1] requires dealing with singular vector potentials. The appropiate mathematical language for describing these vector potentials is that of fiber bundles [2].

To start with, we will consider $U(1)$ bundles on the two sphere S^2. Denoting $H^\pm$ the two hemispheres, with $H^+ \cap H^- = S^1$, the $U(1)$ bundle is defined by

$$g_\pm = e^{i\psi_\pm} \tag{1.6}$$

$U(1)$ valued functions on the two hemispheres and such that on the S^1 equator

$$e^{i\psi_+} = e^{in\varphi}e^{i\psi_-}, \tag{1.7}$$

with φ the equatorial angle, and n some integer number characterizing the $U(1)$ bundle. Notice that n defines the winding number of the map

$$e^{in\varphi} : S^1 \longrightarrow U(1), \tag{1.8}$$

classified under the first homotopy group

$$\Pi_1(U(1)) \simeq \Pi_1(S^1) \simeq \mathbf{Z}. \tag{1.9}$$

Using the $U(1)$ valued functions $g_\pm$, we can define pure gauge connections, $A^\pm_\mu$, on $H^\pm$ as follows:

$$A^\pm_\mu = g^{-1}_\pm \partial_\mu g_\pm. \tag{1.10}$$

From (1.7) we easily get, on the equator,

$$A^+ = A^- + n\varphi, \tag{1.11}$$

and, through Stokes theorem, we get

$$\int_{S^2} F = \frac{1}{2\pi}[\int_{H^+} dA^+ + \int_{H^-} dA^-] = \frac{1}{2\pi}\int_{S^1} A^+ - A^- = n \tag{1.12}$$

identifying the winding number n with the magnetic charge of the monopole.

In quantum mechanics, the presence of a magnetic charge implies a quantization rule for the electric charge. In fact, as we require that the Schrödinger wave function, for an electric field in a monopole background, be single valued, we get

$$\exp \frac{ie}{2\pi\hbar} \oint_\Gamma A = 1, \tag{1.13}$$

with Γ a non contractible loop. In the presence of a magnetic charge $m \equiv \frac{1}{2\pi}\oint_\Gamma A$ we get Dirac's quantization rule [1],

$$em = nh. \tag{1.14}$$

Notice that the quantization rule (1.14) is equivalent to the definition (1.12) of the magnetic charge as a winding number or, more precisely, as minus the first Chern class of a $U(1)$ principal bundle on S^2. In fact, the single valuedness of the Schrödinger wave function is equivalent to condition (1.7), where we have required n to be integer for the transition function, in order to get a manifold. The gauge connection used in (1.12) was implicitely defined as eA, with A standing for the physical gauge configuration appearing in the Schrödinger equation. From now on, we will use units with $\hbar = 1$.

The main problem with Dirac monopoles is that they are not part of the spectrum of standard QED. In order to use the idea of duality as a dynamical symmetry, we need to search for more general gauge theories, containing in the spectrum magnetically charged particles [3, 4, 5].

1.2 The 't Hooft-Polyakov Monopole.

Let us consider the Georgi-Glashow model [6] for $SU(2)$,

$$\mathcal{L} = -\frac{1}{4}F_a^{\mu\nu}F_{a\mu\nu} + \frac{1}{2}\mathcal{D}^\mu\phi \cdot \mathcal{D}_\mu\phi - V(\phi), \quad a = 1,2,3, \tag{1.15}$$

with the Higgs field in the adjoint representation, $\mathcal{D}^\mu\phi_a \equiv \partial^\mu\phi_a - g\epsilon_{abc}A_b^\mu\phi_c$ the covariant derivative, and $V(\phi)$ the Higgs potential,

$$V(\phi) = \frac{1}{4}\lambda(\phi^2 - a^2)^2. \tag{1.16}$$

with $\lambda > 0$ and a arbitrary constants.

A classical vacuum configuration is given by

$$\phi_a = a\delta_{a3}, \qquad A_\mu^a = 0. \tag{1.17}$$

We can now define the vacuum manifold $\mathcal{V}$ as

$$\mathcal{V} = \{\phi, V(\phi) = 0\}, \tag{1.18}$$

which in this case is a 2-sphere of radius equal a. A necessary condition for a finite energy configuration is that at infinity, the Higgs field ϕ takes values in the vacuum manifold $\mathcal{V}$,

$$\phi : S_\infty^2 \longrightarrow \mathcal{V} \tag{1.19}$$

and that $\mathcal{D}_\mu\phi|_{S_\infty^2} = 0$. Maps of the type (1.19) are classified by the second homotopy group, $\Pi_2(\mathcal{V})$, which for the Georgi-Glashow model (with $\mathcal{V} = S^2$) is non trivial, and equal to the set of integer numbers. These maps are characterized by their winding number,

$$N = \frac{1}{4\pi a^3}\int_{S_\infty^2} dS^i \frac{1}{2}\epsilon_{ijk}\phi \cdot (\partial^j\phi \wedge \partial^k\phi). \tag{1.20}$$

Once we impose the finite energy condition $\mathcal{D}_\mu\phi|_{S_\infty^2} = 0$, the gauge field at infinity is given by

$$A^\mu = \frac{1}{a^2 g}\phi \wedge \partial^\mu\phi + \frac{1}{a}\phi f^\mu, \tag{1.21}$$

where f_μ is an arbitrary function. The corresponding stress tensor is given by

$$F_a^{\mu\nu} = \frac{1}{a}\phi_a F^{\mu\nu} = \frac{1}{a}\phi_a\left(\frac{1}{a^2 g}\phi \cdot (\partial^\mu\phi \wedge \partial^\nu\phi) + \partial^\mu f^\nu - \partial^\nu f^\mu\right), \tag{1.22}$$

which implies that the magnetic charge

$$m = -\frac{1}{2ga^3}\int_{S_\infty^2} \epsilon_{ijk}\phi \cdot (\partial^j\phi \wedge \partial^k\phi)dS^i, \tag{1.23}$$

for a finite energy configuration is given in terms of the winding number (1.20) as [7]

$$m = -\frac{4\pi N}{g}. \tag{1.24}$$

In order to combine (1.24) with Dirac's quantization rule we should define the $U(1)$ electric charge. The $U(1)$ photon field is defined by

$$A^{\mu} = (A^{\mu} \cdot \phi)\frac{1}{a}. \tag{1.25}$$

Thus, the electric charge of a field of isotopic spin j is given by

$$e = g\,j. \tag{1.26}$$

From (1.24) and (1.26) we recover, for $j = \frac{1}{2}$, Dirac's quantization rule.

For a generic Higgs model, with gauge group G spontaneously broken to H, the vacuum manifold $\mathcal{V}$ is given by

$$\mathcal{V} = G/H, \tag{1.27}$$

with

$$\Pi_2(G/H) \simeq \Pi_1(H)_G, \tag{1.28}$$

where $\Pi_1(H)_G$ is the set of paths in H that can be contracted to a point in G, which again contains Dirac's condition in the form (1.9).

The mass of the monopole is given by

$$M = \int d^3x \frac{1}{2}[(E_a^i)^2 + (B_a^i)^2 + (D^0\phi_a)^2 + (D^i\phi_a)^2] + V(\phi). \tag{1.29}$$

For a static monopole, the mass becomes

$$M = \int d^3x \frac{1}{2}[(B_a^i)^2 + (D^i\phi_a)^2] + V(\phi); \tag{1.30}$$

then, in the Prasad-Sommerfeld [8] limit $\lambda = 0$ (see equation (1.16)), we get

$$M = \int d^3x \frac{1}{2}[((B_a^i + D^i\phi_a)^2 - 2B_a^i D^i\phi_a], \tag{1.31}$$

which implies the Bogomolny [9] bound $M \geq am$. The Bogomolny bound is saturated if $B_a^k = D^k\phi_a$, which are known as the Bogomolny equations.

1.3 Instantons.

Let us now consider pure $SU(N)$ Yang-Mills theory,

$$\mathcal{L} = -\frac{1}{4} F^{a\mu\nu} F^a_{\mu\nu}. \tag{1.32}$$

In euclidean spacetime $\mathbf{R}^4$, the region at infinity can be identified with the 3-sphere S^3. A necessary condition for finite euclidean action of configurations is

$$F^a_{\mu\nu}|_{S^3_\infty} = 0, \tag{1.33}$$

or, equivalently, that the gauge configuration A^μ at infinity is a pure gauge,

$$A^\mu|_{S^3_\infty} = g(x)^{-1}\partial^\mu g(x). \tag{1.34}$$

Hence, finite euclidean action configurations are associated with maps

$$g : S^3 \to SU(N), \tag{1.35}$$

which are topologically classified in terms of the third homotopy group,

$$\Pi_3(SU(N)) \simeq \mathbf{Z}. \tag{1.36}$$

The winding number of the map g defined by (1.35) is given by

$$n = \frac{1}{24\pi^2} \int_{S^3} d^3x \epsilon_{ijk} tr[g^{-1}\nabla_i g(x) g^{-1}\nabla_j g(x) g^{-1}\nabla_k g(x)]. \tag{1.37}$$

As for the Dirac monopole construction, we can use the map g in order to define $SU(N)$ bundles on S^4. In this case, g defines the transition function on the equator. So, for the simplest group, $SU(2)$, we will have different bundles, depending on the value of n; in particular, for $n = 1$, we obtain the Hopft bundle

$$S^7 \longrightarrow S^4. \tag{1.38}$$

Interpreting S^4 as the compactification of euclidean space $\mathbf{R}^4$, we can define a gauge configuration on S^4 such that on the equator, which now has the topology of S^3, we have

$$A^+_\mu = g A^-_\mu g^{-1} + g^{-1}\partial_\mu g, \tag{1.39}$$

with A^+ and A^- the gauge configurations on the two hemispheres. Using now the relation

$$\mathrm{tr}(F_{\mu\nu}\tilde{F}^{\mu\nu}) = d\mathrm{tr}(F \wedge A - \frac{1}{3} A \wedge A \wedge A), \tag{1.40}$$

we get

$$-\frac{1}{8\pi^2} \int_{S^4} \mathrm{tr}(F_{\mu\nu}\tilde{F}^{\mu\nu}) = \frac{1}{24\pi^2} \int_{S^3} \epsilon_{ijk} \mathrm{tr}[g^{-1}\partial_i g g^{-1}\partial_j g g^{-1}\partial_k g] = n, \tag{1.41}$$

which is the generalization to S^4 of the relation we have derived above between the magnetic charge of the monopole and the winding number of the transition function defining the $U(1)$ bundle on S^2. The topological charge defined by (1.41) is a bound for the total euclidean action. In fact,

$$\frac{1}{4}\int F^{a\mu\nu}F^{a}_{\mu\nu} \equiv \frac{1}{2}\int \mathrm{tr}(F^{\mu\nu}F_{\mu\nu}) \geq \left|\frac{1}{2}\int \mathrm{tr}(F^{\mu\nu}\tilde{F}_{\mu\nu})\right|. \tag{1.42}$$

The instanton configuration will be defined by the gauge field saturating the bound (1.42),

$$F_{\mu\nu} = \tilde{F}_{\mu\nu}, \tag{1.43}$$

and with topological charge equal one. Bianchi identity, $DF = 0$, together with the field equations, implies $D\tilde{F} = 0$; in fact, the self duality condition (1.43) can be related to the Bogomolny equation. If we start with euclidean Yang-Mills, and reduce dimensionally to three dimensions through the definition $A_4 \equiv \phi$, we get the three dimensional Yang-Mills-Higgs lagrangian. Then, the self duality relation (1.43) becomes the Bogomolny equation.

A solution to (1.43) for $SU(2)$ was discovered by Belavin et al [10]. Including the explicit dependence on the bare coupling constant g,

$$F_{\mu\nu} = \partial_\mu A_\nu - \partial_\nu A_\mu + g[A_\mu, A_\nu], \tag{1.44}$$

the BPST solution to (1.43) is given by

$$\begin{aligned} A^a_\mu &= -\frac{2i}{g}\frac{\eta_{a\mu\nu}x^\nu}{x^2+\rho^2}, \\ F^a_{\mu\nu} &= \frac{4i}{g}\frac{\eta_{a\mu\nu}\rho^2}{(x^2+\rho^2)^2}, \end{aligned} \tag{1.45}$$

with $\eta_{a\mu\nu}$ satisfying $\eta_{a\mu\nu} = \eta_{aij} = \epsilon_{aij}$, $\eta_{ai0} = \delta_{ai}$, $\eta_{a\mu\nu} = -\eta_{a\mu\nu}$, and $\bar{\eta}_{a\mu\nu} = (-1)^{\delta_{\mu 0}+\delta_{\nu 0}}\eta_{a\mu\nu}$, where a, i, j take values $1, 2, 3$.

The value of the action for this configuration is

$$S = \frac{8\pi^2}{g^2}, \tag{1.46}$$

with Pontryagin number

$$\frac{g^2}{32\pi^2}\int F^a_{\mu\nu}\tilde{F}^{a\mu\nu}d^4x = 1. \tag{1.47}$$

Notice that the instanton solution (1.45) depends on a free parameter ρ, that can be interpreted as the classical size of the configuration. In particular, we can consider the gauge zero modes of the instanton solution, i. e., small self dual fluctuations around the instanton solution. From (1.45), it is clear that the action is invariant under changes of the size ρ, and under translations $x^\mu \to x^\mu + a^\mu$. This means that we will have five independent gauge zero modes. The number of gauge zero modes is called, in the mathematical literature, the dimension of the moduli space of self dual solutions. This number can be computed [11, 12, 13] using index theorems [14]; the result for $SU(N)$ instantons on S^4 is

$$\text{dim Instanton Moduli} = 4nk - n^2 + 1, \tag{1.48}$$

with k the Pontryagin number of the instanton[1]. For $k = 1$ and $n = 2$ we recover the five zero modes corresponding to translations and dilatations of the solution (1.45)[2]. The generalization of equation (1.48) to instantons on a manifold $\mathcal{M}$ is

$$\dim = 4nk - \frac{1}{2}(N^2 - 1)[\chi - \tau], \tag{1.49}$$

with χ and τ the Euler number and the signature of the manifold $\mathcal{M}$.

In order to get a clear physical interpretation of instantons, it is convenient to work in the $A^0 = 0$ temporal gauge [15, 16, 17]. If we compactify $\mathbf{R}^3$ to S^3 by impossing the boundary condition

$$A_i(\mathbf{r})|_{|\mathbf{r}|\to\infty} \to 0, \tag{1.50}$$

the vacuum configurations in this gauge are pure gauge configurations, $A_\mu = g^{-1}\partial_\mu g$, with g a map from S^3 into the gauge group $SU(N)$. We can now define different vacuum states $|n>$, characterized by the winding number of the corresponding map g. In the temporal gauge, an instanton configuration of Pontryagin number equal one satisfies the following boundary conditions:

$$\begin{aligned} A_i(t = -\infty) &= 0, \\ A_i(t = +\infty) &= g_1^{-1}\partial_i g_1, \end{aligned} \tag{1.51}$$

with g_1 a map from S^3 into $SU(N)$, of winding number equal one. We can now interpret the instanton configuration (1.51) as defining a tunnelling process between the $|0>$ and $|1>$ vacua.

Moreover, the vacuum states $|n>$ are not invariant under gauge transformations with non vanishing winding number. A vacuum state invariant under all gauge transformations would be defined by the coherent state

$$|\theta> = \sum_n e^{in\theta}|n>, \tag{1.52}$$

with θ a free parameter taking values in the interval $[0, 2\pi]$. Under gauge transformations of winding number m, the vacuum states $|n>$ transform as

$$\mathcal{U}(g_m)|n> = |n+m>, \tag{1.53}$$

[1] k must satisfy the irreducibility condition $k \geq \frac{n}{2}$. This condition must hold if we require the gauge configuration to be irreducible, i. e., that the connection can not be obtained by embedding the connection of a smaller group.

[2] Observe that the total number of gauge zero modes is 4, and that $n^2 - 1$ are simply gauge rotations of the instanton configuration.

and therefore the θ-vacua will transform as

$$\mathcal{U}(g_m)|\theta >= e^{im\theta}|\theta >, \tag{1.54}$$

which means invariance in the projective sense, i. e., on the Hilbert space of rays.

The generating functional now becomes

$$<\theta|\theta>=\sum_n <0|n> e^{in\theta} = \int dA \exp - \left(i \int \mathcal{L}(A) \right), \tag{1.55}$$

with the Yang-Mills lagrangian

$$\mathcal{L} = -\frac{1}{4} F^{a\mu\nu} F^a_{\mu\nu} + \frac{\theta g^2}{32\pi^2} F^{a\mu\nu} \tilde{F}^a_{\mu\nu}. \tag{1.56}$$

The θ-topological term in (1.56) breaks explicitely the CP invariance of the lagrangian. Notice that if we consider the euclidean functional integral

$$\int dA \exp - \left(\int -\frac{1}{4} F^{a\mu\nu} F^a_{\mu\nu} + \frac{i\theta g^2}{32\pi^2} F\tilde{F} \right) d^4x, \tag{1.57}$$

the instanton euclidean action becomes

$$S = \frac{8\pi^2}{g^2} + i\theta. \tag{1.58}$$

1.4 Dyon Effect.

Let us now add the topological θ-term of (1.56) to the Georgi-Glashow model (1.15). At this level, we are simply considering the θ-angle as an extra coupling constant, multiplying the topological density $F\tilde{F}$. In order to define the $U(1)$ electric charge, we can simply apply Noether's theorem for a gauge transformation in the unbroken $U(1)$ direction [18]. An infinitesimal gauge transformation in the ϕ direction would be defined by

$$\begin{aligned} \delta A_\mu &= \frac{1}{ag} \mathcal{D}_\mu \phi, \\ \delta\phi &= 0. \end{aligned} \tag{1.59}$$

The corresponding Noether charge,

$$N = \frac{\delta \mathcal{L}}{\delta \partial_0 A} \cdot \delta A + \frac{\delta \mathcal{L}}{\delta \partial_0 \phi} \cdot \phi, \tag{1.60}$$

will be given, after the θ-term is included, by

$$N = \frac{1}{ag} \int d^3x \partial_i (\phi \cdot F_{0i}) + \frac{\theta g}{8\pi^2 a} \int d^3x \partial_i (\phi \frac{1}{2} \epsilon_{ijk} F_{jk}) \tag{1.61}$$

or, in terms of the electric charge, as

$$N = \frac{e}{g} + \frac{\theta g}{8\pi^2} m. \tag{1.62}$$

Notice from (1.61) that the θ-term only contributes to N in the background of the monopole field. If we now require invariance under a $U(1)$ rotation, we get

$$e^{2\pi i N} = e^{2\pi i (\frac{e}{g} + \frac{\theta g}{8\pi^2} m)} = 1, \tag{1.63}$$

and the electric charge becomes equal to [18]

$$e = ng - \frac{\theta g^2}{8\pi^2} m, \tag{1.64}$$

which implies that a magnetic monopole of charge m becomes a dyon with electric charge $-\frac{\theta g^2}{8\pi^2} m$.

We can reach the same result, (1.63), without incuding a θ term in the lagrangian, if we require, for the monopole state,

$$e^{2\pi i N} |m >= e^{i\theta} |m >, \tag{1.65}$$

for $N = \frac{e}{g}$. Equation (1.65) implies that the monopole state transforms under $e^{2\pi i N}$ as the θ vacua with respect to gauge transformations of non vanishing winding number. However, $e^{2\pi i N}$ can be continously connected with the identity which, in physical terms, means that the induced electric charge of the monopole is independent of instantons, and is not suppresed by a tunnelling factor of the order of $\exp -\frac{8\pi^2}{g^2}$ [18, 19].

1.5 Yang-Mills Theory on T^4.

We will consider now $SU(N)$ pure Yang-Mills on a 4-box [20], with sides of length a_0, a_1, a_2, a_3. Let us impose periodic boundary conditions for gauge invariant quantities,

$$\begin{aligned} A^\mu(x_0 + a_0, x_1, x_2, x_3) &= \Omega_0 A^\mu(x_0, x_1, x_2, x_3), \\ A^\mu(x_0, x_1 + a_1, x_2, x_3) &= \Omega_1 A^\mu(x_0, x_1, x_2, x_3), \\ A^\mu(x_0, x_1, x_2 + a_2, x_3) &= \Omega_2 A^\mu(x_0, x_1, x_2, x_3), \\ A^\mu(x_0, x_1, x_2, x_3 + a_3) &= \Omega_3 A^\mu(x_0, x_1, x_2, x_3), \end{aligned} \tag{1.66}$$

where

$$\Omega_\rho A^\mu \equiv \Omega_\rho A^\mu \Omega_\rho^{-1} + \Omega_\rho^{-1} \partial^\mu \Omega_\rho. \tag{1.67}$$

As the gauge field transforms in the adjoint representation, we can allow the existence of $\mathbf{Z}(N)$ twists,

$$\Omega_\mu \Omega_\nu = \Omega_\nu \Omega_\mu e^{2\pi i n_{\mu\nu}/N}, \tag{1.68}$$

and therefore we can characterize different configurations in T^4 by the topological numbers $n_{\mu\nu}$. Three of these numbers, n_{12}, n_{13} and n_{23}, can be interpreted as magnetic fluxes in the 3, 2 and 1 directions, respectively. In order to characterize these magnetic fluxes, we introduce the numbers

$$m_i = \epsilon_{ijk} n_{jk}, \tag{1.69}$$

These magnetic fluxes carry $\mathbf{Z}(N)$ charge, and their topological stability is due to the fact that

$$\Pi_1(SU(N)/\mathbf{Z}(N)) \simeq \mathbf{Z}(N). \tag{1.70}$$

In order to characterize the physical Hilbert space of the theory, let us again work in the temporal gauge $A^0 = 0$. For the three dimensional box T^3, we impose twisted boundary conditions, corresponding to magnetic flow $\mathbf{m} = (m_1, m_2, m_3)$. The residual gauge symmetry is defined by the set of gauge transformations preserving these boundary conditions. We may distinguish the following different types of gauge transformations:

i) Periodic gauge transformations, which as usual are characterized by their winding number in $\Pi_3(SU(N)) \simeq \mathbf{Z}$.

ii) Gauge transformations periodic, up to elements in the center:

$$\begin{aligned} \Omega(x_1 + a_1, x_2, x_3) &= \Omega(x_1, x_2, x_3) e^{2\pi i k_1/N}, \\ \Omega(x_1, x_2 + a_2, x_3) &= \Omega(x_1, x_2, x_3) e^{2\pi i k_2/N}, \\ \Omega(x_1, x_2, x_3 + a_3) &= \Omega(x_1, x_2, x_3) e^{2\pi i k_3/N}. \end{aligned} \tag{1.71}$$

These transformations are characterized by the vector

$$\mathbf{k} = (k_1, k_2, k_3), \tag{1.72}$$

and will be denoted by $\Omega_{\mathbf{k}}(\mathbf{x})$. Among this type of transformations we can extract an extra classification:

ii-1) Those such that $(\Omega_{\mathbf{k}}(\mathbf{x}))^N$ is periodic, with vanishing Pontryagin number.

ii-2) Those such that $(\Omega_{\mathbf{k}}(\mathbf{x}))^N$ is periodic, with non vanishing Pontryagin number.

In the temporal gauge, we can represent the transformations in ii-2) in terms of unitary operators. Let $|\Psi>$ be a state in the Hilbert space $\mathcal{H}(\mathbf{m})$; then, we get

$$\Omega_{\mathbf{k}}(\mathbf{x})|\Psi> = e^{2\pi i \frac{\mathbf{e}\cdot\mathbf{k}}{N}} e^{i\theta \frac{\mathbf{k}\cdot\mathbf{m}}{N}} |\Psi>, \tag{1.73}$$

where $\mathbf{e}$ and θ are free parameters. Notice that the second term in (1.73) is equivalent, for $\mathbf{Z}(N)$ magnetic vortices, to the Witten dyon effect described in the previous section. In fact, we can write (1.73) in terms of an effective $\mathbf{e}_{eff}$,

$$\mathbf{e}_{eff} = \mathbf{e} + \frac{\theta \mathbf{m}}{2\pi}. \tag{1.74}$$

Moreover, as $\theta \to \theta + 2\pi$, we change $\mathbf{e}_{eff} \to \mathbf{e}_{eff} + \mathbf{m}$. On the other hand, the Pontryagin number of a gauge field configuration with twisted boundary conditions, determined by a set $n_{\mu\nu}$, is given by [21]

$$\frac{g^2}{16\pi^2} \int \operatorname{tr}(F^{\mu\nu} \tilde{F}_{\mu\nu}) d^4x = k - \frac{n}{N}, \tag{1.75}$$

where $n \equiv \frac{1}{4} n_{\mu\nu} n_{\mu\nu}$. A simple way to understand the origin of the fractional piece in the above expression is noticing that, for instance, a twist n_{12} corresponds to magnetic flux in the 3-direction, with value $\frac{2\pi n_{12}}{N}$, which can be formally described by $F_{12} \sim \frac{2\pi n_{12}}{N a_1 a_2}$, and a twist n_{03}, which corresponds to an electric field in the 3-direction, is described by $F_{03} \sim \frac{2\pi n_{03}}{N a_0 a_3}$. Using now the integral representation of the Pontryagin number we easily get the fractional piece, with the right dependence on the twist coefficients (see section 1.5.2). Moreover, $(\Omega_{\mathbf{k}}(\mathbf{x}))^N$ acting on the state $|\Psi>$ produces

$$(\Omega_{\mathbf{k}}(\mathbf{x}))^N |\Psi> = e^{i\theta \mathbf{k}\cdot\mathbf{m}} |\Psi>, \tag{1.76}$$

which means that $\mathbf{k}\cdot\mathbf{m}$ is the Pontryagin number of the periodic gauge configuration $(\Omega_{\mathbf{k}}(\mathbf{x}))^N$. For a generic gauge configuration with Pontryagin number n we will get, as usual,

$$\Omega(\mathbf{x}; n)|\Psi> = e^{in\theta}|\Psi> . \tag{1.77}$$

Using (1.71), it is easy to see that the $\mathbf{k}$'s characterizing the residual gauge transformations are nothing else but the n_{0i} twists. The physical interpretation of the parameter $\mathbf{e}$ introduced in (1.73), in the very same way as the θ-term, is that of an electric flux. In fact, we can define the Wilson loop

$$A(C) = \frac{1}{N} \operatorname{tr} \exp \int_C igA(\xi) d\xi, \tag{1.78}$$

with C a path in the 3-direction. Under $\Omega_{\mathbf{k}}(\mathbf{x})$, $A(C)$ transforms as

$$A(C) \to e^{2\pi i k_3/N} A(C); \tag{1.79}$$

therefore, we get

$$\Omega_{\mathbf{k}}(\mathbf{x}) A(C)|\Psi> = e^{2\pi i \frac{\mathbf{e}\cdot\mathbf{k}}{N}} A(C)|\Psi>, \tag{1.80}$$

which means that $A(C)$ creates a unit of electric flux in the 3-direction.

1.5.1 The Toron Vortex.. We will now consider a vacuum configuration with non vanishing magnetic flux. It may a priori come as a surprise that we can have magnetic flux for a classical vauum configuration. What we need, in order to achieve this goal, is to find two constant matrices in the gauge group, such that [21]

$$PQ = QP\mathcal{Z} \tag{1.81}$$

with $\mathcal{Z}$ a non trivial element in the center of the group. If such matrices exist, we can use them to define twisted boundary conditions in two directions in the box. The trivial configuration $A = 0$ automatically satisfies these boundary conditions, and we will get a classical vacuum with a non vanishing magnetic flux, characterized by the center element $\mathcal{Z}$ in (1.81). For the gauge group $SU(N)$ those matrices exist; they are

$$\begin{aligned} P &= \begin{pmatrix} 0 & 1 & & & \\ & 0 & 1 & & \\ & & & & \\ & & & & 1 \\ 1 & & & & 0 \end{pmatrix}, \\ Q &= \begin{pmatrix} 1 & & & \\ & e^{2\pi i/N} & & \\ & & & \\ & & & e^{2\pi i(N-1)/N} \end{pmatrix} \cdot e^{\pi i(1-N)/N}, \end{aligned} \tag{1.82}$$

satisfying $PQ = QPe^{2\pi i/N}$. If we impose twisted boundary conditions,

$$\begin{aligned} A(x_1 + a_1, x_2, x_3) &= PA(x_1, x_2, x_3)P^{-1}, \\ A(x_1, x_2 + a_2, x_3) &= QA(x_1, x_2, x_3)Q^{-1}, \\ A(x_1, x_2, x_3 + a_3) &= A(x_1, x_2, x_3), \end{aligned} \tag{1.83}$$

in the temporal gauge $A^0 = 0$, then the classical vacuum $A = 0$ is in the sector with non vanishing magnetic flux, $m_3 = 1$.

Classical vacuum configurations, $A_i(\mathbf{x}) = g^{-1}(\mathbf{x})\partial_i g(\mathbf{x})$, satisfying (1.83), would be defined by gauge transformations $g(\mathbf{x})$ satisfying

$$\begin{aligned} g(x_1 + a_1, x_2, x_3) &= Pg(x_1, x_2, x_3)P^{-1}e^{2\pi i k_1/N}, \\ g(x_1, x_2 + a_2, x_3) &= Qg(x_1, x_2, x_3)Q^{-1}e^{2\pi i k_2/N}, \\ g(x_1, x_2, x_3 + a_3) &= g(x_1, x_2, x_3)e^{2\pi i k_3/N}, \end{aligned} \tag{1.84}$$

for generic (k_1, k_2, k_3). Now, any gauge transformation satisfying (1.84) can be written as

$$g = T_1^{k_1} T_2^{k_2} T_3^{k_3} \tilde{g}, \tag{1.85}$$

with

$$T_1 = Q, \qquad T_2 = P^{-1}, \tag{1.86}$$

and $\tilde{g}$ satisfying (1.84), with $k_1 = k_2 = k_3 = 0$. Acting on the vacuum $|A_i = 0>$, we get, from (1.86),

$$\begin{aligned} T_1|A_i = 0> &= |A_i = 0>, \\ T_2|A_i = 0> &= |A_i = 0>, \end{aligned} \tag{1.87}$$

which implies, using (1.73), that the different vacua have $e_1 = e_2 = 0$. On the other hand, we get, acting with T_3,

$$T_3^{k_3}|A_i = 0> \equiv |A_i = 0; k_3>, \tag{1.88}$$

and, therefore, we get N different vacua defined by

$$|e_3> \equiv \frac{1}{N}\sum_{k_3} e^{2\pi i \frac{k_3 e_3}{N}}|A_i = 0; k_3>, \tag{1.89}$$

with $e_3 = 0, \ldots, N-1$. Acting now with $T_3^{k_3}$ on $|e_3>$, we get

$$T_3^{k_3}|e_3> = e^{2\pi i \frac{k_3 e_3}{N}} e^{i\theta \frac{k_3 m_3}{N}}|e_3>, \tag{1.90}$$

from which we observe that

$$T_3^N|e_3> = e^{i\theta}|e_3>, \tag{1.91}$$

i. e., T_3^N is periodic, with winding number equal one. Notice that in the definition of $|e_3>$ we have included the θ-parameter and the magnetic flux $m_3 = 1$, associated with the boundary conditions (1.83).

From the previous discussion we learn two basic things: first, that we can get zero energy states, with both electric and magnetic flux, provided both fluxes are parallel; secondly, that the number of vacuum states with twisted boundary conditions (1.83) is equal to N. In fact, what has been computed above is the well known Witten index, tr $(-1)^F$ [22].

1.5.2 't Hooft's Toron Configurations.. We will now try to find configurations on T^4 with fractional Pontryagin number, satisfying the equations of motion. Configurations of this type were initially discovered by 't Hooft for $SU(N)$ [21]. In order to describe this configurations, we first choose a subgroup $SU(k) \times SU(l) \times U(1)$ of $SU(N)$, with $k + l = N$. Let ω be the matrix corresponding to the $U(1)$ generators of $SU(k) \times SU(l) \times U(1)$,

$$\omega = 2\pi \begin{pmatrix} l & & & & \\ & \ddots & & & \\ & & l & & \\ & & & -k & \\ & & & & \ddots \\ & & & & & -k \end{pmatrix}, \tag{1.92}$$

with $\operatorname{tr}\omega = 0$. The toron configuration is defined by

$$A_\mu(x) = -\omega \sum_\lambda \frac{\alpha_{\mu\lambda} x_\lambda}{a_\lambda a_\mu}, \tag{1.93}$$

with

$$\alpha_{\mu\nu} - \alpha_{\nu\mu} = \frac{n^{(2)}_{\mu\nu}}{Nl} - \frac{n^{(1)}_{\mu\nu}}{Nk}, \tag{1.94}$$

and

$$n_{\mu\nu} = n^{(1)}_{\mu\nu} + n^{(2)}_{\mu\nu}. \tag{1.95}$$

The stress tensor for configuration (1.93) is given by

$$F_{\mu\nu} = \omega \frac{\alpha_{\mu\nu} - \alpha_{\nu\mu}}{a_\mu a_\nu}. \tag{1.96}$$

If we consider the simplest case, $n_{12} = n^{(1)}_{12} = 1$, and $n_{30} = n^{(2)}_{30} = 1$, we will be led to

$$\begin{aligned} F_{12} &= +\omega \frac{1}{Nka_1 a_2}, \\ F_{30} &= -\omega \frac{1}{Nla_3 a_4}, \end{aligned} \tag{1.97}$$

and therefore

$$\frac{g^2}{16\pi^2} \int \operatorname{tr}(F_{\mu\nu} \tilde{F}^{\mu\nu}) = -\frac{1}{N}. \tag{1.98}$$

If we now impose the self duality condition, we get

$$\frac{a_1 a_2}{a_3 a_4} = \frac{l}{k} = \frac{N-k}{k}, \tag{1.99}$$

which constrains the relative sizes of the box.

The gauge zero modes for the toron configuration (1.93) can be derived from the general relation (1.49), with $k = \tau = 0$ for T^4. Thus, for Pontryagin number equal $\frac{1}{N}$, we only get four translational zero modes for gauge group $SU(N)$. In this sense, we can think of the toron as having a size equal to the size of the box.

The toron of Pontryagin number equal $\frac{1}{N}$ can be interpreted, as we did for the instanton, as a tunnelling process between states $|m_3 = 1, \mathbf{k}>$ and $|m_3 = 1, \mathbf{k} + (0,0,1)>$.

Let us fix a concrete distribution of electric and magnetic fluxes, characterized by $\mathbf{e}$ and $\mathbf{m}$. The functional integral for this background is given by [21]

$$<\mathbf{e}, \mathbf{m}|\mathbf{e}, \mathbf{m}> = \sum_{\mathbf{k}} e^{2\pi i \frac{\mathbf{k}\cdot\mathbf{e}}{N}} W(\mathbf{k}, \mathbf{m}), \tag{1.100}$$

where

$$W(\mathbf{k},\mathbf{m}) = \int [dA]_{\mathbf{k},\mathbf{m}} \exp - \int \mathcal{L}(A), \tag{1.101}$$

with the integral in (1.101) over gauge field configurations satisfying the twisted boundary conditions defined by the twists $(\mathbf{k},\mathbf{m})$. We can consider the particular case $\mathbf{m} = (0,0,1)$ to define the effective action for the toron configuration,

$$S = \frac{8\pi^2}{g^2 N} + \frac{2\pi i e_3}{N}. \tag{1.102}$$

A possible generalization is obtained when using configurations with Pontryagin number equal $\frac{1}{N}$, but with $\mathbf{k} = (k_1, k_2, 1)$. In this case, the action (1.102) becomes

$$S = \frac{8\pi^2}{g^2 N} + \frac{2\pi i(\mathbf{k}\cdot\mathbf{e})}{N}. \tag{1.103}$$

It must be noticed that we have not included in (1.102) the effect of θ, which contributes to the action with a factor $\frac{i\theta}{N}$.

1.6 Instanton Effective Vertex.

Next, we will consider the effect of instantons on fermions [15, 11]. For the time being, we will work on compactified euclidean spacetime, S^4. The Dirac matrices satisfy

$$\{\gamma^\mu, \gamma^\nu\} = -2\delta^{\mu\nu}, \tag{1.104}$$

and the chiral operator γ_5,

$$\gamma_5 \equiv \gamma^0\gamma^1\gamma^2\gamma^3 = \begin{pmatrix} \mathbf{1} & \mathbf{0} \\ \mathbf{0} & \mathbf{1} \end{pmatrix}. \tag{1.105}$$

The space of Dirac fermions splits into two spaces of opposite chirality,

$$\gamma_5 \psi_\pm = \pm\psi_\pm. \tag{1.106}$$

Let us work with massless Dirac fermions coupled to an instanton gauge configuration. We consider normalized solutions to Dirac's equation,

$$\gamma^\mu \mathcal{D}_\mu(A)\psi = 0. \tag{1.107}$$

As a consequence of the index theorem, the number ν_+ of solutions to (1.107) with positive chirality, minus the number of solutions with negative chirality, ν_-, is given by

$$\nu_+ - \nu_- = \frac{g^2 N_f}{32\pi^2} \int F^a_{\mu\nu} \tilde{F}^{\mu\nu a} d^4x, \tag{1.108}$$

i. e., by the topological charge of the instanton gauge configuration. Thus, the change of chirality induced by an instanton configuration is given by

$$\Delta Q_5 = 2N_f k, \tag{1.109}$$

with k the Pontryagin number, and N_f the number of different massless Dirac fermions, transforming in the fundamental representation of the gauge group. We can generalize equation (1.108) to work with instanton configurations on a generic four dimensional euclidean manifold $\mathcal{M}$. The index theorem then becomes

$$\nu_+ - \nu_- = \frac{N}{24 \cdot 8\pi^2} \int_{\mathcal{M}} \mathrm{tr}(R \wedge R) - \frac{g^2 N_f}{32\pi^2} \int_{\mathcal{M}} F^a_{\mu\nu} \tilde{F}^{\mu\nu a} d^4x, \tag{1.110}$$

where again we consider fermions in the fundamental representation of $SU(N)$. Equation (1.109) implies that instanton configurations induce effective vertices, with change of chirality given by (1.109). In order to compute these effective vertices, we will use a semiclassical approximation to the generating functional,

$$Z(J, \bar{J}) = \int [dA][d\bar{\psi}][d\psi] \exp - \int \mathcal{L}(A, \bar{\psi}, \psi) + J\bar{\psi} + \psi\bar{J}, \tag{1.111}$$

around the instanton configuration. Let us first perform the gaussian integration of fermions in (1.111):

$$Z(J, \bar{J}) = \int [dA] \det{}' \not{D}(A) \exp \int \bar{J}(x) G(x, y; A) J(y) dx dy \cdot \exp - \int \mathcal{L}(A) \cdot$$

$$\prod_{n(A)} \int \bar{\psi}_0^{(n)}(x) J(x) d^4x \int \bar{J}(y) \psi_0^{(n)}(y) d^4y, \tag{1.112}$$

where $\psi_0^{(n)}$ are the fermionic zero modes for the configuration A, $\det' \not{D}(A)$ is the regularized determinant, and $G(x, y; A)$ is the regularized Green's function,

$$\not{D}(A) G(x, y; A) = -\delta(x - y) + \sum_n \psi_0^n(x) \bar{\psi}_0^n(y). \tag{1.113}$$

In semiclassical approximation around the instanton, we get

$$Z(J, \bar{J}) = \int [dQ] \det{}' \not{D}(A_{inst}) \exp - \frac{8\pi^2}{g^2} \cdot \exp \int \bar{J}(x) G(x, y; A) J(y) d^4x d^4y \cdot$$

$$\exp \int \mathcal{L}_0''(A_{inst}) Q^2 \cdot \prod_{i=1}^{m} \int \bar{\psi}_0^i(x) J(x) d^4x \int \bar{J}(y) \psi_0^i(y) d^4y, \tag{1.114}$$

where

$$\mathcal{L}_0''(A_{inst}) = \left(\frac{\delta^2 \mathcal{L}_0}{\delta A \delta A} \right)_{A = A_{inst}}, \tag{1.115}$$

for $\mathcal{L}_0 = -\frac{1}{4} F^{a\mu\nu} F^a_{\mu\nu}$, and Q the small fluctuation. It is clear from (1.115) that the only non vanishing amplitudes are those with

$$\left(\frac{\delta^{2m} Z(J, \bar{J})}{\delta J(x_1) \delta \bar{J}(x_1) \ldots \delta J(x_m) \delta \bar{J}(x_m)} \right)_{J = \bar{J} = 0}, \tag{1.116}$$

for $m = \nu^+ + \nu^-$. In order to perform the integration over Q, we need to consider the gauge zero modes. Each gauge zero mode contributes with a factor $\frac{1}{g}$. So, as we have $4N$ zero modes, we get

$$< \psi(x_1)\bar{\psi}(x_1)\ldots\psi(x_m)\bar{\psi}(x_m) >=$$

$$C\int\left(\frac{1}{g}\right)^{4N}\frac{1}{\rho^5}\rho^{3N_f}\exp-\frac{8\pi^2}{g^2(\mu)}[\mu]^{\beta_1}\prod_{i=1}^{N_f}(\bar{\psi}_0^i\psi_0^i)d^4z d\rho, \tag{1.117}$$

where β_1 is the coefficient of the β-function, in such a way that the result (1.117) is independent of the renormalization point μ. It must be stressed that $d^4z d\rho\rho^{-5}$ is a translation and dilatation invariant measure. The factor ρ^{3N_f} comes from the fermionic zero modes[3],

$$\psi_0^i = \frac{\rho^{3/2}}{(x^2+\rho^2)^{3/2}}\left(\frac{2}{\pi^2}\right)^{1/2}\omega. \tag{1.118}$$

(A chiral symmetry breaking condensate is obtained in the $N_f = 1$ case). The proportionality factor C in (1.117) comes from the determinants for fermions, gauge bosons and Faddeev-Popov ghosts.

The previous computation was carried out for $\theta = 0$. The effect of including θ is simply

$$< \psi\bar{\psi}(x_1)\ldots\psi\bar{\psi}(x_m) >_\theta=<\ldots>_{\theta=0}\cdot e^{i\theta} \tag{1.119}$$

It is important to stress that the integration over the instanton size in (1.117) is infrared divergent; thus, in order to get finite instanton contributions, we should cut off the integration size, something that can be implemented if we work with a Higgs model. The so defined instantons are known as constrained instantons [11].

1.7 Three Dimensional Instantons.

An instanton in three dimensions is a finite euclidean action configuration. This necesarily implies, in order to have topological stability, that the second homotopy group of the vacuum manifold is different from zero. This can not be realized for pure gauge theories, as $\Pi_2(SU(N)) \simeq 0$, so we will consider a Higgs model with spontaneous symmetry breaking from the G gauge group to a subgroup H, such that $\Pi_2(G/H)) \neq 0$. Think of $G = SU(N)$ and $H = U(1)^{N-1}$, then $\Pi_2 = \mathbf{Z}^{N-1}$. Thus, we see that three dimensional instantons are nothing but 't Hooft-Polyakov monopoles (see table).

The first thing to be noticed in three dimensions is that the dual to the photon is a scalar field,

---

[3] In fact, ρ^{3N_f} is the factor that appears in the fermionic Berezin measure for the fermionic zero modes.

Dimension	Energy Density	Energy	Action
$1+1$		Π_0	Π_1
$2+1$	Π_0	Π_1	Π_2
$3+1$	Π_1	Π_2	Π_3
Name	Vortex	Monopole	Instanton

$$
\begin{aligned}
H_\mu &= *F_{\rho\sigma} \equiv \frac{1}{2}\epsilon_{\mu\rho\sigma}F^{\rho\sigma}, \\
H_\mu &= \partial_\mu \chi .
\end{aligned}
\tag{1.120}
$$

In the weak coupling regime, we can describe the dilute gas of instantons and anti-instantons as a Coulomb gas. The partition function is given by [23]

$$
Z = \sum_n \int \prod_{i=1}^{n\pm} \frac{dx_i^+ dx_i^-}{n^+!\, n^-!} [\exp - S_0]^{n_+ + n_-} .
$$

$$
\exp - \frac{1}{2}\left(\frac{4\pi}{g}\right)^2 \int \rho(x)\left(-\frac{1}{\partial^2}\right)\rho(y) d^3x d^3y, \tag{1.121}
$$

with $n^+ + n^- = n$, S_0 the instanton action, and ρ the instanton density,

$$
\rho(x) = \sum_i \delta(x - x_i^+) - \sum_i \delta(x - x_i^-). \tag{1.122}
$$

The Coulomb interaction term admits the following gaussian representation, in terms of the dual photon [23]:

$$
\exp - \frac{1}{2}\left(\frac{4\pi}{g}\right)^2 \int d^3x d^3y \rho(x) - \frac{1}{\partial^2}\rho(y) = \int [d\chi] \exp - \int \frac{1}{2}(\partial\chi)^2 + \frac{4\pi i \chi \rho}{g} . \tag{1.123}
$$

When we sum up the instanton and anti-instanton contributions, we get the effective lagrangian for χ,

$$
\mathcal{L}_{eff}(\chi) = \frac{1}{2}(\partial\chi)^2 + e^{-S_0} \cos\frac{4\pi\chi}{e}, \tag{1.124}
$$

which implies a mass for the dual photon χ equal to e^{-S_0}. That χ is the dual photon becomes clear from the $\chi - \rho$ coupling in (1.123), between χ and the magnetic density ρ. The generation of a mass for the dual photon in a dilute gas of instantons is a nice example of confinement in the sense of dual Higgs phenomena.

The inclusion of massless fermions will drastically change the physical picture. In particular, as will be shown, the photon will become a massless Goldstone boson [24]. This will be due to the existence of effective fermionic vertices induced by the three dimensional instanton, of similar type to the ones studied in previous section. In order to analyze instanton induced effective interactions in three dimensions, we should first consider the problem of fermionic zero modes in the background of a monopole.

1.7.1 Callias Index Theorem.. Consider Dirac matrices in euclidean three dimensional spacetime,

$$\gamma^i \gamma^j + \gamma^j \gamma^i = 2\delta^{ij}. \tag{1.125}$$

We can get a representation of (1.125) using constant 2×2 matrices. In general, for euclidean space of dimension n, the corresponding γ^i are constant $2^{(n-1)/2}$ matrices.

Now, we define the Dirac operator,

$$L = i\gamma^i \partial_i + \gamma^i A_i + i\Phi(x), \tag{1.126}$$

with $A_i = gT^a A_i^a$, and $\Phi(x) = \phi^a(x)T^a$, for T^a the generators of the gauge group in some particular representation. We can now consider a Dirac fermion in Minkowski $3+1$ spacetime. This is a four component spinor,

$$\psi = \begin{pmatrix} \psi_+ \\ \psi_- \end{pmatrix}. \tag{1.127}$$

Then, Dirac's equation in $3+1$ dimensions becomes

$$\begin{pmatrix} \mathbf{0} & L \\ L^+ & \mathbf{0} \end{pmatrix} \begin{pmatrix} \psi_+ \\ \psi_- \end{pmatrix} = E \begin{pmatrix} \psi_+ \\ \psi_- \end{pmatrix}, \tag{1.128}$$

for fermion fields $\psi(x,t) = \psi(x)e^{iEt}$, and where L^+ is the adjoint of L. If we consider solutions to (1.128) with $E = 0$, we get

$$\begin{aligned} L\psi_- &= 0, \\ L^+\psi_+ &= 0, \end{aligned} \tag{1.129}$$

i. e., ψ_- and ψ_+ are zero modes of the euclidean Dirac equation in three dimensions, defined by (1.126).

Now, we can define the index

$$I(L) = k_- - k_+, \tag{1.130}$$

where k_- and k_+ are, respectively, the dimensions fo $\mathrm{Ker}(L)$ and $\mathrm{Ker}(L^+)$. By generalizing the Atiyah-Singer index theorem, Callias [25] got the following formula for $I(L)$:

$$I(L) = \frac{1}{2\left(\frac{n-1}{2}\right)} \left(\frac{i}{8\pi}\right)^{\frac{n-1}{2}} \int_{S_\infty^{n-1}} \mathrm{tr}[U(x)(dU(x))^{n-1}], \tag{1.131}$$

with n the dimension of euclidean spacetime, and

$$U(x) \equiv |\Phi(x)|^{-1}\Phi(x). \tag{1.132}$$

In our case, $n = 3$. In terms of the magnetic charge of the monopole, (1.20),

$$N = \frac{1}{8\pi} \int \epsilon_{ijk} \phi^i \partial \phi^j \partial \phi^k, \tag{1.133}$$

where we have normalized $a = 1$ in equation (1.20), and using (1.127) for Φ we get, for $SU(2)$,

$$I(L) = 2N, \tag{1.134}$$

for fermions in the adjoint representation. Notice that in odd dimensions, the index is zero for compact spaces. The contribution in (1.131) appears because we are working in a non compact space, with special boundary conditions at infinity, which are the ones defining the monopole configuration. We can also consider the more general case of massive fermions replacing (1.127) by

$$\Phi = \phi^a T^a + m. \tag{1.135}$$

In this case, we get, from (1.131),

$$I(L) = (j(j+1) - \{m\}(\{m\}+1))N, \tag{1.136}$$

with $\{m\}$ the largest eigenvalue of $\phi^a T^a$ smaller than m or, if there is no such eigenvalue, the smallest minus one. Thus, for massless fermions in the fundamental representation we have $\{m\} = -\frac{1}{2}$, and $I(L) = N$. It is important to observe that by changing the bare mass, the index also changes (we are using the normalization $a = 1$). Thus, for $m > \frac{1}{2}$, and fermions in the fundamental representation, we get $I(L) = 0$.

1.7.2 The Dual Photon as Goldstone Boson.. We will consider the $SU(2)$ lagrangian

$$\mathcal{L} = -\frac{1}{4} F_{\mu\nu}^2 + \frac{1}{2}(\mathcal{D}_\mu \phi)^2 + V(\phi) + \psi_+ (i\not{\mathcal{D}} + g\phi)\psi_-, \tag{1.137}$$

where we have used notation (1.127), and the Dirac operator (1.126). Lagrangian (1.137) is invariant under the $U(1)$ trasnformation

$$\begin{aligned} \psi_- &\rightarrow e^{i\theta}\psi_-, \\ \psi_+ &\rightarrow e^{i\theta}\psi_+. \end{aligned} \tag{1.138}$$

We will assume that the $\psi_\pm$ transform in the adjoint representation of $SU(2)$. Using (1.134), the induced instanton couple ψ_- fermions to $\psi_-^T \gamma_0$, through an instanton, while ψ_+ is coupled to $\psi_+^T \gamma$ in the anti-instanton case (the number $\psi_+(\psi_-)$ of zero modes for spherically symmetric monopoles in the instanton (anti-instanton) configuration is zero, and the two zero modes are $\psi_\pm$ and $\psi_\pm^T \gamma_0$. These vertices induce effective mass terms for fermions with mass $\mathcal{O}(e^{-S_0})$[4].

Now, we should include the Coulomb interaction between instantons; then, the effective lagrangian becomes

[4] These mass terms clearly break the $U(1)$ symmetry (1.138).

$$\mathcal{L} = \frac{1}{2}(\partial\chi)^2 + m\psi_-^T\gamma_0 e^{\frac{4\pi i\chi}{g^2}}\psi_- + m\psi_+\gamma_0 e^{-\frac{4\pi i\chi}{g^2}}\psi_+ + \cdots, \tag{1.139}$$

so that now the old vertices coupling $\psi_\pm$ to $\psi_\pm^T\gamma_0$ become vertices where the instanton or anti-instanton couple $\psi_\pm$ and $\psi_\pm^T\gamma_0$ to the dual photon χ[5]. From (1.139) it is now clear that χ becomes a Goldstone boson for the $U(1)$ symmetry [24] (1.138). In fact, $\mathcal{L}$ is invariant under (1.138) if

$$\chi \to \chi + \frac{g^2\theta}{2\pi}. \tag{1.140}$$

Notice that now χ is massless, and that no potential for χ is generated by instanton effects. It is also important to stress that the symmetry (1.138) is not anomalous in $2+1$ dimensions, which explains, from a different point of view, the Goldstone boson nature of χ.

1.8 $N=1$ Supersymmetric Gauge Theories.

As a first example, we will consider the $N=1$[6] extension of pure Yang-Mills theory. This model is defined in terms of a vector superfield, containing the gluon and the gluino. The gluino will be represented by a real Majorana spinor, transforming in the adjoint representation. The lagrangian is given by

$$\mathcal{L} = -\frac{1}{4}F^{a\mu\nu}F^a_{\mu\nu} + \frac{1}{2}i\bar{\lambda}^a\gamma_\mu\mathcal{D}^\mu(A)\lambda^a + \frac{\theta g^2}{32\pi^2}F^{a\mu\nu}\tilde{F}^a_{\mu\nu}. \tag{1.141}$$

As it can be easily checked, (1.141) is invariant under the supersymmetry transformations

$$\begin{aligned}\delta A^a_\mu &= i\bar{\alpha}\gamma_\mu\lambda^a, \\ \delta\lambda^a &= \frac{1}{4}[\gamma_\mu,\gamma_\nu]\alpha F^{a\mu\nu}, \\ \delta\bar{\lambda}^a &= -\bar{\alpha}\frac{1}{4}[\gamma_\mu,\gamma_\nu]F^{a\mu\nu},\end{aligned} \tag{1.142}$$

with α a Majorana spinor. Notice that, for λ^a in (1.141), we can use either real Majorana or complex Weyl spinors.

We will now study instanton effects for (1.141) [27, 28, 29, 30, 31, 32]. For $SU(N)$ gauge group, the total number of fermionic zero modes is

$$\#\text{zero modes} = 2Nk, \tag{1.143}$$

with k the Pontryagin number of the instanton. For $SU(2)$ and Dirac fermions in the isospin representation, of dimension $2j+1$, the generalization of (1.108) is

[5] The effective lagrangian (1.139) will not be interpreted in the wilsonian sense, but simply as the generating functional of the effective vertices induced by instantons.

[6] For a complete reference on supersymmetry, see [26].

$$\nu_+ - \nu_- = \frac{2}{3}(j+1)(2j+1)k, \tag{1.144}$$

from which we certainly get (1.143) for $j = 1$, using Majorana fermions.

The $2N$ zero modes for $k = 1$ decompose, relative to the $SU(2)$ subgroup where the instanton lies, into

$$\begin{array}{cl} 4 & \text{triplets,} \\ 2(N-2) & \text{doublets.} \end{array} \tag{1.145}$$

The meaning of the 4 triplet zero modes is quite clear from supersymmetry. Namely, two of them are just the result of acting with the supersymmetric charges on the instanton configuration. For $N = 1$ we have four supersymmetric charges, two of which anhilate the instanton configuration. The two other triplets result from superconformal transformations on the instanton. In fact, lagrangian (1.141) is not only invariant under supersymmetry, but also under the superconformal group. Now, we can repeat the computation of section 1.6. The only non vanishing amplitudes will be of the type

$$< \lambda\lambda(x_1) \cdots \lambda\lambda(x_N) > . \tag{1.146}$$

Impossing the instanton measure on collective coordinates to be translation and dilatation invariant, we get

$$\int \frac{d^4 z d\rho \rho^{2N}}{\rho^5}, \tag{1.147}$$

where the factor ρ^{2N} comes from the $2N$ fermionic zero modes, that scale as $\frac{1}{\rho^2}$ (see table). We must include the instanton action, and the renormalization point, μ,

$$\mu^{4N - \frac{2N}{2}} \exp - \frac{8\pi^2}{g(\mu)^2}, \tag{1.148}$$

where the power of μ is given by $+1$ for each gauge zero mode, and $-\frac{1}{2}$ for each Majorana fermionic zero mode. Defining the scale,

$$\Lambda = \mu \exp - \int \frac{dg'}{\beta(g')}, \tag{1.149}$$

and using the β-function for $SU(N)$ supersymmetric Yang-Mills,

$$\beta(g') = -\frac{g^3}{16\pi^2} 3N, \tag{1.150}$$

(1.148) becomes

$$\Lambda^{3N}, \tag{1.151}$$

with

$$\Lambda = \mu \exp - \frac{8\pi^2}{3N g(\mu)^2}. \tag{1.152}$$

Combining all these pieces, we get

$$< \lambda\lambda(x_1) \cdots \lambda\lambda(x_N) >= \int \frac{d^4 z d\rho \rho^{2N}}{\rho^5} \Lambda^{3N} .$$

$$\sum_{\text{permutations}} (-1)^P \text{tr}(\lambda_{i_1} \lambda_{i_2}(x_1)) \cdots \text{tr}(\lambda_{i_{2N-1}} \lambda_{i_{2N}}(x_N)). \quad (1.153)$$

In order to now perform the integration over the collective coordinates, we need the expression for the zero modes given in the table[7].

Supersymmetric triplet	$\sim \rho^2 (f(x))^2$
Superconformal triplet	$\sim \rho x (f(x))^2$
Doblets	$\rho (f(x))^{1/2}$

The fermionic zero modes in the above table are given for the instanton in the singular gauge,

$$A_\mu^{inst} = \frac{2}{g} \frac{\rho^2}{(x-z)^2 + \rho^2} \eta^a_{\mu\nu} (x-z)_\nu , \quad (1.154)$$

with z the instanton position. Using the expressions given above, we can perform the integration over z and ρ, to obtain the result

$$< \lambda\lambda(x_1) \cdots \lambda\lambda(x_N) >\sim \text{constant} \Lambda^{3N} , \quad (1.155)$$

which is a very amusing and, a priori, surprising result. The reason leading to (1.155) is that the integral (1.153) is saturated by instantons with size of the same order as the $|x_1 - x_N|$ distance. If we now use cluster decomposition in (1.155), we get

$$< \lambda\lambda >\sim \text{constant} \Lambda^3 e^{2\pi i n/N} , \quad (1.156)$$

with $n = 0, \ldots, N-1$. Notice that result (1.156) is not generated by instanton configurations, and that we get it assuming clustering or, equivalently, the existence of mass gap in the theory. This map gap should be interpreted as confinement.

A different approach for computing the $< \lambda\lambda >$ condensate starts with massive supersymmetric QCD, and requires a decoupling limit, $m \to \infty$. So, for $SU(2)$ with one flavor of mass m we get, from the instanton computation

$$< \lambda\lambda(x_1)\lambda\lambda(x_2) >\sim \text{ constant } \Lambda^5 m , \quad (1.157)$$

with Λ the scale of the $N = 1$ QCD theory. Relying now upon clustering, we get

[7] The function $f(x)$ is the instanton factor $f(x) = \frac{1}{(x-z)+\rho^2}$.

$$< \lambda\lambda >\sim \text{ constant } \Lambda^{5/2} m^{1/2} e^{2\pi i n/2}. \qquad (1.158)$$

We can now take the $m \to \infty$ limit, and define the scale Λ of pure $N = 1$ supersymmetric Yang-Mills as

$$\Lambda^3 \sim \Lambda^{5/2} m^{1/2}. \qquad (1.159)$$

The only difference with the previous computation is that now we perform cluster decomposition before definig the decoupling limit.

Until now we have consider $< \lambda\lambda >$ condensates for vacuum angle θ equal zero. We will now show the dependence of the condensate on θ, through an argument given by Shifman and Vainshtein. For $SU(N)$ gauge group, the axial anomaly is given by

$$\partial j_\mu^5 = \frac{N}{16\pi^2} F\tilde{F}. \qquad (1.160)$$

This means that under the chiral transformation

$$\lambda \to e^{i\alpha}\lambda, \qquad (1.161)$$

the lagrangian changes as

$$\mathcal{L} \to \mathcal{L} + \frac{\alpha N}{16\pi^2} F\tilde{F}. \qquad (1.162)$$

Thus, $< \lambda\lambda >$ at a non zero value of θ is the same as $< \lambda'\lambda' >_{\theta=0}$, with

$$\lambda' = e^{i\alpha}\lambda, \qquad (1.163)$$

where now

$$2\pi\alpha = \theta. \qquad (1.164)$$

Hence [33],

$$< \lambda\lambda >_{\theta=0} = < \lambda'\lambda' >_{\theta=0} = < \lambda\lambda >_{\theta=0} e^{i\frac{\theta}{n}}. \qquad (1.165)$$

1.9 Instanton Generated Superpotentials in Three Dimensional $N=2$.

To start with, we will consider dimensional reduction of lagrangian (1.141) to three dimensions. In this case, we arrive to the Higgs lagrangian in $2+1$ discussed in section 1.7. We can then define a complex Higgs field, with the real part given by the fourth component of A_μ in $3+1$, and the imaginary part by the photon field χ. If, as was the case in section 1.7, we consider $< \phi >= 0$ for the real Higgs field, then we automatically break superconformal invariance, and for the $SU(2)$ case we will find only two fermionic zero modes in the instanton background ('t Hooft-Polyakov monopole). The action of the three dimensional instanton is

$$S_{\text{inst}} = \frac{4\pi\phi}{g^2}, \qquad (1.166)$$

with ϕ standing for the vacuum expectation value of the Higgs field[8]. The effective lagrangian (1.139) becomes

$$\mathcal{L} = \frac{1}{2}((\partial\chi)^2 + (\partial\phi)^2) + \bar{\psi}i\partial\!\!\!/\psi + me^{-4\pi\phi/g^2}\psi^T\gamma_0\psi e^{i4\pi\chi/g^2} +$$

$$me^{4\pi\phi/g^2}\bar{\psi}\gamma_0\bar{\psi}^T e^{-i4\pi\chi/g^2}, \tag{1.167}$$

where we have included the kinetic term for both the real Higgs field ϕ, and the dual photon χ. In (1.167) we can define a complex Higgs field,

$$\Phi = \phi + i\chi, \tag{1.168}$$

in order to notice that the instanton is certainly generating a Yukawa coupling, which is nothing but the vertex coupling ψ fields to the dual photon χ. In order to write (1.167) as a supersymmetric lagrangian, we need to add a superpotential term of the type [24]

$$W(\Phi)_{\theta\theta} = \exp -\Phi + \text{hc}, \tag{1.169}$$

which induces an effective potential for ϕ of the type

$$V(\phi) = \frac{\partial W}{\partial \Phi}\frac{\partial \bar{W}}{\partial \Phi^*} = \exp -\phi, \tag{1.170}$$

i. e., no potential, as expected for the dual photon field, χ. The minima for the potential (1.170) is at $\phi = \infty$[9].

It is important to stress some aspects of the previous computation: first of all, the superpotential (1.169) is simply given by the instanton action, with the extra term $\frac{i4\pi\chi}{g^2}$ the analog of a topological θ term in four dimensions. Secondly, the fermions appearing in (1.167), the effective lagrangian, are the ones in the hypermultiplet of the $N = 2$ theory. Finally, the superpotential for Φ is defined on a flat direction.

The generalization of the previous picture to the four dimensional case is certainly not straightforward, as in that case we have not flat directions, and the effective lagrangian can not be written in terms of chiral superfields containing the gluino, but the gluino-gluino pair.

---

[8] Notice that the gauge coupling constant, in three dimensions, has length$^{-1/2}$ dimensions.

[9] The reader might be slightly surprised concerning potential (1.170) for the Higgs field. The crucial issue for the correct understanding of this potential requires noticing that the $N = 2$ three dimensional theory has been obtained through dimensional reduction of $N=1$ four dimensional Yang-Mills, which contains a flat direction (in next chapter we will define these flat directions more precisely, as Coulomb branches of moduli of vacua).

1.9.1 A Toron Computation.. A direct way to obtain $<\lambda\lambda>$ condensates in four dimensional $N=1$ Yang-Millsis using self dual gauge configurations, with Pontryagin number $\frac{1}{N}$[10] [34]. In subsection 1.5.2 we have described these configurations. The main point in using these torons is that the number of fermionic zero modes automatically reduces to two, which we can identify with the two triplets defined by supersymmetry transformations of one instanton configurations. We will perform the computation in a box, sending at the end its size to infinity. The size of the box is the size of the toron, but we will avoid the dilatation zero mode and the two triplet zero modes defined by superconformal transformations. The toron measure now becomes, simply,

$$\int d^4z \tag{1.171}$$

for the translation collective coordinate. Now, we have a power of μ, given by the four translation zero modes, and two fermionic zero modes,

$$\mu^3 \exp -\frac{8\pi^2}{g(\mu)^2 N}, \tag{1.172}$$

where we have included the toron action $\frac{8\pi^2}{g^2 N}$. Notice that (1.172) is simply Λ^3. Now, we integrate z over the box of size L. The two fermionic zero modes are obtained by the supersymmetry transformation (1.142) over the toron configuration (1.93), which means that each fermionic zero mode behaves as $\frac{1}{L^2}$, and therefore no powers of L should be included in the measure. The final result is

$$<\lambda\lambda>\sim \text{constant}\Lambda^3 e^{2\pi i e/N}, \quad e=0,1,\ldots,N-1 \tag{1.173}$$

in agreement with the cluster derivation. How should this result be interpreted? First of all, the expectation value (1.173) corresponds to the amplitude

$$< \mathbf{e}, \mathbf{m} = (0,0,1)|\lambda\lambda|\mathbf{e}, \mathbf{m} = (0,0,1) >=$$

$$< \mathbf{k} + (0,0,1), \mathbf{m} = (0,0,1)|\lambda\lambda|\mathbf{k}, \mathbf{m} = (0,0,1) > e^{2\pi i \frac{\mathbf{e}\cdot(0,0,1)}{N}}. \tag{1.174}$$

Then, the e in (1.173) is e_3, and the different values in (1.173) correspond to the set of N different vacua described in subsection 1.5.1.

Notice that a change $\theta \to \theta + 2\pi$ in equation (1.165) produces a change

$$<\lambda\lambda>_\theta \to <\lambda\lambda>_\theta e^{2\pi i/N}, \tag{1.175}$$

i. e., a $\mathbf{Z}(N)$ rotation. In other words, $\theta \to \theta + 2\pi$ exchanges the different vacua. Let us now try the same argument for (1.174). Using (1.74), we observe that

[10] It should already be noticed that topological configurations directly contributing to $<\lambda\lambda>$ are most probably the relevant configurations for confinement, as $<\lambda\lambda>$ was derived through a cluster argument assuming the existence of a mass gap.

$$< \lambda\lambda > \sim \Lambda^3 e^{2\pi i e_{eff}/N} = \Lambda^3 e^{2\pi i e/N} e^{i\theta/N}, \tag{1.176}$$

in agreement with (1.165). So, under $\theta \to \theta + 2\pi$, we go, using (1.74), from $\mathbf{e}_{eff}$ to $\mathbf{e}_{eff} + \mathbf{m}$. Notice that for the toron compuation we are using $\mathbf{m} = \mathbf{1}$.

2. Chapter II

2.1 Moduli of Vacua.

In this part of the lectures, we will consider gauge theories possessing potentials with flat directions. The existence of flat potentials will motivate the definition of moduli of vacua, which we will understand as the quotient manifold

$$\mathcal{M} = \mathcal{V}/\mathcal{G}, \tag{2.1}$$

obtained from the modding of the vacuum manifold $\mathcal{V}$ by gauge symmetries. In the first chapter, an example has already been discussed, namely three dimensional $N = 2$ Yang-Mills, defined as dimensional reduction of $N = 1$ Yang-Mills in four dimensions. Denoting by $\phi^a = A_4^a$ the fourth component of the gauge field, the dimensionally reduced lagrangian is

$$\mathcal{L} = -\frac{1}{4} F_{ij}^a F^{aij} + \frac{1}{2} \mathcal{D}_i \phi^a \mathcal{D}^i \phi^a + i \bar{\chi}^a \gamma_i \mathcal{D}^i \chi + i f_{abc} \bar{\chi}^b \chi^c \phi^a. \tag{2.2}$$

This is the Yang-Mills-Higgs lagrangian in the Prasad-Sommerfeld limit $V(\phi) = 0$. At tree level, the vacuum expectation value for the field ϕ is undetermined; therefore, at the classical level we can define a moduli of (real) dimension one, parametrizing the different values of $< \phi >$. As we already know, in addition to the scalar ϕ we have yet another scalar field, χ, the dual photon field. No potential can be defined for χ, neither classically nor quantum mechanically. If we took into account the action of the Weyl group, $\phi \to -\phi$, $\chi \to -\chi$, the classical moduli manifold should be

$$\mathbf{R} \times S^1/\mathbf{Z}_2. \tag{2.3}$$

The fields ϕ and χ can be combined into a complex scalar, $\Phi = \phi + i\chi$. As discussed in chapter I, instantons generate a superpotential of type $e^{-\Phi}$, which induces a potential for the ϕ fields with its minimum at ∞. This potential lifts the classical degeneracy of vacua. The vacuum expectation value of χ still remains undetermined, but can be changed by just shifting the coefficient of the topological term. The physics of this first example is what we expect from physical grounds: quantum effects breaking the classical vacuum degeneracy. However, there are cases where the amount of supersymmetry prevents, a priori, the generation of superpotential terms; it is in these cases, where we should be able to define the most general concept of quantum moduli [35, 36], where quantum effects will modify the topology and geometry of the classical moduli manifold.

2.2 $N=4$ Three Dimensional Yang-Mills.

$N=4$ three dimensional Yang-Mills will be defined through dimensional reduction of $N=1$ six dimensional Yang-Mills [37]. The three real scalars ϕ_i^a, with $i=1,2,3$, corresponding to the $\mu=3,4,5$ vector components of A_μ^a, are in the adjoint representation of the gauge group, and will transform as a vector with respect to the $SO(3)_R$ group of rotations in the 3,4,5-directions. The fermions in the model will transform, with respect to the $SU(2)_R$ double cover, as doublets, i. e., as spin one half particles. If we now consider the $SU(2)_E$ rotation group of euclidean space, $\mathbf{R}^3$, then fermions transform again as doublets, while scalars, ϕ_i^a, are singlets. By dimensional reduction, we get the following potential for the ϕ_i:

$$V(\phi)=\frac{1}{4g^2}\sum_{i<j}\mathrm{tr}[\phi_i,\phi_j]^2, \tag{2.4}$$

where we have used a six dimensional lagrangian $-\frac{1}{4g^2}F_{\mu\nu}^aF^{a\mu\nu}$. Obviously, the potential (2.4) possesses flat directions, obtained as those whose ϕ_i fields are in the Cartan subalgebra of the gauge group. For $SU(N)$, we will get

$$\phi_i=\begin{pmatrix} a_i^1 & & \\ & \ddots & \\ & & a_i^N \end{pmatrix}, \tag{2.5}$$

with $\sum_{j=1}^N a_i^j=0$, so that $3(N-1)$ parameters characterize a point in the flat directions of (2.4). In the general case of a gauge group of rank r, $3r$ coordinates will be required. A vacuum expectation value like (2.5) breaks $SU(N)$ to $U(1)^{N-1}$. As each $U(1)$ has associated a dual photon field, χ_j, with $j=1,\ldots,N-1$, the classical moduli has a total dimension equal to $3r+r=4r$. The simplest case of $SU(2)$, corresponds to a four dimensional moduli, which classically is

$$\mathcal{M}=(\mathbf{R}^3\times S^1)/\mathbf{Z}_2. \tag{2.6}$$

We have quotiented by the $\mathbf{Z}_2$ Weyl action, and S^1 parametrizes the expectation value of the dual photon field. The group $SU(2)_R$ acts on the $\mathbf{R}^3$ piece of (2.6). In the same sense as for the $N=2$ theory considered in chapter I, the $N=4$ model posseses instanton solutions to the Bogomolny equations which are simply the dimensional reduction to three dimensional euclidean space of four dimensional self dual Yang-Mills equations. These Bogomolny-Prasad-Sommerfeld instantons involve only one scalar field, ϕ_i, out of the three available (we will therefore choose one of them, say ϕ_3), and satisfy the equation

$$F=*\mathcal{D}\phi_3. \tag{2.7}$$

Once we choose a particular vacuum expectation value for the ϕ_i fields, we break $SU(2)_R$ to an $U(1)_R$ subgroup. In particular, we can choose $\phi_1 = \phi_2 = 0$, and ϕ_3 different from zero, with ϕ_3 the field used in the construction of the BPS instanton. The remaining $U(1)_R$ stands for rotations around the ϕ_3 direction.

Now, as discussed above, the BPS instanton can induce effective fermionic vertices. We first will consider the case of no extra hypermultiplets. In this case, we have four fermionic zero modes, corresponding to the four supersymmetry transformations that do not annihilate the instanton solution. In addition, we know, from the results in chapter I, that the effective vertex should behave like

$$\psi\psi\psi\psi \exp -(I + i\chi), \tag{2.8}$$

with χ the dual photon field, and I the classical instanton action, behaving like $\frac{\phi_3}{g^2}$. The term (2.7) breaks the $U(1)_R$ symmetry, as the ψ fermions transform under the $U(1)_R$ subgroup of $SU(2)_R$ like

$$\psi \to e^{i\theta/2}\psi. \tag{2.9}$$

However, the dual photon field plays the role of a Goldstone boson, to compensate this anomalous behaviour; χ transforms under $U(1)_R$ as

$$\chi \to \chi - 4\left(\frac{\theta}{2}\right). \tag{2.10}$$

This is a quantum effect, that will have important consequences on the topology of the classical moduli. In fact, the $U(1)_R$ is not acting only on the $\mathbf{R}^3$ part of $\mathcal{M}$, but also (as expressed by (2.10)) on the S^1 piece.

The topological meaning of (2.10) can be better understood if we work [37] on the boundary of $\mathbf{R}^3 \times S^1$, namely $S^2 \times S^1$. In this region at infinity, (2.10) will define a non trivial S^1 bundle on S^2. In order to see the way this is working, it is useful to use the spinorial notation. Henceforth, let u_1 and u_2 be two complex variables, satisfying

$$|u_1|^2 + |u_2|^2 = 1. \tag{2.11}$$

This defines the sphere S^3. Parametrizing points in S^2 by a vector $\mathbf{n}$, defined as follows:

$$\mathbf{n} = \bar{u}\boldsymbol{\sigma}u, \tag{2.12}$$

with $\boldsymbol{\sigma}$ the Pauli matrices, and using (2.12), we can define a projection,

$$\Phi : S^3 \to S^2, \tag{2.13}$$

associating to u, in S^3, a point $\mathbf{n}$ in S^2. The fiber of the projection (2.13) is S^1. In fact, u_α and $e^{i\theta}u_\alpha$ yield the same value of $\mathbf{n}$; therefore, we conclude

$$S^2 = S^3/U(1), \tag{2.14}$$

with the $U(1)$ action

$$u_\alpha \to e^{i\theta} u_\alpha. \tag{2.15}$$

A $U(1)$ rotation around the point $\mathbf{n}$ preserves $\mathbf{n}$, but changes u as

$$u_\alpha \to e^{i\theta/2} u_\alpha, \tag{2.16}$$

where now θ in (2.16) is the $U(1)_R$ rotation angle. Moreover, it also changes χ to $\chi - 4\left(\frac{\theta}{2}\right)$. So, the infinity of our quantum moduli looks like

$$(S^3 \times S^1)/U(1), \tag{2.17}$$

with the $U(1)$ action

$$u_\alpha \to e^{i\theta} u_\alpha, \quad \chi \to \chi - 4\theta. \tag{2.18}$$

The space (2.17) is the well known Lens space L_{-4}. Generically, L_s spaces can be defined through (2.17), with the $U(1)$ action defined by

$$\chi \to \chi + s\theta. \tag{2.19}$$

The spaces L_s can also be defined as

$$L_s = S^3/\mathbf{Z}_s, \tag{2.20}$$

with the $\mathbf{Z}_s$ action

$$\beta : u_\alpha \to e^{2\pi i/s} u_\alpha. \tag{2.21}$$

To finish the construction of the infinity of the quantum moduli, we need to include the Weyl action, $\mathbf{Z}_2$, so that we get

$$L_s/\mathbf{Z}_2 \tag{2.22}$$

or, equivalently, S^3/Γ_s, where Γ_s is the group generated by β given in (2.21), and the Weyl action

$$\alpha : (u_1, u_2) \to (\bar{u}_2, -u_1), \tag{2.23}$$

which reproduces $\mathbf{n} \to -\mathbf{n}$. The relations defining the group Γ_s are

$$\alpha^2 = \beta^s = 1, \quad \alpha\beta = \beta^{-1}\alpha. \tag{2.24}$$

Moreover, for $s = 2k$, Γ_s is the dihedral group D_{2k}.

Before entering into a more careful analysis of the moduli space for these $N=4$ theories, let us come back to the discussion on the $SU(2)_R$ symmetry, and the physical origin of the parameter s. In order to do so, we will first compactify six dimensional super Yang-Mills down to four dimensions. The resulting $N=2$ supersymmetric Yang-Mills theory has only two scalar fields, ϕ_1 and ϕ_2. The rotation symmetry in the compactified $(4,5)$-plane becomes now a $U(1)$ symmetry for the ϕ_i fields, which is the well known $U(1)_R$ symmetry of $N=2$ supersymmetric four dimensional Yang-Mills. As in the case for the $N=4$ theory in three dimensions, instantons in four dimensions generate effective fermionic vertices which break this $U(1)_R$ symmetry. Following the steps of the instanton computations presented in chapter I, we easily discover a breakdown of $U(1)_R$ to $\mathbf{Z}_8$. In the case of $N=2$ in four dimensions, the potential obtained from dimensional reduction is

$$V(\phi) = \frac{1}{4g^2}\mathrm{tr}[\phi_1, \phi_2]^2, \tag{2.25}$$

with the flat direction, for $SU(2)$,

$$\phi_i = \begin{pmatrix} a_i & \\ & -a_i \end{pmatrix}. \tag{2.26}$$

The moduli coordinate is therefore $u = \mathrm{tr}\phi^2$, where we now define ϕ as a complex field. The $U(1)_R$ transformation of ϕ is given by

$$\phi \to e^{2i\alpha}\phi, \tag{2.27}$$

so that u is $\mathbf{Z}_4$ invariant, and the $\mathbf{Z}_8$ symmetry acts as $\mathbf{Z}_2$ on the moduli space.

The difference with the three dimensional case is that now the instanton contribution does not contain any dual photon field, that can play the role of a Goldstone boson. From the three dimensional point of view, the $U(1)_R$ symmetry of the four dimensional theory is the rotation group acting on fields ϕ_1 and ϕ_2, and can therefore be identified with the $U(1)_R$ part of $SU(2)_R$ fixing the ϕ_3 direction. We should wonder about the relationship between the effect observed in three dimensions, and the breakdown of $U(1)_R$ in four dimensions. A qualitative answer is simple to obtain. The breakdown of $U(1)_R$ in four dimensions can be studied in the weak coupling limit (corresponding to $u \to \infty$), since the theory is assymptotically free, using instantons. Then, we get an effective vertex of the type

$$< \psi\psi\psi\psi >_0 e^{i\theta} =< \psi\psi\psi\psi >_\theta . \tag{2.28}$$

We have only four fermionic zero modes, since for $u \neq 0$ we break the superconformal invariance of the instanton. It is clear, from (2.28), that a $U(1)_R$ transformation is equivalent to the change

$$\theta \to \theta - 4\alpha, \tag{2.29}$$

with α the $U(1)_R$ parameter. Now, this change in θ is, in fact, the perfect analog of transformation rule (2.10), for the dual photon field. This should not come as a surprise; in fact, the four dimensional topological term

$$\frac{i\theta}{32\pi^2} F * F \tag{2.30}$$

produces, by dimensional reduction, the three dimensional topological term

$$\frac{i\theta}{32\pi^2}\epsilon_{ijk}F_{jk}\partial_i\phi_3. \tag{2.31}$$

This is precisely the type of coupling of the dual photon, in three dimensions, with the topological charge, and thus we again recover the result of section 1.7.

From the previous discussion, we can discover something else, specially interesting from a physical point of view. The transformation law of χ was derived counting instanton fermionic zero modes; however, the effect we are describing is a pure perturbative one loop effect, as is the $U(1)_R$ anomaly in four dimensions. Consider the wilsonian [38, 39] effective coupling constant for the $N=2$ theory, without hypermultiplets. Recall that in the wilsonian approach [40], the effective coupling constant is defined in terms of the scale we use to integrate out fluctuations with wave length smaller than that scale (this is the equivalent to the Kadanoff approach for lattice models). In a Higgs model, the natural scale is the vacuum expectation value of the Higgs field. Using the above notation, the wilsonian coupling constant in the four dimensional model is $\frac{1}{g(u)^2}$, with u the moduli parameter defined by $\mathrm{tr}\phi^2$. Now, let us write the lagrangian as follows:

$$\mathcal{L} = \frac{1}{64\pi}\mathrm{Im}\int \tau(F + i*F)^2, \tag{2.32}$$

with τ defined by

$$\tau \equiv \frac{i8\pi}{g^2} + \frac{\theta}{\pi}. \tag{2.33}$$

Using $F^2 = -*F^2$ we get, from (2.32), the standard lagrangian in Minkowski space,

$$\mathcal{L} = \int \frac{1}{4g^2}FF + \frac{\theta}{32\pi^2}F*F. \tag{2.34}$$

Now, we use the one loop effective beta function for the theory,

$$\frac{8\pi}{g(u)^2} = \frac{2}{\pi}\ln u + \mathcal{O}(1). \tag{2.35}$$

In general, if we add n hypermultiplets, we get, for the four dimensional theory,

$$\frac{8\pi}{g(u)^2} = \frac{4-n}{2\pi}\ln u + \mathcal{O}(1), \tag{2.36}$$

recovering the well known result for $N=2$ supersymmetric $SU(2)$ gauge theories in four dimensions, of finiteness of the theory when $n=4$, and infrared freedom for $n>4$. For $n<4$ the theory is assymptotically free, so that the perturbative computation (2.36) is only valid at small distances, for u in the assymptotic infinity.

Now, let us perform a rotation on u,

$$u \to e^{2\pi i}u \tag{2.37}$$

From (2.35) we get, for $n = 0$,

$$\frac{8\pi}{g(u)^2} \to \frac{8\pi}{g(u)^2} + 4i, \tag{2.38}$$

so, using (2.33), we get

$$\theta \to \theta - 4\pi, \tag{2.39}$$

in perfect agreement with equation (2.29). Thus, we observe that the s term (at least for the case without hypermultiplets) that we have discovered above using three dimensional instanton effects, is exactly given by the one loop effect of the four dimensional $N = 2$ theory. But what about higher order loop effects? As the argument we have presented is nothing but the non renormalization theorem [39] in supersymmetric theories, the $U(1)_R$ action on the wilsonian scale u forces the renormalization of the coupling constant to be consistent with the $U(1)$ anomalous behaviour of the lagrangian, which is determined by the Adler-Bardeen theorem [41] to be exact at one loop. What happens as we include hypermultiplets? First of all, and from the point of view of the three dimensional theory, the instanton effect will now be a vertex of type,

$$\psi\psi\psi\psi \prod^{2N_f} \chi e^{-(I+i\chi)}, \tag{2.40}$$

with the $2N_f$ fermionic zero modes appearing as a consequence of Callias index theorem [25], (1.136), for $j = 1/2$ and $\{m\} = -1/2$. From (2.40), we get $s = -4 + 2N_f$, which means a dihedral group Γ_s of type D_{2N_f-4} or, equivalently, a Dynkin diagram of type D_{N_f}. Notice that in deriving this diagram we have already taken into account the Weyl action, $\mathbf{Z}_2$.

The connection between the dihedral group, characterizing the moduli of the three dimensional $N = 4$ theory, and the beta function for the four dimensional $N = 2$ theory can be put on more solid geometrical grounds. The idea is simple. Let us work with the $N = 2$ four dimensional theory on $\mathbf{R}^3 \times S^1$, instead of on euclidean space, $\mathbf{R}^4$. The massless fields, from the three dimensional point of view, contain the fourth component of the photon in four dimensions, and the standard dual photon χ in three dimensions. Requiring, as in a Kaluza-Klein compactification, all fields to be independent of x_4, we still have residual gauge transformations of the type

$$A_4 \to A_4 + \partial_4 \Lambda(x_4), \tag{2.41}$$

with $\Lambda(x_4) = \frac{x_4 b}{\pi R}$, with b an angular variable, $b \in [0, 2\pi]$. This is equivalent to saying that we have non trivial Wilson lines in the S^1 direction, if the gauge field is in $U(1)$, as happens to be the case for a generic value of u[11]. Now, at

[11] We are not impossing, to the magnetic flux through the S^1, to be topologically stable in the sense of $\Pi_1(U(1)) \simeq \mathbf{Z}$. The crucial point is that the value of b at this point of the game is completely undetermined, and in that sense is a moduli parameter.

each point u in the four dimensional $N=2$ moduli, we have a two torus E_u, parametrized by the dual photon field χ, and the field b. This E_u is obtained from the S^1 associated with χ, and the S^1 associated with b. Its volume, in units defined by the three dimensional coupling constant, is of order $\frac{1}{R}$,

$$\text{Vol}\, E \sim \frac{1}{R}. \tag{2.42}$$

In fact, the volume is $\frac{1}{R}\cdot g_3^2$, where g_3^2, the three dimensional coupling constant, is the size of the S^1 associated to the dual photon (notice that the coupling constant g_3^2, in three dimensions, has units of inverse length). Equation (2.42) shows how, in the four dimensional limit, E_u goes to zero volume. Now, we have a picture of the theory in $\mathbf{R}^3 \times S^1_R$, if we keep R finite, namely that of an elliptic fibration over the u-plane, parametrizing the vacuum expectation values of the $N=2$ four dimensional theory.

If we keep ourselves at one particular point u, the torus E_u should be the target space for the effective lagrangian for the fields b and χ. There is a simple way to derive this lagrangian by means of a general procedure, called dualization, that we will now describe. To show the steps to follow, we will consider the four dimensional lagrangian (2.32). In order to add a dual photon field, let us say A_D^μ, we must couple A_D^μ to the monopole charge,

$$\epsilon^{0ijk}\partial_i F_{jk} = 4\pi\delta^{(3)}(x). \tag{2.43}$$

Thus, we add a term

$$\frac{1}{4}\int A_D^\mu \epsilon_{\mu\nu\rho\sigma}\partial^\nu F^{\rho\sigma} \equiv \frac{1}{4\pi}\int *F_D F. \tag{2.44}$$

Using the same notation as in (2.32), we get

$$\frac{1}{4\pi}\int *F_D F = \frac{1}{8\pi}\text{Re}\int (*F_D - iF_D)(F + i*F), \tag{2.45}$$

so that our lagrangian is

$$\mathcal{L} = \frac{1}{64\pi}\text{Im}\int \tau(F+i*F)^2 + \frac{1}{8\pi}\text{Re}\int (*F_D - iF_D)(F+i*F). \tag{2.46}$$

After gaussian integration, we finally get

$$\mathcal{L} = \frac{1}{64\pi}\text{Im}\int \left(\frac{-1}{\tau}\right)(*F_D - iF_D)^2, \tag{2.47}$$

i. e., lagrangian (2.32), with τ replaced by $\frac{-1}{\tau}$. The reader should take into account that these gaussian integrations are rather formal manipulations. Now, we use the same trick to get an effective lagrangian for the fields b and χ. Start with the four dimensional lagrangian,

$$\mathcal{L} = \int \frac{1}{4g^2} FF + \frac{i\theta}{32\pi^2} F * F, \tag{2.48}$$

where we now work in euclidean space. In three dimensions, using $d^4x = 2\pi R d^3x$, we get

$$\mathcal{L} = \int d^3x \frac{1}{\pi R g^2} |db|^2 + \frac{\pi R}{2g^2} F_{ij}^2 + \frac{i\theta}{8\pi^2} \epsilon^{ijk} F_{jk} \partial_k b. \tag{2.49}$$

Now, as we did before, we couple the dual photon field to the monopole charge,

$$\partial_i H^i = 4\pi \delta^{(3)}(x), \tag{2.50}$$

with $H^i \equiv \epsilon^{ijk} F_{jk}$, to get a term

$$\frac{i}{8\pi} \epsilon^{ijk} F_{jk} \partial_i \chi, \tag{2.51}$$

so that we can perform a gaussian integration,

$$\mathcal{L} = \int d^3x \frac{1}{\pi R g^2} |db|^2 + \frac{g^2}{\pi R (8\pi)^2} |d\chi - \frac{\theta}{\pi} db|^2. \tag{2.52}$$

What we get is precisely a target space for χ and b fields, which is the torus of moduli τ given by (2.33). Observe that the complex structure of the torus E_u is given in terms of the four dimensional coupling constant g [37], and the four dimensional θ-parameter, while its volume, (2.42), depends on the three dimensional coupling constant g_3, that acts as unit. When we go to the four dimensional $R \to \infty$ limit, this volume becomes zero, but the complex structure remains the same. The fact that the complex structure of E_u is given by the four dimensional effective coupling will make more transparent the meaning of equation (2.36). In fact, the monodromy around $u = \infty$ for (2.36) is given by a matrix

$$\begin{pmatrix} a & b \\ c & d \end{pmatrix} = \begin{pmatrix} -1 & -n+4 \\ 0 & -1 \end{pmatrix}, \tag{2.53}$$

with τ transforming as

$$\tau \to \frac{a\tau + b}{c\tau + d}, \tag{2.54}$$

so that for $n = 0$ we get transformation (2.39). Next, we will see that transformation (2.53) is precisely what we need, in order to match the dihedral group characterization of the $N=4$ three dimensional moduli space; however, in order to do that we need a few words on Atiyah-Hitchin spaces [42].

2.3 Atiyah-Hitchin Spaces.

Atiyah-Hitchin spaces appear in the study of moduli spaces for static multi-monopole configurations. Static solutions are defined by the BPS equations, (2.7), which are simply the dimensional reduction to $\mathbf{R}^3$ of euclidean self-dual equations for instantons. Next, we simply summarize some of the relevant results on Atiyah-Hitchin spaces for our problem (we refer the interested reader to the book by M. Atiyah and N. Hitchin, [42]). First of all, the Atiyah-Hitchin spaces are hyperkähler manifolds of dimension $4r$, on which a rotation $SO(3)$ is acting in a specific way. This is part of what we need to define the moduli space of $N=4$ three dimensional Yang-Mills theory for gauge group of rank r. In fact, in order to define $N=4$ supersymmetry on this space, interpreted as a σ-model target space of the low energy effective lagrangian, we have to require hyperkähler structure. Recall here that hyperkähler simply means that we have three different complex structures, I, J anf K, and therefore three different Kähler forms, ω_i, ω_j and ω_k, which are closed. Following the notation used by Atiyah and Hitchin, we define N_k as the moduli space of a k monopole configuration. The dimension of N_k is $4k-1$, so for $k=1$ we get dimension 3, corresponding to the position of the monopole center. If we mode out by the translation of the center of mass, we get the space

$$M_k^0 = N_k/\mathbf{R}^3, \tag{2.55}$$

of dimension $4(k-1)$. For two monopoles, we get $\dim M_2^0 = 4$. Now, the spaces M_k^0 are generically non simply connected,

$$\Pi_1(M_k^0) \simeq \mathbf{Z}_k, \tag{2.56}$$

so we can define its k-fold covering $\tilde{M}_k^0$. The known results, for $k=2$, are the following: the spaces M_2^0 and $\tilde{M}_2^0$ are, at infinity, respectively of type $L_{-4}/\mathbf{Z}_2$, and $L_{-2}/\mathbf{Z}_2$, which strongly indicates that M_2^0 is a good candidate for the moduli of the $N_f=0$ case, and $\tilde{M}_2^0$ is the adequate for the $N_f=1$ case. Moreover, the spaces $\tilde{M}_2^0$ can be represented by a surface in $\mathbf{C}^3$, defined by

$$y^2 = x^2 v + 1. \tag{2.57}$$

The space $M_2^0 = \tilde{M}_2^0/\mathbf{Z}_2$, can be obtained using variables $\chi = x^2$ and $y = x$, so that we get

$$y^2 = \chi^2 v + \chi. \tag{2.58}$$

The spaces $\tilde{M}_2^0$, defined by (2.55), can be interpreted as a limit of the family of spaces

$$y^2 = x^2 v + v^l, \tag{2.59}$$

where l should, in our case, be identified with $N_f - 1$. Surfaces (2.59) are well known in singularity theory; they give rise to the type of singularities obtained from $\mathbf{C}^3/\Gamma$, with Γ a discrete subgroup of $SO(3)$, and are classified according to the following table [43],

Γ	Name	Singularity
$\mathbf{Z}_n$	A_{n-1}	$v^n + xy = 0$
$\mathbf{D}_{2n}$	D_{n+2}	$vx^2 - v^{n+1} + y^2 = 0$
$\mathbf{T}_{12}$	E_6	$v^4 + x^3 + y^2 = 0$
$\mathbf{O}_{24}$	E_7	$v^3 + vx^3 + y^2 = 0$
$\mathbf{I}_{60}$	E_8	$v^5 + x^3 + y^2 = 0$

As can be seen from this table, the manifold (2.59) corresponds to a D_{n+2} singularity, with $n = N_f - 2$, and dihedral group D_{2N_f-4}, i. e., the group Γ we have discussed in the previous section.

It is important to stress that the type of singularities we are describing in the above table are the so called rational singularities [44]. The geometrical meaning of the associated Dynkin diagram is given by the resolution of the corresponding singularity as the intersection matrix of the irreducible components obtained by blowing up the singularity. In this interpretation, each mode of the diagram corresponds to an irreducible component, which is a rational curve X_i, with self dual intersection $X_i.X_i = -2$, and each line to the intersection $X_i.X_j$ between different irreducible components.

In the previous section we have modelled the $N = 4$ moduli space as an elliptic fibration, with fiber E_u of volume $\frac{1}{R}$, and moduli τ given by the coupling constant of the four dimensional $N = 2$ gauge theory. Next, we will try to connect the dihedral group, characterizing the Atiyah-Hitchin space describing the $N = 4$ moduli, with the monodromy at infinity of the elliptic modulus of E_u. But before doing this, we will briefly review Kodaira's theory on elliptic singularities.

2.4 Kodaira's Classification of Elliptic Fibrations.

According to Kodaira's notation [45], we define an elliptic fibration V onto Δ, where Δ will be chosen as a compact Riemann surface. In general, we take Δ to be of genus equal zero. The elliptic fibration,

$$\Phi : V \rightarrow \Delta, \tag{2.60}$$

will be singular at some discrete set of points, a_ρ. The singular fibers, $C_{a\rho}$, are given by

$$C_{a\rho} = \sum n_s \Theta_{\rho s}, \tag{2.61}$$

with $\Theta_{\rho s}$ irreducible curves. According to Kodaira's theorem (see section 4.7 for more details), all posible types of singular curves are of the following types:

–

$$I_{n+1}: \ C_{a\rho} = \Theta_0 + \Theta_1 + \cdots + \Theta_n, \quad n+1 \geq 3, \qquad (2.62)$$

where Θ_i are non singular rational curves with intersections $(\Theta_0, \Theta_1) = (\Theta_1, \Theta_2) = \cdots = (\Theta_n, \Theta_0) = 1$.
The A_n affine Dynkin diagram can be associated to I_{n+1}. Different cases are
i) I_0, with $C_\rho = \Theta_0$ and Θ_0 elliptic and non singular.
ii) I_1, with $C_\rho = \Theta_0$ and Θ_0 a rational curve, with one ordinary double point.
iii) I_2, with $C_\rho = \Theta_0 + \Theta_1$ and Θ_0 and Θ_1 non singular rational points, with intersection $(\Theta_0, \Theta_1) = p_1 + p_2$, i. e., two points.
Notice that I_1 and I_2 correspond to diagrams A_0 and A_1, respectively.
– Singularities of type I^*_{n-4} are characterized by

$$I^*_{n-4}: \ C_\rho = \Theta_0 + \Theta_1 + \Theta_2 + \Theta_3 + 2\Theta_4 + 2\Theta_5 + \cdots + 2\Theta_n, \qquad (2.63)$$

with intersections $(\Theta_0, \Theta_4) = (\Theta_1, \Theta_4) = (\Theta_2, \Theta_4) = (\Theta_3, \Theta_4) = (\Theta_4, \Theta_5) = (\Theta_5, \Theta_6) = \cdots = 1$, these singularities correspond to the D_n Dynkin diagram.
– Singularities of type II^*, III^* and IV^* correspond to types E_6, E_7 and E_8.
In addition to these singularities, we have also the types
– $II: \ C_\rho = \Theta_0$, with Θ_0 a rational curve with a cusp.
– $III: \ C_\rho = \Theta_0 + \Theta_1$, with Θ_0 and Θ_1 non singular rational curves, with intersection $(\Theta_0, \Theta_1) = 2p$.
– $IV: \ C_\rho = \Theta_0 + \Theta_1 + \Theta_2$, with Θ_0, Θ_1 and Θ_2 non singular rational curves, with intersections $(\Theta_0, \Theta_1) = (\Theta_1, \Theta_2) = (\Theta_2, \Theta_0) = p$.

In contrast to the singularities described in last section (the rational ones), these singularities are associated to affine Dynkin diagrams. Observe that for all these singularities we have

$$C.C = 0, \qquad (2.64)$$

while in the rational case the corresponding maximal cycle satisfies

$$C.C = -2. \qquad (2.65)$$

The origin for the affinization of the Dynkin diagram is the elliptic fibration structure. In fact, we can think of a rational singularity of ADE type in a surface, and get the affinization of the Dynkin diagram whenever there is a singular curve passing through the singularity. In the case of an elliptic fibration, this curve is the elliptic fiber itself. So, the extra node in the Dynkin diagram can be interpreted as the elliptic fiber. This can be seen more clearly as we compute the Picard of the surface. In fact, for the elliptic fibration the contribution to the Picard comes from the fiber, the basis, and the contribution from each singularity. Now, in the contribution to Picard from each singularity, we should not count the extra node, since this has already been taken into account when we count the fiber as an element in the Picard.

The previous discussion is already telling us what happens when we go to the $R = 0$ limit, i. e., to the three dimensional $N = 4$ gauge theory. In this limit, the elliptic fiber E_u becomes of infinite volume, and therefore we can not consider it anymore as a compact torus, i. e., as an elliptic curve. Thus, in this limit the corresponding singularity should become rational, and the Dynkin diagram is not affine.

However, before entering that discussion, let us work out the monodromies for the elliptic fibrations of Kodaira's classification.

We will then define

$$\tau(u) = \frac{\int_{u\times\gamma_1}\varphi(u)}{\int_{u\times\gamma_2}\varphi(u)}, \tag{2.66}$$

with $\varphi(u)$ the holomorphic one form on C_u. From (2.66), it follows that $\tau(u)$ is a holomorphic function of u. Next, we define the elliptic modular function, $j(\tau(u))$, on the upper half plane,

$$j(\tau(u)) = \frac{1728 g_2^3}{\Delta}, \tag{2.67}$$

where Δ is the discriminant $\Delta = g_2^3 - 27g_3^2$, with the modular functions

$$\begin{aligned} g_2 &= 60 \sum_{n_1,n_2\in\mathbf{Z}} \frac{1}{(n_1+n_2\tau)^4}, \\ g_3 &= 140 \sum_{n_1,n_2\in\mathbf{Z}} \frac{1}{(n_1+n_2\tau)^6}. \end{aligned} \tag{2.68}$$

Defining $\mathcal{F}(u) \equiv j(w(u))$ as a function of Δ, it turns out to be a meromorphic function. To each pole a_ρ, and each non contractible path γ_ρ in Δ (that is, an element in $\Pi_1(\Delta)$), we want to associate a monodromy matrix, A_ρ,

$$A_\rho\tau \to \frac{a\tau+b}{c\tau+d}. \tag{2.69}$$

If a_ρ is a pole of $\mathcal{F}(u)$, of order b_ρ, then it can be proved that A_ρ is of type

$$A_\rho\tau \to \tau + b_\rho, \tag{2.70}$$

for some b_ρ. The matrix A_ρ, of finite order, $A_\rho^m = 1$, for some m, corresponds to singularities which can be removed. Moreover, if A_ρ is of infinite order, then it is always possible to find numbers p, q, r and s such that

$$\begin{pmatrix} s & -r \\ -q & p \end{pmatrix}\begin{pmatrix} a & b \\ c & d \end{pmatrix}\begin{pmatrix} p & r \\ q & s \end{pmatrix} = \begin{pmatrix} 1 & b_\rho \\ 0 & 1 \end{pmatrix}, \tag{2.71}$$

with $ps - qr = 1$. Next, we relate matrices A_ρ with the different types of singularities. The classification, according to Kodaira's work, is as shown in the table below.

Matrix	Type of singularity
$\begin{pmatrix} 1 & 1 \\ 0 & 1 \end{pmatrix}$	I_1
$\begin{pmatrix} 1 & b \\ 0 & 1 \end{pmatrix}$	I_b
$\begin{pmatrix} -1 & -b \\ 0 & -1 \end{pmatrix}$	I_b^*
$\begin{pmatrix} 1 & 1 \\ -1 & 0 \end{pmatrix}$	II
$\begin{pmatrix} 0 & 1 \\ -1 & 0 \end{pmatrix}$	III
$\begin{pmatrix} 0 & 1 \\ -1 & -1 \end{pmatrix}$	IV

Now, we can compare the monodromy (2.53) with the ones in the table. It corresponds to the one associated with a singularity of type I_b^*, with $b = n-4$, i. e., a Dynkin diagram of type D_n. In the rational case, this corresponds to a dihedral group D_{2n-4}. In (2.53), n represents the number of flavors, so that we get the dihedral group of the corresponding Atiyah-Hitchin space.

Summarizing, we get that the dihedral group of $N = 4$ in three dimensions is the one associated with the type of elliptic singularity at infinity of the elliptic fibration defined by the $N = 2$ four dimensional theory. In other words, the picture we get is the following: in the $R \to 0$ three dimensional limit we have, at infinity, a rational singularity of type $\mathbf{C}^3/D_{2N_f-4}$. When we go to the $R \to \infty$ limit we get, at infinity, an elliptic singularity with Dynkin diagram D_{N_f}. Both types of singularities describe, respectively, one loop effects in three dimensional $N = 4$ and four dimensional $N = 2$.

2.5 The Moduli Space of the Four Dimensional $N=2$ Supersymmetric Yang-Mills Theory. The Seiberg-Witten Solution.

From our previous discussion, we have observed that the complex structure of the moduli space of three dimensional $N=4$ supersymmetric Yang-Mills theory is given by the elliptic fibration on the moduli space of the four dimensional $N=2$ theory, where the elliptic modulus is identified with the effective complexified coupling constant τ, as defined in (2.33). This result will in practice mean that the complete solution to the four dimensional $N=2$ theory can be directly read out form the complex structure of the Atiyah-Hitchin spaces (2.59), with $l = N_f - 1$. In previous sections, we have already done part of this job, comparing the monodromy of τ around $u = \infty$, i. e., in the assymptotic freedom regime, with the dihedral group characterizing the infinity of the three dimensional $N=4$ moduli space. In this section, we will briefly review the Seiberg-Witten solution [35, 36, 46, 47, 48, 49, 50, 51, 52, 53] for four dimensional $N=2$ Yang-Mills theory, and compare the result with the complex structure of Atiyah-Hitchin spaces. Recall that the Atiyah-Hitchin spaces are hyperkähler, and therefore possess three different complex structures. The complex structure determined by the four dimensional $N=2$ solution is one of these complex structures, namely the one where the Atiyah-Hitchin space becomes elliptically fibered. The analysis of Seiberg and Witten was originally based on the following argument: the moduli space parametrized by u should be compactified to a sphere (we will first of all consider the $N_f = 0$ case, for $SU(2)$ gauge group). According to Kodaira's notation, Δ is taken to be of genus equal zero. Next, the behaviour of τ at $u = \infty$ is directly obtained from the one loop beta function (see equation (2.36)); this leads to a monodromy around infinity of the type (2.53). Next, if $\tau(u)$ is a holomorphic function of u, which is clear from the elliptic fibration mathematical point of view (see equation (2.66)), and is a direct consequence of $N=2$ supersymmetry, then the real and imaginary parts are harmonic functions. As the coupling constant is the imaginary part of the complex structure $\tau(u)$, which is on physical grounds always positive, we are dealing with an elliptic fibration, so we already know all posible types of singularities. That some extra singularities should exist, in addition to the one at infinity, is clear form the harmonic properties of $\mathrm{Im}\tau(u)$, and the fact that it is positive, but in principle we do not how many of them we should expect, and of what type. The answer to this question can not, in principle, be derived from Kodaira's theory. In fact, all what we can obtain from Kodaira's approach, using the adjunction formula, is a relation between the canonical bundle K of the elliptic fibration, the K of the base space, which we can take as $\mathbb{P}^1$, and the type of singularities,

$$K_V = \Pi^*(K_\Delta + \sum a_i P_i), \tag{2.72}$$

where the a_i, for each type of singularity, are given below.

Singularity	a_i
I_1	$1/12$
I_b^*	$1/2 + b/12$
I_b	$b/12$
II	$1/6$
III	$1/4$
IV	$1/3$
II^*	$5/6$
III^*	$3/4$
IV^*	$2/3$

However, (2.72) is not useful, at this point, since we do not know the V manifold, which is what we are looking for. We will therefore proceed according to physical arguments.

The singularities we are looking for are singularities in the strong coupling regime of the moduli space of the theory, so it is hopeless to try to use a naive perturbative analysis; instead, we can rely on a duality approach. In dual variables (see equation (2.47)), the effective coupling constant behaves like $\frac{-1}{\tau}$, i. e., we have performed an S transformation, with

$$S = \begin{pmatrix} 0 & -1 \\ 1 & 0 \end{pmatrix}. \tag{2.73}$$

Thinking of $\frac{-1}{\tau}$ as the effective magnetic coupling, τ^{mag}, we can reduce our analysis to looking for perturbative monodromies of type

$$\tau^{mag} \to \tau^{mag} + b. \tag{2.74}$$

Indeed, we know that any singularity of Kodaira's type is related to a monodromy of type (2.74), up to a unitary transformation, (see equation (2.71)).

Now, and on physical grounds, we can expect a transformation of the type (2.74) as the monodromy singularity for the effective coupling constant of an effective $U(1)$ theory, with b equal to the number of massless hypermultiplets. In fact, the beta function for the $U(1)$ $N=2$ theory, with n hypermultiplets, is given by

$$\tau^{mag}(u) = -\frac{ik}{2\pi} \ln(u), \tag{2.75}$$

with k the number of massless hypermultiplets. This yields the monodromy

$$\begin{pmatrix} 1 & k \\ 0 & 1 \end{pmatrix}, \tag{2.76}$$

or, in Kodaira's notation, a monodromy of type A_{k-1}. Notice that the difference in sign between the type D, and the type A monodromies, reflects that we are obtaining type A for infrared free theories, and type D (that is D_0, D_1, D_2, and D_3) for assymtotically free theories (notice the sign in (2.75)) [54]. Now, we should wonder about the meaning of (2.75). Recall that our analysis relies upon the wilsonian coupling constant, so the meaning of u in (2.75) must be related to the scale in the $U(1)$ theory, i. e. the vacuum expectation value for the scalar field in the photon multiplet or, more properly, in the dual photon multiplet. This vacuum expectation value gives a mass to the hypermultiplets through the standard Yukawa coupling, so the singularity of (2.75) should be expected at $u = 0$, with u proportional to the mass of the hypermultiplet. Fortunately, we do know which hypermultiplet we should consider: the one defined by the monopole of the theory. In fact, we should rewrite (2.75) as

$$\tau^{mag}(u) = -\frac{ik}{2\pi} \ln(M(u)), \tag{2.77}$$

with $M(u)$ the mass of the monopole, and consider (2.77) perturbatively around the point u_0, where

$$M(u_0) = 0. \tag{2.78}$$

Therefore, we conclude that a singularity of A_0 type will appear whenever the mass of the monopole equals zero. The nature of the point u_0 is quite clear from a physical point of view: the magnetic effective coupling constant is zero, as can be seen from (2.77), so that the dual electric coupling should become infinity. But the point where the coupling constant is infinity is by definition the scale Λ of the theory; then, $u_0 = \Lambda$.

Now, it remains to discover how many singularities of A_0 type are there. In principle, a single point where the monopole becomes massless should be expected (the $u_0 = \Lambda$ point); however, as mentioned in section 2.2, the $U(1)_R$ symmetry is acting on the moduli space as a $\mathbf{Z}_2$ transformation. Therefore, in order to implement this symmetry, an extra singularity of A_0 type must exist. The simplest solution for the $N_f = 0$ theory, with $SU(2)$ gauge group, corresponds to an elliptic fibration over $\mathbb{P}^1$, the compactified u-plane, with three singular points, of type

$$D_0;\ A_0,\ A_0, \tag{2.79}$$

with D_0 the singularity at infinity, and the two A_0 singularities at the points $\pm\Lambda$, with Λ the scale of the theory.

What about the inclusion of flavors? In this case, we know that D_0 in (2.79) is replaced by D_{N_f}. The case $N_f = 2$ should be clear, as D_2 is equivalent to two A_1 singularities and therefore, we should expect

$$D_2;\ A_1,\ A_1. \tag{2.80}$$

The singularities of A_1 type indicate that two hypermultiplets become massless. Another simple case is that with $N_f = 4$, where there is a trivial monodromy D_4, which is now the monodromy around the origin. The two other cases of assymptotically free theories can be obtained through decoupling arguments, and taking into account the residual $U(1)_R$ symmetry. The results are [54]

$$\begin{aligned} D_1 \quad &; \quad A_0,\ A_0,\ A_0, \\ D_3 \quad &; \quad A_0,\ A_3. \end{aligned} \tag{2.81}$$

Now, with these elliptic fibrations, we shoud consider the complex structure. As we know from Kodaira's argument for the $N_f = 0$ case, the A_0 singularities correspond to a rational curve with a double singular point; as we know that this double singularity appears at $u = \pm\Lambda$, the simplest guess for the corresponding complex structure is, with $\Lambda = 1$,

$$y^2 = x^3 - x^2 u + x. \tag{2.82}$$

The curve (2.83), for generic u, does not have singular points. Recall that for a curve defined by $f(x,y;u)=0$, the singular points are those such that

$$\begin{aligned} F &= 0, \\ F_x = F_y &= 0, \end{aligned} \tag{2.83}$$

with F_x and F_y the derivatives with respect to x and y, respectively. The genus of the curve can be obtained using Riemann's theorem,

$$g = \frac{(n-1)(n-2)}{2} - \sum_p \frac{r_p(r_p-1)}{2}, \tag{2.84}$$

where the sum is over singular points, r_p is the order of the singularity, and n in (2.84) is the degree of the polynomial F, defining the curve. So, for generic u, we get, for (2.82), $g=1$.

Now, for $u=\pm 2$, we have a singular point satisfying (2.83), namely

$$\begin{aligned} y &= 0, \\ x &= \frac{u}{2}. \end{aligned} \tag{2.85}$$

This is a double point and therefore, using (2.84), we get $g=0$. From Kodaira's classification, we know that at this points we get two singularities of A_0 type. Notice also that at the origin, $u=0$, we have the curve $y^2=x^3+x$, which is of genus one, since there are no singular points. Moreover, if we take $\Lambda=0$, we get the curve

$$y^2 = x^3 - x^2 u. \tag{2.86}$$

This curve has a double point at $x=y=0$ for generic u. Using (2.84), we now get genus equal zero. Thus, the curve (2.82) satisfies all the properties derived above.

The curve (2.82) has a point at $x=y=\infty$. In order to compactify the curve, we must add the point at infinity. This can be done going to the projective curve

$$zy^2 = x^3 - zx^2u + z^2x. \tag{2.87}$$

The region at infinity of this curve is defined by $z=0$. The curve, in the three dimensional $R\to 0$ limit, can be described by (2.82), but with $\mathrm{Vol}(E_u)=\infty$. Next, we will see that this limit is equivalent to deleting the points at infinity of (2.87), i. e., the points with $z=0$. In fact, for $z\neq 0$ we cab define a new variable,

$$v = x - zu, \tag{2.88}$$

and write (2.87) as

$$zy^2 = x^2v + z^2x. \tag{2.89}$$

We can interpret (2.89) as defining a surface in the projective space $\mathbb{P}^3$, but (2.89) is in fact the Atiyah-Hitchin space in homogeneous coordinates. Thus, we conclude that the $R \to 0$ limit is equivalent to deleting the points at infinity of the curves E_u defined by (2.82).

We can see this phenomena in a different way as follows. The representation (2.82) of the Atiyah-Hitchin space is as an elliptic fibration, so that we have selected one complex structure. However, we can yet rotate in the space of complex structures, preserving the one selected by the elliptic fibration. This defines a $U(1)$ action. This $U(1)$ action must act on E_u; however, this is impossible if E_u is a compact torus. But when we delete the point at infinity, and pass to the projective curve (2.89), we have a well defined $U(1)$ action [37],

$$\begin{aligned} x &\to \lambda^2 x, \\ y &\to \lambda y, \\ v &\to \lambda^{-2} v. \end{aligned} \tag{2.90}$$

Only a $\mathbf{Z}_2$ subgroup of this action survives on u:

$$u \to \lambda^2 x - \lambda^{-2} v \equiv \hat{\lambda} u, \tag{2.91}$$

which means

$$\lambda^2 = \lambda^{-2} = \hat{\lambda}, \tag{2.92}$$

i. e., $\lambda^4 = 1$ or $\hat{\lambda}^2 = 1$. This $\mathbf{Z}_2$ action moves $u \to -u$, and is the only part of the $U(1)$ action surviving when we work in the four dimensional limit. More simply, at $z = 0$, i. e., at infinity, in the projective sense, $v = x$, and we get $\lambda^2 = \lambda^{-2}$, and the $\mathbf{Z}_4$ symmetry of (2.82) becomes

$$\begin{aligned} y &\to iy, \\ x &\to -x, \\ u &\to -u. \end{aligned} \tag{2.93}$$

Notice also the relation between Λ and the breaking of $U(1)$. In fact, for $\Lambda = 0$ we have

$$y^2 = x^3 - x^2 u, \tag{2.94}$$

which is invariant under

$$\begin{aligned} y &\to \lambda^3 y, \\ x &\to \lambda^2 x, \\ u &\to \lambda^2 u. \end{aligned} \tag{2.95}$$

2.6 Effective Superpotentials.

Maybe the most spectacular result derived from the Seiberg-Witten solution to $N = 2$ supersymmetric theories is the first dynamical proof of electric confinement. In order to properly understand this proof, we need first to go through the recent history of confinement. The simplest physical picture of confinement is that of dual BCS superconductivity theory [23, 55, 56]. In that picture, a confining vacua is to be represented as the dual of the standard superconducting vacua, which is characterized by the condensation of Cooper pairs. In ordinary superconductivity we find, under the name of Meisner effect, the mechanism for magnetic confinement. In a superconducting vacua, a monopole-antimonopole pair creates a magnetic flux tube that confines them. The relativistic Landau-Ginzburg description of superconductivity was first introduced by Nielsen and Olesen [57], where vortices in the Higgs phase are interpreted as Meisner magnetic flux tubes. The order parameter of the phase is the standard vacuum expectation value of the Higgs field; in this model, simply a scalar coupled to the $U(1)$ electric-magnetic field. The confined monopoles would be $U(1)$ Dirac monopoles, and the magnetic string is characterized by the Higgs mass of the photon. The dual version of this picture is in fact easy to imagine. We simply consider a dual photon, or dual $U(1)$ theory, now coupled to magnetic Higgs matter, a field representing the magnetic monopoles with magnetic $U(1)$ charge, and we look for a dual Higgs mechanism that, by a vacuum expectation value of the monopole field, will induce a Higgs mass for the dual magnetic photon. This mass gap will characterize the confinement phase. As the reader may realize, this whole picture of confinement is based on Higgs, or dual Higgs mechanisms for abelian gauge theories; however, in standard QCD, we expect confinement to be related to the very non abelian nature of the gauge groups. Indeed, only non abelian gauge theories are assymptotically free, and would possess the infrared slavery, or confinement, phenomena. Moreover, in a pure non abelian gauge theory, we do not have the right topology to define stable 't Hooft-Polyakov abelian monopoles, so the extesion of the superconductivity picture to the $N = 0$ pure Yang-Mills theory, or standard QCD, is far from being direct. Along the last two decades, with 't Hooft and Polyakov as leaders, some pictures for confinement have been sugested. Perhaps, the main steps in the story are

i) $2 + 1$ Polyakov quantum electrodynamics [23].
ii)'t Hooft $\mathbf{Z}(N)$ duality relations [56].
iii)'t Hooft twisted boundary conditions [20].
iv)'t Hooft abelian projection gauge [58].

Concerning *i*), we have already described the relevant dynamics in chapter I. Let us therefore now consider the other points. Concerning *ii*), the general idea is dealing with the topology underlying pure $SU(N)$ Yang-Mills theory, namely

$$\Pi_1(SU(N)/U(1)) \simeq \mathbf{Z}(N). \tag{2.96}$$

This is the condition for the existence of magnetic $\mathbf{Z}(N)$ vortices. The 't Hooft loop $B(C)$ is the magnetic analog of the Wilson loop $A(C)$, and was defined for creating a $\mathbf{Z}(N)$ magnetic flux tube along the path C. The Wilson criteria for confinement, $A(C)$ going like the area, has now its dual in $B(C)$ behaving like the perimeter, reproducing again the picture that dual Higgs is equivalent to confinement. The duality relations established by 't Hooft reduce to

$$A(C)B(C') = e^{2\pi i\nu(C,C')/N}B(C')A(C), \tag{2.97}$$

where $\nu(C,C')$ is the link number between the loops C and C'. From (2.96), the different posible phases compatible with duality were obtained. A way to make more quantitative the previous picture was also introduced by 't Hooft, by means of twisted boundary conditions in a box. Some of the main ingredients were already introduced in chapter I, but we will come back to them later on. In what follows of this section we will mainly be interested in the abelian projection gauge.

The idea of the abelian projection gauge was originally that of defining a unitary gauge, i. e., a gauge absent of ghosts. The simplest way to do it is first reducing the theory to an abelian one, and then fixing the gauge, which is (in the abelian theory) a certainly easier task. Using a formal notation, if G is the non abelian gauge group, and L is its maximal abelian subgroup, then the non abelian part is simply given by G/L, so that we can take, as the degrees of freedom for the abelian gauge theory, the space $R/(G/L)$, where R generically represents the whole space of gauge configurations. Now, the theory defined by $R/(G/L)$, is an abelian theory, and we can fix the gauge, going finally to the unitary gauge, characterized by $R/G = L\backslash R/(G/L)$. Now, two questions arise, concerning the content of the intermediate abelian theory, $R/(G/L)$, and the more important point of how such a theory should be defined. In order to fix the non abelian part of the gauge group, i. e., the piece G/L, 't Hooft used the following trick [58]: let X be a field that we can think of as a functional of A, $X(A)$, or an extra field that will be decoupled at the end. For the time being, we simply think of X as a functional, $X(A)$. We will require $X(A)$ to transform under the adjoint representation, i. e.,

$$X(A) \to gX(A)g^{-1}. \tag{2.98}$$

Now, the gauge condition that fixes the non abelian part of the gauge group is

$$X(A) = \begin{pmatrix} \lambda_l & & \\ & \ddots & \\ & & \lambda_N \end{pmatrix}. \tag{2.99}$$

Indeed, if $X(A)$ is diagonal, the residual group is just the maximal abelian subgroup. Notice that $X(A)$ is playing a similar role to a Higgs field in the adjoint representation, and (2.99) is what we will interpret as a vacuum expectation value, breaking the G symmetry to its maximal abelian subgroup. As in the standard Higgs mechanism, now the degrees of freedom are the diagonal parts of the gauge field, $A_\mu^{(ij)}$, that transform as $U(1)$ charged particles. In addition, we have the N scalars fields λ, appearing in (2.99). Summarizing, the particle content we get in the maximal abelian gauge is

i) $N-1$ photons, $A^{(ii)}$.
ii) $\frac{1}{2}N(N-1)$ charged particles, $A^{(ij)}$.
iii) N scalar fields, λ_i.

Notice that (2.99) does not require the λ_i to be constant; in fact, λ_i are fields depending on the spacetime position. Another important aspect of (2.99) is that, by means of this maximal abelian gauge we are not introducing, in principle, any form of potential for the λ_i fields, so that their expectation values are a priori undetermined. Concerning the previous spectrum, charged particles of type *ii*) can be considered formally massive, with the mass being proportional to $\lambda_i - \lambda_j$, as is the case in the standard Higgs mechanism.

The spectrum *i*), *ii*) and *iii*) is not complete. Extra spectrum, corresponding to singularities of the maximal abelian gauge, (2.99), is also allowed. These singularities correspond to points in spacetime, where $\lambda_i(x) = \lambda_{i+1}(x)$, i. e., where two eigenvalues coincide. We have impossed that $\lambda_i > \lambda_{i+1}$, i. e., the eigenvalues of (2.99) are ordered. These singularities are point-like in three dimensions, and $d-3$ dimensional for spaces of dimension d. It is easy to see that these singularities of the gauge (2.99) are 't Hooft-Polyakov monopoles. Once we have this set of degrees of freedom to describe the non abelian theory, we may proceed to consider the phenomenum of confinement, following in essence the same philosophy as in abelian superconductors. 't Hooft's rules of construction are:

R1 Eliminate the electric charges. This means constructing an effective lagrangian, where the "massive" electric particles $A^{(ii)}$ have been integrated out inside loops.

R2 Perform duality transformations on the effective lagrangian obtained upon the above integration of the electric charges, going to dual photons. These dual photons should interact with the charged monopoles by ordinary vertices, coupling the dual photon to two monopoles. The interaction between monopoles is certainly not reduced to the the single exchange of dual photons; there is in practice a missing link connecting the dual photon-monopole vertices, and the effective lagrangian, and which is played by the λ-fields: the Γ_{eff} action depends also on the λ-fields, that have Yukawa coupling with the charged $A^{(ij)}$ particles, running inside the loop. As we dualize, we should also take into account duality on these fields λ_i. In fact, this should be the most relevant part of our story, as it is the potential interaction between monopoles and the dual λ_i fields what naturally leads to next rule.

R3The expectation value $< M >$, for the theory obtained in R2, must be computed. In fact, this vacuum expectation value should be obtained after minimizing the theory with respect to the λ_i field values.

In spite of the beatiful physical structure underlying ‘t Hooft's approach, this program is far from being of practical use in standard QCD or pure Yang-Mills theory. However, progress in lattice computations is being made at present.

After this introduction to ‘t Hooft's abelian projection gauge, let us come back to the simpler example of $N=2$ pure Yang-Mills theory to find out the validity of the above rules. The careful reader wil have already found some similarities in our discussion and the way the Seiberg-Witten solution for $N=2$ supersymmetric Yang-Mills has been presented. In fact, in the $N=2$ theory, the X field can simply be interpreted as the Higgs field in the adjoint, breaking $SU(2)$ to $U(1)$ on generic points of the moduli (for a group of higher rank, r, the breaking is down to $U(1)^r$). Moreover, we also have the spectrum of ‘t Hooft-Polyakov monopoles and, according to degrees of freedom, we are certainly quite close to the abelian projection picture; however, we should be careful at this point. In ‘t Hooft's abelian projection, it was not assumed at any moment that we must be at a Higgs phase with well defined massive monopoles. The type of monopoles we find in the abelian projection gauge are not massive in the usual sense and, moreover, they have not finite size but are simply point like singularities.

Rule R1 is almost accomplished through the Seiberg-Witten solution [35, 36]. In fact, we can consider the effective lagrangian obtained from $\Gamma_{eff}(A_\mu^0, a)$, where A_μ^0 represents the photon, and a is the scalar field in the $N=2$ hypermultiplet (notice that this effective lagrangian is constrained to be $N = 2$ invariant). For each value of $u = \frac{1}{2} < \mathrm{tr}\phi^2 >$, the vacuum expectation value of the field a, in the perturbative regime, is simply

$$a(u) = \sqrt{2u}. \tag{2.100}$$

The effective lagrangian contains only one loop logarithmic contributions (see equation (2.36)), and instanton effects. The instanton and multiinstanton contributions contribute each with four fermionic zero modes, as we kill the four zero modes associated with superconformal transformations. The expansion of the effective lagrangian in perturbative and non perturbative effects can be done in the weak coupling regime and, if we know how to perform the duality trasnformation, we can start obtaining non trivial information on the strong coupling regime. Let us formally denote through $\Gamma_{eff}(A^D, a_D)$ the dual effective lagrangian. In the dual perturbative regime, the effective lagrangian is an expansion in one loop terms, corresponding to light magnetic monopoles, and non perturbative higher order terms. From the moduli space point of view, the dual perturbative expansion should appear as a good description of the infrared region, i. e., for values of u such that the electric constant is large, which are points at the neighbourhood of $u \simeq \Lambda$, with Λ the dynamically generated scale. To complete the equivalent dual description, the equivalent to expression (2.100) for the dual variable a_D should be constructed; impossing $N=2$ supersymmetry, we obtain that the dual theory has a coupling

$$a_D M\tilde{M}, \tag{2.101}$$

of Yukawa type for monopoles. Then, a_D is the mass of the monopole, in the very same way as the mass of $W^{\pm}$ particles, in the standard Higgs mechanism, is given by a. We can now write a general formula for electrically and magnetically charged particles,

$$M(n_e, n_m) = |n_e a + n_m a_D|. \tag{2.102}$$

Here, we have only motivated equation (2.102) from physical arguments but, as we will see, the mathematical and supersymmetric meaning of (2.102) goes far beyond the scope of the simple argument we have used.

Coming back to our problem of discovering $a_D(u)$, a proper description will require some results on Kähler geometry. In fact, we know that the metric on the u moduli, is certainly Kähler with respect to the complex structure distinguished by the elliptic fibration representation of the $N=4$ three dimensional moduli space. If it has a Kähler structure, the corresponding Kähler potential can be defined through

$$g_{u\bar{u}} = \mathrm{Im}\left(\frac{\partial^2 K}{\partial u \partial \bar{u}}\right). \tag{2.103}$$

This Kähler potential can be read out from the effective $N=2$ low energy action. In fact, as a general statement, the metric on the moduli space is given by the quadratic terms of the effective low energy lagrangian. Now, for $N=2$ the lagrangian can be written in terms of the so called prepotential as follows:

$$\mathcal{L} = \int d^4\theta \mathcal{F}(\mathcal{A}), \tag{2.104}$$

where $\mathcal{A}$ is an $N=2$ superfield, which is holomorphic or, in supersymmetric language, depends only on chiral fields. The Kähler potential is derived from $\mathcal{F}$ as

$$K = \mathrm{Im}\left(\frac{\partial \mathcal{F}}{\partial \mathcal{A}} \cdot \bar{\mathcal{A}}\right), \tag{2.105}$$

from which (2.103) becomes

$$g_{u\bar{u}} = \mathrm{Im}\left(\frac{\partial a_D}{\partial u}\frac{\partial \bar{a}}{\partial \bar{u}}\right), \tag{2.106}$$

where we have defined

$$a_D \equiv \frac{\partial \mathcal{F}}{\partial a}, \tag{2.107}$$

in the sense of lower components. Using (2.104) and (2.105) we get, for the metric,

$$ds^2 = \mathrm{Im}\frac{\partial^2 \mathcal{F}}{\partial a^2} da d\bar{a}, \tag{2.108}$$

and therefore we can identify $\frac{\partial^2 \mathcal{F}}{\partial a^2}$ with $\tau(u)$ or, equivalently,

$$\tau(u) = \frac{da_D}{da}. \tag{2.109}$$

Notice that equation (2.109) is perfectly consistent with what we expect for the definition of a_D, as it provides the mass of the monopole. In the perturbative regime, we know that it behaves like $\mathrm{Im}\tau \cdot a \simeq \frac{a}{g^2}$. Therefore, (2.107) is the right generalization. Fortunatelly, thanks to (2.109) and relation (2.66), we get a definite representation of $a_D(u)$ in terms of the Seiberg-Witten elliptic fibration solution,

$$\frac{da_D}{du} = \oint_\gamma \varphi(z;u) dz, \tag{2.110}$$

where φ is the holomorphic differential of E_u, which is given by

$$\varphi(z;u) = \frac{dx}{y}. \tag{2.111}$$

Now that we have a candidate for $a_D(u)$, we can continue our analysis following 't Hooft's rules (in fact, $a_D(u)$ or, equivalently, how to define the dual scalar field λ_i^D, was a missing part in 't Hooft's program). Next, we want to work out the dynamics of the monopoles. Until now, we have used $N=2$ dynamics, so that the fields a and a_D are part of our original lagrangian, and not a gauge artifact, as in 't Hooft's abelian projection gauge. However, if we softly break $N=2$ to $N=1$ [35] adding a mass term for the scalar fields,

$$m\mathrm{tr}\Phi^2, \tag{2.112}$$

then for large enough m the low energy theory is $N=1$, where the interpretation of the fields a and a_D should become closer and closer to the fields of the abelian projection. The soft breaking term (2.112) should reproduce 't Hooft's hidden dynamics governing the λ-fields. In fact, there is a simple procedure, discovered by Seiberg and Witten, to do that. The effect of (2.112) on the low energy description of the theory is to add a superfield $\mathcal{U}$, with lower component u, such that $< u > = < \mathrm{tr}\phi^2 >$, with superpotential

$$W = m\,\mathcal{U}. \tag{2.113}$$

This extra term contains in fact the dynamics about a_D fields we are looking for, so we can write (2.113) as

$$W = m\,\mathcal{U}(a_D), \tag{2.114}$$

and interpret it as a lagrangian term for a_D. The monopole dynamics is then controlled by a superpotential of type

$$W = a_D M \tilde{M} + m\,\mathcal{U}(a_D), \tag{2.115}$$

where the first term is the $N=2$ Yukawa coupling. Now, in order to fulfill rule R3, we only need to minimize the superpotential (2.115). Clearly, we get two minima with monopole vacuum expectation value given by

$$< M >= \pm \left(\frac{\partial \mathcal{U}}{\partial a_D}\right)^{1/2} m, \tag{2.116}$$

which is the desired proof of confinement. 't Hooft's program is then completed. In order to extend this approach to non supersymmetric theories, we can still use the the trick of adding a mass term for the X field; however, because of the lack of holomorphy, no translation of such procedure in the form of (2.114) is possible.

Instead of using the relation for $\mathcal{U}(a_D)$, we can try to get a more direct geometrical interpretation of (1.103): let us work with the curve (2.82), and consider the points A and B with $y=0$,

$$x^2 - xu + \Lambda^2 = 0. \tag{2.117}$$

Now, we can define the function

$$\mathcal{U}(x) = x + \frac{\Lambda^2}{x}. \tag{2.118}$$

The purpose of this function is giving a value of $\mathcal{U}$, such that x is one of the crossing points. Obviously, $U(x)$ posseses two minima, at

$$x = \pm\Lambda, \tag{2.119}$$

and therefore the superpotential $m\,\mathcal{U}$ has two minina, at $\pm\Lambda_1$, with Λ_1 the scale of the $N=1$ theory. Of course, the minima of $\mathcal{U}(x)$ take place when the tow points A and B coincide, i. e., at the singular nodal curves. Now, we can use the following heuristic argument to find out what happens in the three dimensional $R \to 0$ limit. In projective coordinates, the region at infinity of (2.82) is

$$zy^2 = x^3 - zx^2u + \Lambda^2 xz^2, \tag{2.120}$$

at $z=0$. If we delete the infinity point, i. e., the intersection of the projective curve C defined by (2.120) and $H_\infty = \{(x,y,0)\}$, and we then put $x^3=0$ in (2.120) we get, instead of (2.118) [37],

$$\mathcal{U}_{3D}(x) = \frac{\Lambda^2_{N=2}}{x}, \tag{2.121}$$

with $\Lambda^2_{N=2}$ the $N=2$ three dimensional scale.

3. Chapter III

Taking into account the enormous amount of good reviews and books [59, 61, 60, 62] in string theory, we will reduce ourselves in this section to simply stablishing some notation and motivating fundamental relations as mass formulas.

3.1 Bosonic String.

3.1.1 Classical Theory.. Let us start considering classical bosonic string theory in flat Minkowski spacetime. This physical system is characterized by the lagrangian

$$\mathcal{L} = -\frac{T}{2}\int d^2\sigma\sqrt{h}h^{\alpha\beta}\partial_\alpha X\partial_\beta X, \tag{3.1}$$

where $h^{\alpha\beta}$ is the worldsheet metric. The equations of motion, with respect to $h^{\alpha\beta}$, imply that

$$T_{\alpha\beta} = -\frac{2}{T}\frac{1}{\sqrt{h}}\frac{\delta S}{\delta h^{\alpha\beta}} = 0. \tag{3.2}$$

The parameter T in (3.1) has units of squared mass, and can be identified with the string tension,

$$T = \frac{1}{2\pi\alpha'}. \tag{3.3}$$

Using the Weyl invariance of (3.1), the gauge

$$h_{\alpha\beta} = \eta_{\alpha\beta} = \begin{pmatrix} -1 & 0 \\ 0 & 1 \end{pmatrix} \tag{3.4}$$

can be chosen. In this gauge, the equations of motion for (3.1) become

$$\Box X = 0. \tag{3.5}$$

Defining light cone coordinates,

$$\begin{aligned} \sigma^- &= \tau - \sigma, \\ \sigma^+ &= \tau + \sigma, \end{aligned} \tag{3.6}$$

the generic solution to (3.5) can be written as

$$X^\mu = X^\mu_R(\sigma^-) + X^\mu_L(\sigma^+). \tag{3.7}$$

Now, we will introduce open and closed strings. We will first work out the case of the closed bosonic string; in this case, we impose periodic boundary conditions,

$$X^\mu(\tau,\sigma) = X^\mu(\tau,\sigma+\pi). \tag{3.8}$$

The solution to (3.5), compatible with these boundary conditions, becomes

$$X_R^\mu = \frac{1}{2}x^\mu + \frac{1}{2}(2\alpha')p^\mu(\tau-\sigma) + i\sqrt{\frac{\alpha'}{2}}\sum_{n\neq 0}\frac{1}{n}\alpha_n^\mu e^{-2in(\tau-\sigma)},$$

$$X_L^\mu = \frac{1}{2}x^\mu + \frac{1}{2}(2\alpha')p^\mu(\tau+\sigma) + i\sqrt{\frac{\alpha'}{2}}\sum_{n\neq 0}\frac{1}{n}\tilde{\alpha}_n^\mu e^{-2in(\tau+\sigma)}. \tag{3.9}$$

Using this Fourier decomposition we get, for the hamiltonian,

$$H = \frac{1}{2}\left[\sum_{-\infty}^{\infty}\alpha_{m-n}\alpha_n + \sum_{-\infty}^{\infty}\tilde{\alpha}_{m-n}\tilde{\alpha}_n\right], \tag{3.10}$$

where we have used the notation

$$\alpha_0^\mu = \sqrt{\frac{\alpha'}{2}}p^\mu. \tag{3.11}$$

Using now (3.2), we get the classical mass formula

$$M^2 = \frac{2}{\alpha'}\sum_{n=1}^{\infty}(\alpha_{-n}\alpha_n + \tilde{\alpha}_{-n}\tilde{\alpha}_n). \tag{3.12}$$

The constraint (3.2) also implies that the left and right contributions to (3.12) are equal. Using the standard quantization rules,

$$\begin{aligned}
[\alpha_m^\mu, \tilde{\alpha}_n^\nu] &= 0, \\
[\alpha_m^\mu, \alpha_n^\nu] &= m\delta_{m+n}\eta^{\mu\nu}, \\
[\tilde{\alpha}_m^\mu, \tilde{\alpha}_n^\nu] &= m\delta_{m+n}\eta^{\mu\nu}, \\
[x^\mu, p^\nu] &= i\eta^{\mu\nu},
\end{aligned} \tag{3.13}$$

and taking into account the normal ordering factors we get, for $\alpha' = \frac{1}{2}$,

$$M^2 = -8a + 8\sum_{n=1}^{\infty}\tilde{\alpha}_{-n}\tilde{\alpha}_n = -8a + 8\sum_{n=1}^{\infty}\alpha_{-n}\alpha_n. \tag{3.14}$$

Two things are left free in deriving (3.14), the constant a, defining the zero point energy, and the number of dimensions of the target space. The classical way to fix these constants is impossing Lorenz invariance in the light cone gauge, where physical degrees of freedom are reduced to transversal oscillations. The result, for the closed bosonic string, is that a should equal one and the number of dimensions should be 26.

From (3.14), we can easily deduce the spectrum of massless states. First of all, we have a tachyon with no oscillator modes, and squared mass negative (-8). The massless modes are of the type

$$\alpha_{-1}^\mu\alpha_{-1}^\mu|0> . \tag{3.15}$$

To discover the meaning of these modes, we can see the way they transform under $SO(24)$ in the light cone gauge; then, we get three different types of particles: gravitons for the symmetric and traceless part, a dilaton for the trace part and, finally, the antisymmetric part.

3.1.2 Background Fields.. The simplest generalization of the worldsheet lagrangian (3.1) corresponds to including background fields. The obvious is the $G^{\mu\nu}$ metric of the target spacetime,

$$S_1 = -\frac{T}{2}\int d^2\sigma\sqrt{h}h_{\alpha\beta}G^{\mu\nu}(X)\partial_\alpha X_\mu \partial_\beta X_\nu . \tag{3.16}$$

However, not any background $G^{\mu\nu}$ is allowed, since we want to preserve Weyl invariance on the worldsheet. Scale invariance, for the two dimensional system defined by (3.16) is equivalent, from the quantum field theory point of view, to requiring a vanishing β-function. At one loop, the β-function for (3.16) is given by

$$\beta = -\frac{1}{2\pi}R, \tag{3.17}$$

for $\alpha' = \frac{1}{2}$, and with R the Ricci tensor of the target spacetime. Therefore, the first condition we require on allowed spacetime backgrounds is to be Ricci flat manifolds. We will allow the addition of extra manifolds to (3.16), namely the spectrum of massless particles of the bosonic closed string,

$$S = S_1 - \frac{T}{2}\int d^2\sigma \epsilon^{\alpha\beta}\partial_\alpha X^\mu \partial_\beta X^\nu B_{\mu\nu}(X) + \frac{1}{4}\int d^2\sigma\sqrt{h}\Phi(X)R^{(2)}, \tag{3.18}$$

where $R^{(2)}$ in (3.18) is the worldsheet curvature. α' does not appear in the last term due to dimensional reasons (the first two terms in (3.18) contain the X^μ field, which has length units).

Notice that for a constant dilaton field, the last term in (3.18) is simply

$$\chi \cdot \Phi, \tag{3.19}$$

with χ the Euler number; in terms of the genus, g, for a generic Riemann surface the Euler number is simply given by

$$\chi = 2 - 2g. \tag{3.20}$$

Thus, the powers of Φ in the partition function behave like $2 - 2g$. This topological number possesses a nice meaning in string theory: it is equal to the number of vertices joining three closed strings, needed to build up a Riemann surface of genus g. This naturally leads to a precise physical meaning of the dilaton background field: it is the string coupling constant,

$$g = e^{\Phi}. \tag{3.21}$$

Once the background fields in (3.18) have been added, the condition of Weyl invariance generalizes to vanishing β-funtions for G, B and Φ. At one loop, they are

$$\begin{aligned} R_{\mu\nu} + \frac{1}{4}H_\mu^{\lambda\rho}H_{\nu\lambda\rho} - 2D_\mu D_\nu \Phi &= 0, \\ D_\lambda H^\lambda_{\mu\nu} - 2(D_\lambda\Phi)H^\lambda_{\mu\nu} &= 0, \\ 4(D_\mu\Phi)^2 - 4D_\mu D^\mu\Phi + R + \frac{1}{12}H_{\mu\nu\rho}H^{\mu\nu\rho} + (D-26) &= 0, \end{aligned} \tag{3.22}$$

where $H_{\mu\nu\rho} = \partial_\mu B_{\nu\rho} + \partial_\rho B_{\mu\nu} + \partial_\nu B_{\rho\mu}$.

3.1.3 World Sheet Symmetries.. Before ending this quick survey on the bosonic string, let us mention an aspect of worldsheet symmetries. Worldsheet parity acts exchanging left and right oscillators,

$$\Omega : \alpha_n^\mu \leftrightarrow \tilde{\alpha}_n^\mu. \tag{3.23}$$

Among massless states (3.15), only the symmetric part (the graviton) is invariant under this transformation. We can now reduce the Hilbert space to states invariant under Ω. The inmediate effect of this on the worldsheet geometry is that a one loop surface can be defined in two ways: the opposite S^1 boundaries of a cylinder can be glued preserving orientation, to generate a torus, or up to an Ω trasnformation, giving rise to a Klein bottle.

3.1.4 Toroidal Compactifications.. A torus is a Ricci flat manifold that can be used as target spacetime. Let us consider the simplest case, $\mathbf{R}^{25} \times S^1$, where the compact S^1 dimension is taken to be of radius R. Then, the coordinate x^{25}, living on this S^1, must satisfy

$$x^{25} \equiv x^{25} + 2\pi n R. \tag{3.24}$$

If we now include the identification (3.24) in the mode expansion (3.9) we get, for the right and left momenta,

$$\begin{aligned} p_L &= \frac{m}{2R} - nR, \\ p_R &= \frac{m}{2R} + nR, \end{aligned} \tag{3.25}$$

while the mass formula becomes

$$M^2 = 4\left(\frac{m}{2R} - nR\right)^2 + 8(N-1) = 4\left(\frac{m}{2R} + nR\right)^2 + 8(\bar{N}-1), \tag{3.26}$$

with N and $\bar{N}$ the total level of left and right moving excitations, respectively. The first thing to be noticed, from (3.25), is the invariance under the transformation

$$\begin{aligned} T : R &\rightarrow \frac{1}{2R}, \\ m &\rightarrow n. \end{aligned} \tag{3.27}$$

A nice way to represent (3.25) is using a lattice of $(1,1)$ type, which will be referred to as $\Gamma^{1,1}$. This is an even lattice, as can be observed from (3.25),

$$p_L^2 - p_R^2 = 2mn. \tag{3.28}$$

If Π is the spacelike 1-plane where p_L lives, then $p_R \in \Pi^\perp$. In fact, p_L froms a θ[12] angle with the positive axis of the $\Gamma^{1,1}$ lattice, while p_R forms a negative angle, $-\theta$, and changes in R, which are simply changes in θ (or

[12] θ is the coordinate parametrizing the radius of the compact dimension.

Lorentz rotations in the $\Gamma^{1,1}$ hyperbolic space), are changes in the target space preserving the $\beta = 0$ condition, and therefore are what can be called the *moduli of the σ-model* (3.16). Of course, no change arises in the spectrum upon rotations of the Π and $\Pi^{\perp}$ planes. We have now obtained a good characterization of the moduli space for the string σ-model on a simple S^1 torus. However, in addition to rotations in Π and $\Pi^{\perp}$, we should also take into account the symmetry (3.27), representing rotations of the $\Gamma^{1,1}$ lattice.

The previous discussion can be generalized to compactifications on higher dimensional tori, T^d (i. e., working in a background spacetime $\mathbf{R}^{26-d} \times T^d$). In this case, (p_L, p_R) will belong to a lattice $\Gamma^{d,d}$, and the moduli space will be given by [64]

$$O(d,d;\mathbf{Z})\backslash O(d,d)/O(d) \times O(d), \tag{3.29}$$

where the $O(d,d;\mathbf{Z})$ piece generalizes the T-transformations (3.27) to T^d. From now we will call these transformations T-duality [65]. Notice also that the dimension of the moduli (3.29) is $d \cdot d$, which is the number of massles degrees of freedom that have been used to define the background fields of the σ-model (3.18). The manifold (3.29) is the first example of moduli of a σ-model we find; these moduli spaces will be compared, in next section, to the $K3$ moduli described.

3.1.5 σ-Model $K3$ Geometry. A First Look at Quantum Cohomology.. The concept of moduli space introduced in previous paragraph, for the σ-model (3.18), when the target space is a T^d torus, leading to manifold (3.29), can be generalized to more complicated spacetime geometries satisfying the constraints derived from conformal invariance, namely Ricci flat manifolds. This is a physical way to approach the theory of moduli spaces where, instead of working out the cohomology of the manifold, a string is forced to move on it, which allows to wonder about the moduli of the so defined conformal field theory. In order to properly use this approach, let us first review some facts about $K3$ geometry.

Let us first recall the relation between supersymmetry and the number of complex structures. Let us think of a σ-model, with target space $\mathcal{M}$. Now, we want this σ-model to be invariant under some supersymmetry transformations. It turns out that in order to make the σ-model, whose bosonic part is given by

$$\eta^{\mu\nu} g_{ij}(\phi(x))\partial_\mu \phi^i \partial_\nu \phi^j, \tag{3.30}$$

with η the metric on spacetime, and g the metric on the target, invariant under $N=2$ supersymmetry we have to require the manifold to be Kähler and, in order to be $N=4$, to be hyperkähler.

Let us now enter the description of the $K3$ manifold [66, 67, 68]. To characterize topologically $K3$, we will first obtain its Hodge diamond. The first property of $K3$ is that the canonical class,

$$K \equiv -c_1(T), \tag{3.31}$$

with $c_1(T)$ the first Chern class of the tangent bundle, T, is zero,

$$K = 0. \tag{3.32}$$

Equation (3.32) implies that there exists a holomorphic 2-form Ω, everywhere non vanishing. Using the fact that only constant holomorphic functions are globally defined, we easily derive, from (3.32), that

$$\dim H^{2,0} = h^{2,0} = 1. \tag{3.33}$$

In fact, if there are two different 2-forms Ω_1 and Ω_2, then Ω_1/Ω_2 will be holomorphic and globally defined, and therefore constant.

The second important property characterizing $K3$ is

$$\Pi_1 = 0, \tag{3.34}$$

so that

$$h^{1,0} = h^{0,1} = 0, \tag{3.35}$$

as $b_1 = h^{1,0} = h^{0,1} = 0$, because of (3.34).

The Euler number can be now derived using Noether-Riemann theorem, and property (3.32), and it turns out to be 24. Using now the decomposition of the Euler number as an alternating sum of Betti numbers, we can complete the Hodge diamond,

$$24 = b_0 - b_1 + b_2 - b_3 + b_4 = 1 - 0 + b_2 - 0 + 1, \tag{3.36}$$

which implies that

$$\dim H^2 = 22, \tag{3.37}$$

and therefore, from (3.33), we get

$$\dim H^{1,1} = h^{1,1} = 20, \tag{3.38}$$

leading to the Hodge diamond

$$\begin{array}{ccccc} & & 1 & & \\ & 0 & & 0 & \\ 1 & & 20 & & 1 \\ & 0 & & 0 & \\ & & 1 & & \end{array} \tag{3.39}$$

Using Hirzebuch's pairing, we can give an inner product to the 22 dimensional space H^2. In homology terms, we have

$$\alpha_1 \cdot \alpha_2 = \#(\alpha_1 \cap \alpha_2), \tag{3.40}$$

with $\alpha_1, \alpha_2 \in H^2(X, Z)$, and $\#(\alpha_1 \cap \alpha_2)$ the number of oriented intersections. From the signature complex,

$$\tau = \int_X \frac{1}{3}(c_1^2 - 2c_2) = -\frac{2}{3}\int_X c_2 = -\frac{2 \cdot 24}{3} = -16, \tag{3.41}$$

we know that $H^2(X, Z)$ is a lattice of signature $(3, 19)$. The lattice turns out to be self dual, i. e., there exits a basis α_i^* such that

$$\alpha_i \cdot \alpha_j^* = \delta_{ij}, \tag{3.42}$$

and even,

$$\alpha \cdot \alpha \in 2\mathbf{Z}, \qquad \forall \alpha \in H^2(X, \mathbf{Z}). \tag{3.43}$$

Fortunatelly, lattices with these characteristics are unique up to isometries. In fact, the $(3, 19)$ lattice can be represented as

$$E_8 \perp E_8 \perp \mathcal{U} \perp \mathcal{U} \perp \mathcal{U}, \tag{3.44}$$

with $\mathcal{U}$ the hyperbolic plane, with lattice $(1, 1)$, and E_8 the lattice of $(0, 8)$ signature, defined by the Cartan algebra of E_8. The appearance of E_8 in $K3$ will be at the very core of future relations between $K3$ and string theory, mainly in connection with the heterotic string.

Next, we should separetely characterize the complex structure and the metric of $K3$. Recall that this is exactly what we did in our study of the moduli of $N=4$ supersymmetric three dimensional Yang-Mills theories. Concerning the complex structure, the proper tool to be used is Torelli's theorem, that stablishes that the complex structure of a $K3$ marked surface[13] is completely determined by the periods of the holomorphic 2-form, Ω. Thus, the complex structure is fixed by

i) The holomorphic form Ω.
ii) A marking.

To characterize $\Omega \in H^{2,0}(X, \mathbf{C})$, we can write

$$\Omega = x + iy, \tag{3.45}$$

with x and y in $H^2(X, \mathbf{R})$, that we identify with the space $\mathbf{R}^{3,19}$. Now, we know that

$$\begin{aligned} \int_X \Omega \wedge \Omega &= 0, \\ \int_X \Omega \wedge \bar{\Omega} &> 0, \end{aligned} \tag{3.46}$$

and we derive

$$\begin{aligned} x \cdot y &= 0, \\ x \cdot x &= y \cdot y. \end{aligned} \tag{3.47}$$

[13] By a marked $K3$ surface we mean a specific map of $H^2(X, \mathbf{Z})$ into the lattice (3.44), that we will denote, from now on, $\Gamma_{3,19}$.

Therefore, associated with Ω, we define a plane of vectors $v = nx + my$ which, due to (3.46), is space-like, i. e.,

$$v \cdot v > 0. \tag{3.48}$$

The choice of (3.45) fixes an orientation of the two plane, that changes upon complex conjugation. Thus, the moduli space of complex structures of $K3$, will reduce to simply the space of oriented space-like 2-planes in $\mathbf{R}^{3,19}$. To describe this space, we can use a Grassmanian [67],

$$Gr = \frac{(O(3,19))^+}{(O(2) \times O(1,19))^+}, \tag{3.49}$$

where $(\ )^+$ stands for the part of the group preserving orientation. If, instead of working with the particular marking we have been using, we change it, the result turns out to be an isometry of the $\Gamma^{3,19}$ lattice; let us refer to this group by $O(\Gamma^{3,19})$. The moduli then becomes

$$\mathcal{M}^C = Gr/O^+(\Gamma^{3,19}). \tag{3.50}$$

The group $O(\Gamma^{3,19})$ is the analog to the modular group, when we work out the moduli space of complex structures for a Riemann surface ($Sl(2,\mathbf{Z})$ for a torus).

Let us now make some comments on the distinguished complex structure we have used in the study of the moduli of the three dimensional $N = 4$ theories. This complex structure is such that the the elliptic curve is a $(1,1)$-form, and is characterized by the 2-form

$$\Omega = du \wedge \frac{dx}{y}, \tag{3.51}$$

with $\frac{dx}{y}$ the holomorphic differential on the elliptic fiber. However, before entering a more detailed discussion on this issue, let us consider the question of metrics. Once a complex structure has been introduced, we have a Hodge decomposition of H^2, as

$$H^2 = H^{2,0} \oplus H^{1,1} \oplus H^{0,2}. \tag{3.52}$$

Thus, relative to a complex structure characterized by Ω, the Kähler form J in $H^{1,1}$ is orthogonal to Ω, and such that

$$\mathrm{Vol} = \int_X J \wedge J > 0, \tag{3.53}$$

which means that J is represented by a space-like vector in $\mathbf{R}^{3,19}$ and, therefore, together with Ω, spans the whole three dimensional space-like subspace of $\mathbf{R}^{3,19}$. Yau's theorem now shows how the metric is completely determined by J and Ω, i. e., by a space-like 3-plane in $\mathbf{R}^{3,19}$. Thus, we are in a smilar position to the characterization of the moduli space of complex structures, and we end up with a Grassmannian manifold of three space-like planes in $\mathbf{R}^{3,19}$,

$$Gr = O(3,19)/O(3) \times O(19). \tag{3.54}$$

Now, we need to complete Gr with two extra ingredients. One is the volume of the manifold, that can change by dilatations, and the other is again the modular part, corresponding to isometries of $\Gamma^{3,19}$, so that finally we get

$$\mathcal{M}^M = O(\Gamma_{3,19}) \backslash Gr \times \mathbf{R}^+. \tag{3.55}$$

Hence, the moduli of the σ-model (3.18), defined on a $K3$ surface, will contain the moduli of Einstein metrics on $K3$ (see equations (3.54) and (3.55)). Now, the dimension of manifold (3.26) is 58. For the σ-model (3.18) we must also take into account the moduli of B-backgrounds. In the string action, what we have is the integral, $\int B$, over the worldsheet, which now becomes a 2-cycle of $K3$; thus, the moduli of B-backgrounds is given by the second Betti number of the $K3$ manifold, which is 22. Finally, the dilaton field Φ has to be taken into account in (3.18). As mentioned, if Φ is constant, as we will require, it counts the number of loops in the perturbation series, so we will not consider it as an extra moduli. More precisely, we will probe the $K3$ geometry working at tree level in string theory. Under these conditions, the σ moduli space is of dimension [69]

$$58 + 22 = 80, \tag{3.56}$$

and the natural guess is the manifold

$$\mathcal{M}^\sigma = O(4,20)/O(4) \times O(20). \tag{3.57}$$

Naturally, this is not the final answer, as we have not divided yet by the equivalent to the T-duality trasnformations in the toroidal case, which are, for $K3$, isometries of the $H^2(X;\mathbf{Z})$ lattice, i. e.,

$$O(\Gamma^{3,19}). \tag{3.58}$$

However, the final answer is not the quotient of (3.57) by (3.58), as an important symmetry from the point of view of conformal field theory is yet being missed: mirror symmetry. In order to get a geometrical understanding of mirror symmetry [70], we need first to define the Picard lattice.

Let us consider curves inside the $K3$ manifold. The Picard lattice is defined as

$$\mathrm{Pic}(X) = H^{1,1}(X) \cap H^2(S,\mathbf{Z}), \tag{3.59}$$

which means curves (i. e., 2-cycles) holomorphically embedded in X. By definition (3.59), $\mathrm{Pic}(X)$ defines a sublattice of $H^2(S;\mathbf{Z})$. This Picard lattice has signature $(18,t)$. Let us consider, as an example, an elliptic fibration where the base is a 2-cycle B, and F is the fiber. The Picard lattice defined by these two 2-cycles is given by

$$\begin{aligned} B \cdot B &= -2, \\ B \cdot F &= 1, \\ F \cdot F &= 0, \end{aligned} \tag{3.60}$$

which is a lattice of $(1,1)$ type. Self intersections are given by the general expression

$$C \cdot C = 2(g-1), \tag{3.61}$$

where g is the genus, so that for $g = 0$, the base space, we get -2, and for the elliptic fiber, with $g = 1$, we get 0 for the intersection. The intersection between the base and the fiber, $B \cdot F$, reflects the nature of the fibration. Notice that expression (3.61) is consistent with the even nature of the lattice $\Gamma_{3,19}$. Now, from (3.59), it is clear that the number of curves we have in $\mathrm{Pic}(X)$ depends on the complex structure. Taking this fact into account, we can ask ourselves about the moduli space of complex structures preserving a given Picard sublattice; for instance, we can be interested in the moduli space of elliptic fibrations preserving the structure of the fibration. As $\mathrm{Pic}(X)$ are elements in $H^{1,1}(X)$, they should be orthogonal to Ω, so the moduli we are looking for will be defined in terms of the Grassmannian of space-like 2-planes in $\mathbf{R}^{2,19-t}$, i. e.,

$$Gr^P = O(2, 19-t)/O(2) \times O(19-t), \tag{3.62}$$

where we should again quotient by the corresponding modular group. This modular group will be given by isometries of the lattice Λ, called the transcendental lattice, and is simply defined as the orthogonal complement to the Picard lattice. Thus, Λ is of $\Gamma^{2,19-t}$ type, and the moduli preserving the Picard group is

$$\mathcal{M}^P = Gr^P/O(\Lambda). \tag{3.63}$$

As is clear from (3.62), the dimension of the moduli space of complex structures preserving the Picard group, reduces in an amount given by the value of t for the Picard lattice. At this point of the discussion, a question at the core of mirror symmetry comes naturally to our mind, concerning the posibility to define a manifold X^* whose Picard group is the transcendental lattice Λ of X [71]. In these terms, the answer is clearly negative, as the Picard lattice is of signature $(1,t)$, and Λ is of signature $(2, 19-t)$, so that we need either passing from Λ to a $(1,t')$ lattice, or generalize the concept of Picard lattice, admiting lattices of signature $(2,t)$. It turns out that both approaches are equivalent, but the second has a more physical flavor; in order to get from Λ a Picard lattice, what we can do is to introduce an isotropic vector f in Λ, and define the new lattice through

$$f^{\perp}/f, \tag{3.64}$$

which is of $(1, 18-t)$ type; now, the mirror manifold X^* is defined as the manifold possesing as Picard lattice the one defined by (3.64). The moduli space of the mirror manifold is therefore given by the equivalent to expression (3.62),

$$Gr^{*P} = O(2, t+1)/O(2) \times O(t+1). \tag{3.65}$$

Then, we observe that the dimension of the two moduli spaces sums up to 20, and that the dimension of the moduli space of the mirror manifold is exactly given by the rank $t+1$ of the Picard of the original moduli space.

A different approach will consist in definig the so called quantum Picard lattice. Given a Picard lattice of signature $(1, t)$, we define its quantum analog as the lattice of signature $(2, t+1)$, obtained after multiplying by the hyperbolic lattice $\Gamma^{1,1}$. So, the question of mirror will be that of given a manifold X, with transcendental lattice Λ, finding a manifold X^* such that its quantum Picard lattice is precisely Λ. Now, we observe that the quantum Picard lattices of X and X^* produce a lattice of signature $(4, 20)$. The automorphisms $O(\Gamma^{4,20})$ will result of compossing the T-duality transformations and mirror symmetry. Coming back to (3.57), and including mirror symmetry, we get, as moduli space of the σ-model on $K3$,

$$O(4, 20; \mathbf{Z})\backslash O(4, 20)/O(4) \times O(20). \tag{3.66}$$

This concludes our analysis of σ-models on $K3$.

3.1.6 Elliptically Fibered $K3$ and Mirror Symmetry.. We are now going to consider singularities in the $K3$ manifold. Let C be a rational curve in the $K3$ manifold; then, by equation (3.61), $C \cdot C = -2$. If the curve C is holomorphically embedded it will be an element of the Picard lattice. Its volume is defined as

$$\mathrm{Vol}(C) = J \cdot C, \tag{3.67}$$

with J the Kähler class. A singularity will appear whenever the volume of C goes zero, i. e., whenever the Kähler class J is orthogonal to C. Notice that this implies that C should be orthogonal to the whole 3-plane defined by Ω and J, as C is in fact $(1, 1)$, and therefore orthogonal to Ω.

Now, we can define the process of blowing up or down a curve C in X. In fact, a way to blow up is simply changing the moduli space of metrics J, until $J \cdot C$ becomes different from zero. The opposite is the blow down of the curve. The other way to get rid off the singularity is simply changing the complex structure in such a way that the curve is not in $H^{1,1}$, i. e., the curve does not exist anymore.

We can have different types of singularities, according to how many rational curves C_i are orthogonal to J. The type of singularity will be given by the lattice generated by these C_i curves. Again, these lattices would be characterized by Dynkin diagrams.

Let us now consider an elliptically fibered $K3$ manifold,

$$E \to X \to B. \tag{3.68}$$

Now, we can come back to Kodaira's analysis on elliptic fibrations, as presented in chapter II. Elliptic singularities of Kodaira type are characterized by the set of irreducible components X_i of the corresponding singularities. The Picard lattice for these elliptic fibrations contains the $\Gamma^{1,1}$ lattice generated by the fiber and the base, and the contribution of each singularity as given by the Shioda-Tate formula [71]. Defining the Picard number $\rho(X)$ as $1+t$ for a Picard lattice of type $(1,t)$ we get

$$\rho(X) = 2 + \sum_{\nu} \sigma(F_{\nu}), \tag{3.69}$$

where the sum is over the set of singularities, and where σ is given by $\sigma(A_{n-1}) = n-1$, $\sigma(D_{n+4}) = n+4$, $\sigma(E_6) = 6$, $\sigma(E_7) = 7$, $\sigma(E_8) = 8$, $\sigma(IV) = 2$, $\sigma(III) = 1$, $\sigma(II) = 0$. Equation (3.69) is true provided the Mordell-Weyl group of sections is trivial.

As described in the previous section, the mirror map goes from a manifold X, with Picard lattice of type $(1,t)$, to X^*, with Picard lattice $(1, 18-t)$ or, equivalently,

$$\rho(X) + \rho(X^*) = 20. \tag{3.70}$$

Through mirror, we can then pass from an elliptically fibered $K3$ surface, with Picard number $\rho(X) = 2$, which should for instance have all its singularities of type A_0, to a $K3$ surface of Picard number $\rho(X^*) = 18$, which should have 16 singularities of A_1 type, or some other combination of singularities.

3.1.7 The Open Bosonic String.. Repeating previous comments on closed strings for the open case is straightforward. The only crucial point is deciding the type of boundary conditions to be imposed. From (3.1), we get boundary terms of the form

$$\frac{T}{2} \int \partial X^{\mu} \partial_n X_{\mu}, \tag{3.71}$$

with ∂_n the normal boundary derivative. In order to avoid momentum flow away form the string, it is natural to imposse Neumann boundary conditions,

$$\partial_n X_{\mu} = 0. \tag{3.72}$$

Using these boundary conditions the mode expansion (3.9) becomes, for the open string,

$$X^{\mu}(\sigma, \tau) = x^{\mu} + 2\alpha' p^{\mu} \tau + i\sqrt{2\alpha'} \sum_{n \neq 0} \frac{1}{n} \alpha_n^{\mu} e^{-in\tau} \cos n\sigma, \tag{3.73}$$

and the quantum mass formula (3.14) is, for $\alpha' = \frac{1}{2}$,

$$M^2 = -2 + 2 \sum_{n=1}^{\infty} \alpha_{-n} \alpha_n. \tag{3.74}$$

Now, the first surprise arises when trying to generalize the T-duality symmetry, (3.27), to the open string case.

3.1.8 D-Branes.. By introducing the complex coordinate

$$z = \sigma^2 + i\sigma, \tag{3.75}$$

with $\sigma^2 \equiv i\tau$, (3.73) can be rewritten as

$$X^\mu(\sigma,\tau) = x^\mu - i\alpha' p^\mu \ln(z\bar{z}) + i\sqrt{\frac{\alpha'}{2}}\sum_{n\neq 0}\frac{1}{n}\alpha_n^\mu(z^{-n}+\bar{z}^{-n}). \tag{3.76}$$

Let us now consider the open string moving in $\mathbf{R}^{25}\times S^1$. Neumann boundary conditions in the compactified direction are

$$\partial_n X^{25} = 0. \tag{3.77}$$

Now, we will work out the way these boundary conditions modify under the $R \to \frac{1}{R}$ transformation [72]. To visualize the answer, we will consider the cylinder swept out by a time evolving closed string, both from the closed and open string pictures (in the open string picture the cylinder can be understood as an open string with both ends at the S^1 edges of the cylinder). In fact, from the open string point of view, the propagation of the string is at tree level, while the open string approach is a one loop effect. We will now assume that the S^1 boundary circles of the cylinder are in the 25 direction. Recalling then what happens in the closed string case, under change (3.27), the mode expansion (3.9) turns (3.27) equivalent to the change

$$\tilde{\alpha}_n^{25} \to -\tilde{\alpha}_n^{25}. \tag{3.78}$$

In the $n=0$ case we get, from (3.11) and (3.25) (with $\alpha' = \frac{1}{2}$),

$$\alpha_0^{25} = \frac{m}{2R} - nR \to nR - \frac{m}{2R} = -\tilde{\alpha}_0^{25}. \tag{3.79}$$

What this means is that the theory in the dual circle of radius $\frac{1}{2R}$ is equivalent to a theory on a circle of radius R, but written in terms of a new space coordinate Y^{25}, defined from X^{25} by the change (3.78). Now, it easy to see that

$$\partial_\alpha Y^{25} = \epsilon_{\alpha\beta}\partial^\beta X^{25}. \tag{3.80}$$

Returning now to the cylinder image described above, let us consider boundary conditions in the open string picture. From the closed string approach, they will be represented as

$$\partial_\tau X^{25} = 0. \tag{3.81}$$

Now, after performing the duality transformation (3.27), equation (3.80) implies

$$\partial_\sigma Y^{25} = 0, \tag{3.82}$$

that, from the open string point of view, looks as Dirichlet boundary conditions, so that the extreme points of the open string do not move in time in the 25 direction. Summarizing, we observe that under $R \to \frac{\alpha'}{R}$, Neumann and Dirichlet boundary conditions for the open string are exchanged. Besides, the picture we get if the end points of the open string do not move in the 25 direction is that of D-brane hypersurfaces, with fixed 25 coordinate, where the open string should end.

For a better understanding of the dynamical nature of these D-brane hypersurfaces, and their physical meaning, the above approach must be generalized to include several D-brane hypersurfaces; the tool needed comes from the old fashioned primitive string theory, interpreted as a meson model: the Chan-Paton factors [73].

3.1.9 Chan-Paton Factors and Wilson Lines.. Chan-paton factors are simply defined encoding the end points of the open string with labels i, j, with $i, j = 1, \ldots, N$. The corresponding string states will be defined as $|k; i, j>$. Let us now define a set of $N \times N$ matrices, $\lambda^a_{N\times N}$, hermitian and unitary, which define the adjoint representation of $U(N)$. We can now define the open string state $|k; a>$ as

$$|k; a> = \sum_{i,j} \lambda^a_{i,j} |k; i, j> . \tag{3.83}$$

The string states $|i, j>$ can now be easily interpreted in the language of gauge theories. In order to do that, we will again use the abelian projection introduced in previous chapter. In the abelian projection gauge, states $|i, i>$ correspond to $U(1)$ photons, while $|i, j>$ states (non diagonal components of the gauge field) correspond to charged massive particles. The way they transform under the abelian $U(1)^N$ group is

$$|i, j> \to e^{i(\alpha_j - \alpha_i)} |i, j>, \tag{3.84}$$

for the abelian transformation

$$\begin{pmatrix} e^{i\alpha_1} & & \\ & \ddots & \\ & & e^{i\alpha_N} \end{pmatrix} . \tag{3.85}$$

As discussed in chapter II, to define an abelian projection gauge, a field X must be chosen to transform in the adjoint representation; then, the gauge is fixed through imposing X to be diagonal. A simple example of field X is a Wilson line. So, let us assume we are working in $\mathbf{R}^{25} \times S^1$, and define X as the Wilson line in the 25 compactified direction. Choosing X diagonal means taking A^{25} in the abelian group $U(1)^N$; a diagonal Wilson line is obtained from

$$A^{25} = \frac{1}{2\pi R} \begin{pmatrix} \theta_1 & & \\ & \ddots & \\ & & \theta_N \end{pmatrix} , \tag{3.86}$$

corresponding to a pure gauge

$$A^{25} = \partial_{25}\Lambda = \partial_{25}\frac{X^{25}}{2\pi R}\begin{pmatrix} \theta_1 & & \\ & \ddots & \\ & & \theta_N \end{pmatrix}. \qquad (3.87)$$

Now, $\{\theta_1, \dots, \theta_N\}$ are the analogs to $\{\lambda_1, \dots, \lambda_N\}$, used in the standard abelian projection. The effect of the Wilson line (3.86) on a charged state $|i,j>$ is transforming it in the way (3.84) defines, which in particular means that the p^{25} momentum of the $|i,j>$ state becomes

$$p^{25} = \frac{n}{R} + \frac{\theta_j - \theta_i}{2\pi R}. \qquad (3.88)$$

When moving from R to $R' = \frac{1}{2R}$, the momentum (3.88) turns into a winding,

$$2nR' + (\theta_j R' - \theta_i R')\frac{1}{\Gamma}. \qquad (3.89)$$

The geometrical meaning of (3.89) is quite clear: the open string can wind around the dual circle of radius R' any number of times, but its end points are fixed, as expected after the $R \to R'$ duality transformation, to be in $\theta_j R'$ and $\theta_i R'$ positions. Thus, the picture we get is that of several D-brane hypersurfaces fixed in the dual circle to be at positions $\theta_1 R', \dots, \theta_N R'$, and the string states of type $|i,j>$ are now living between the ith and j^{th} D-brane hypersurface.

Using mass formula (3.26), and equation (3.88) for the momentum, we observe that only $\alpha_{-1}^{\mu}|i,i>$ states can be massless (the $U(1)$ photons), and the mass of the $\alpha_{-1}^{\mu}|i,j>$ states goes like $\left(\frac{(\theta_i - \theta_j)R'}{\Gamma}\right)^2$. Both of these states have the kinematical index μ in the uncompactified directions. We can also consider the massless Kaluza-Klein states, $\alpha_{-1}^{25}|i,i>$, which can be interpreted as scalars living on the 24 dimensional space defined by the D-brane hypersurface. However, this spectrum is the abelian projected gauge spectrum for a $U(N)$ gauge theory, now defined on the D-brane hypersurface. Therefore, two complementary pictures arise,

- The distribution of D-branes represents a new type of background for string theory, where a $U(N)$ Wilson line has been introduced in the internal or compactified S^1.
- The distribution of D-branes provides, for the massless spectrum, a geometrical representation of a gauge theory living on the worldvolume of the D-brane. Moreover, the spectrum is presented as the abelian projection spectrum.

Of course, this second approach only takes into account, as is usual in string theory, low energy degrees of freedom. Properly speaking, what we are doing is embedding the gauge theory into string theory in a new way.

To end this first contact with D-branes (for more details see, for instamce, [60], and references therein) we should, at least qualitatively, answer the question possed above on the dynamical nature of D-branes. The simplest answer will be obtained analizing the gravitational interactions through the computation of the mass density, leading to the tension of the D-brane hypersurface. A graviton, which is a closed string state can couple a D-brane, defining an interaction vertex. The disc coupling the graviton to the D-brane can be interpreted in terms of open strings ending on its circle boundary. Without performing any computation, we already know something on the order of magnitude of the process: it is a process determined by the topology of a disc, with half the Euler number of a sphere, so the order in the string coupling constant, defined in (3.21), is $O(\frac{1}{g})$.

A more detailed discussion on D-branes needs the use of more general string theories (superstring theories), which is what we will discuss in next section.

3.2 Superstring Theories.

Superstrings correspond to the supersymmetric generalization of the σ-model (3.1). This is done adding the fermionic term

$$S_F = \int d^2\sigma i\bar{\psi}^\mu \rho^\alpha \partial_\alpha \psi_\mu, \tag{3.90}$$

where ψ^μ are spinors, relative to the worldsheet, and vectors with respect to the spacetime Lorentz group, $SO(1, D-1)$. Spinors in (3.90) are real Majorana spinors, and the Dirac matrices ρ^α, $\alpha = 0, 1$, are defined by

$$\begin{aligned} \rho^0 &= \begin{pmatrix} 0 & -i \\ i & 0 \end{pmatrix}, \\ \rho^1 &= \begin{pmatrix} 0 & i \\ i & 0 \end{pmatrix}, \end{aligned} \tag{3.91}$$

satisfying

$$\{\rho^\alpha, \rho^\beta\} = -2\eta^{\alpha\beta}. \tag{3.92}$$

The supersymmetry transformations are defined by

$$\begin{aligned} \delta x^\mu &= \bar{\epsilon}\psi^\mu, \\ \delta\psi^\mu &= -i\rho^\alpha \partial_\alpha x^\mu \epsilon, \end{aligned} \tag{3.93}$$

with ϵ a constant anticonmuting spinor. Defining the components

$$\psi^\mu = \begin{pmatrix} \psi^\mu_- \\ \psi^\mu_+ \end{pmatrix}, \tag{3.94}$$

the fermionic lagrangian (3.90) can be written as

$$S_F = \int d^2\sigma (\psi_-^\mu \partial_+ \psi_-^\mu + \psi_+^\mu \partial_- \psi_+^\mu), \tag{3.95}$$

with $\partial_\pm \equiv \frac{1}{2}(\partial_\tau \pm \partial_\sigma)$. As was the case for the bosonic string, we need now to specify the boundary conditions for the fermion fields, both in the open and closed string case. For open strings, there are two posibilities:

$$\begin{aligned} \text{Ramond} &: \quad \psi_+^\mu(\pi,\tau) = \psi_-^\mu(\pi,\tau), \\ \text{Neveu-Schwarz} &: \quad \psi_+^\mu(\pi,\tau) = -\psi_-^\mu(\pi,\tau), \end{aligned} \tag{3.96}$$

which produce the mode expansions

$$\begin{aligned} \text{Ramond} &: \quad \psi_\mp^\mu = \frac{1}{\sqrt{2}} \sum_{z} d_n^\mu e^{-in(\tau \mp \sigma)}, \\ \text{Neveu-Schwarz} &: \quad \psi_\mp^\mu = \frac{1}{\sqrt{2}} \sum_{z+\frac{1}{2}} b_n^\mu e^{-in(\tau \mp \sigma)}. \end{aligned} \tag{3.97}$$

In the case of closed strings, we can impose either periodic or antiperiodic boundary conditions for the fermions, obtaining Ramond (R) or Neveu-Schwarz (NS) for both $\psi_\pm$ fields. After quantization we get, following similar steps to those in the bosonic case, that the critical dimension is 10, and that the mass formulas and normal ordering correlators are given by

$$M^2 = 2(N_L - \delta_L) = 2(N_R - \delta_R), \tag{3.98}$$

with $\delta = \frac{1}{2}$ in the NS sector, and $\delta = 0$ in the R sector. Using this formula, and the GSO projection, we easily get the massless spectrum. For the closed string we get

$$\begin{aligned} \text{NS-NS sector} &: \quad b_{-1/2}^\mu b_{-1/2}^\nu |0>, \\ \text{NS-R sector} &: \quad b_{-1/2}^\mu |S>, \\ \text{R-R sector} &: \quad |S> \otimes |S> . \end{aligned} \tag{3.99}$$

The state $|S>$ corresponds to the Ramond vacua (recall $\delta = 0$ in the Ramond sector).

The d_0^μ oscillators in (3.97) define a Clifford algebra,

$$\{d_0^\mu, d_0^\nu\} = \eta^{\mu\nu}, \tag{3.100}$$

and therefore the $|S>$ vacua can be one of the two $\mathbf{8}_S$, $\mathbf{8}_{S'}$ spinorial representations of $SO(8)$. Depending on what is the spinorial representation chosen we get, from (3.99), two different superstring theories. In the chiral case, we choose the same chirality for the two fermionic states in the NS-R and R-NS sectors. This will lead to two gravitinos of equal chirality. Moreover, in the R-R sector we get, for same chirality,

$$\mathbf{8}_S \times \mathbf{8}_S = \mathbf{1} \oplus \mathbf{28} \oplus \mathbf{35}_S, \tag{3.101}$$

corresponding to a scalar field being identified with the axion, an antisymmetric field, and a 4-form field. We will call this superstring theory type IIB. In case we choose different chiralities for the spinor representations associated with the Ramond vacua, what we get is type IIA superstring theory, which is also an $N=2$ theory, but this time with two gravitinos of different chirality; now, the R-R sector contains

$$\mathbf{8}_S \otimes \mathbf{8}_{S'} = \mathbf{8}_V \oplus \mathbf{56}_V, \tag{3.102}$$

i. e., a vector field and a 3-form. These are the first two types of superstring theories that we will consider.

3.2.1 Toroidal Compactification of Type IIA and Type IIB Theories. *U*-duality.. Before considering different compactifications of superstring theories, we will first review some general results on the maximum number of allowed supersymmetry, depending on the spacetime dimension.

Spinors should be considered as representations of $SO(1,d-1)$. Irreducible representations have dimension

$$2^{[\frac{d+1}{2}]-1}, \tag{3.103}$$

where [] stands for the integer part. Depending on the dimension, the larger spinor can be real, complex or quaternionic,

$$\begin{aligned} \mathbf{R}, \text{ if } d &= 1,2,3 \bmod 8, \\ \mathbf{C}, \text{ if } d &= 0 \bmod 4, \\ \mathbf{H}, \text{ if } d &= 5,6,7 \bmod 8. \end{aligned} \tag{3.104}$$

Using (3.103) and (3.104), we get the number of supersymmetries listed in the table below[14].

Dimension	N	Irreducible Representation
11	1	$\mathbf{R}^{32}$
10	2	$\mathbf{R}^{16}$
9	2	$\mathbf{R}^{16}$
8	2	$\mathbf{C}^{8}$
7	2	$\mathbf{H}^{8}$
6	4	$\mathbf{H}^{4}$
5	4	$\mathbf{H}^{4}$
4	8	$\mathbf{C}^{2}$
3	16	$\mathbf{R}^{2}$

[14] This table is constrained by the physical requirement that particles with spin > 2 do not appear.

The maximum number of supersymmetries in three dimensions is then 16. From the table it is also clear that through standard Kaluza-Klein compactification, starting with six dimensional $N = 1$ supersymmetry leads to four dimensional $N = 2$, and three dimensional $N = 4$ supersymmetry. We can also notice that ten dimensional $N=1$ leads to $N=4$ supersymmetry in four dimensions.

It must be stressed that the counting of supersymmetries after dimensional reduction is slightly more subtle if we compactify on manifolds with non trivial topology. Here, the adequate concept is the holonomy of the internal manifold; let us therefore recall some facts on the concept of holonomy. Given a Riemannian manifold $\mathcal{M}$, the holonomy group $H_{\mathcal{M}}$ is defined as the set of transformations M_γ associated with paths γ in $\mathcal{M}$, defined by parallel transport of vectors in the tangent bundle. The connection used in this definition is the Levi-Civita connection. In general, for a vector budle $E \to \mathcal{M}$, the holonomy group $H_{\mathcal{M}}$ is defined by the paralell transport of v in the fiber, with respect to the connection on E. The Ambrose-Singer theorem shows how the holonomy is generated by the curvature.

Manifolds can be classified according to its holonomy group. Therefore, we get [74]

- $H_{\mathcal{M}} = O(d)$, for real manifolds of dimension d.
- $H_{\mathcal{M}} = U(\frac{d}{2})$, for Kähler manifolds.
- $H_{\mathcal{M}} = SU(\frac{d}{2})$, for Ricci flat Kähler manifolds.
- $H_{\mathcal{M}} = Sp(\frac{d}{4})$, for hyperkähler manifolds[15].

The answer to the question of what the role of holonomy is in the counting of the number of supersymetries surviving after compactification is quite simple: let us suppose we are in dimension d, so that the spinors are in $SO(1, d-1)$. Now, the theory is compactified on a manifold of dimension d_1, down to $d_2 = d - d_1$. Supersymmetries in d_2 are associated with representations of $SO(1, d_2 - 1)$, so we need to decompose an irreducible representation of $SO(1, d-1)$, into $SO(1, d_2 - 1) \times SO(d_1)$. Now, the holonomy group of the internal manifold $H_{\mathcal{M}_{d_1}}$ will be part of $SO(d_1)$. Good spinors in d_2 dimensions would be associated with singlets of the holonomy group of the internal manifold. Let us consider the simplest case, with $d_1 = 4$; then,

$$SO(4) = SU(2) \otimes SU(2) \tag{3.105}$$

and, if our manifold is Ricci flat and Kähler, the holonomy will be one of these $SU(2)$ factors. Therefore, we will need a singlet with respect to this $SU(2)$. As an example, let us consider the spinor in ten dimensions, with $N=1$; as we can see from the above table, it is a **16**, that we can decompose with respect to $SO(1,5) \times SU(2) \times SU(2)$ as

$$\mathbf{16} = (\mathbf{4}, \mathbf{2}, \mathbf{1}) \otimes (\mathbf{4}, \mathbf{1}, \mathbf{2}). \tag{3.106}$$

[15] Notice that any hyperkähler manifold is always Ricci flat.

Therefore, we only get one surviving supersymmetry in six dimensions. This is a general result: if we compactify a ten dimensional theory on a manifold of dimension four, with $SU(2)$ holonomy, we will get a six dimensional theory with only one supersymmetry. However, if the compactification is on a torus with trivial holonomy, two supersymmetries are obtained (the maximum number of supersymmetries available).

As the first contact with type IIA string theory we will then consider its compactification on a d-dimensional torus, T^d. To start with, let us work in the particular case $d = 4$. From the above table, we learn that the number of supersymmetries in six dimensions is 4, as the holonomy of T^4 is trivial. If we do not take into account the R-R fields, the moduli of the string σ-model is exactly the one described in section 3.1,

$$O(4,4;\mathbf{Z})\backslash O(4,4)/O(4)\times O(4), \tag{3.107}$$

with the T-duality $O(4,4;\mathbf{Z})$ corresponding to changes of the type $R_i \to \frac{\alpha'}{R_i}$, for the four S^1 cycles compossing the torus. The situation becomes different if we allow R-R background fields. In such a case, we should take into account the possiblity of including Wilson lines for the A_μ field (the $\mathbf{8}_V$ in (3.102)), and also a background for the 3-form $A_{\mu\nu\rho}$ (the $\mathbf{56}_V$ of (3.102)). The number of Wilson lines is certainly 4, one for each non contractible loop in T^d, so we need to add 4 dimensions to the 16-dimensional space (3.107). Concerning an $A_{\mu\nu\rho}$ background, the corresponding moduli is determined by $H_3(T^4)$, which implies 4 extra parameters. Finally, the dimension equals

$$16+4+4=24. \tag{3.108}$$

Now, a new extra dimension coming form the dilaton field must be added. It is important here to stress this fact: in the approach in previous section to σ-model moduli space the dilaton moduli has not been considered. This corresponds to interpreting the dilaton as a string coupling constant, and allowing changes only in the string. Anyway, this differentiation is rather cumbersome. Adding the dilaton moduli to (3.108), we get a moduli space of dimension (3.25), that can be written as

$$O(5,5;\mathbf{Z})\backslash O(5,5)/O(5)\times O(5). \tag{3.109}$$

The proposal of moduli (3.109) for type IIA on T^4 already contains a lot of novelties. First of all, the modular group $O(5,5;\mathbf{Z})$ now acts on the dilaton and the resting Ramond fields. In fact, relative to the $O(4,4;\mathbf{Z})$ T-duality of toroidal compactifications, we have now an extra symmetry which is S-duality [5, 75, 76, 77, 78, 79, 80, 81, 82, 83, 84, 85, 86],

$$g \to \frac{1}{g}, \tag{3.110}$$

with g the string coupling constant. This new modular symmetry is called in the physics literature U-duality [78]. The phenomena found here resembles very much what arises from mirror symmetry in the analysis of $K3$. There, the "classical" modular group was $O(\Gamma^{3,19};\mathbf{Z})$, and quantum mirror symmetry creates the enhancement to $O(\Gamma^{4,20};\mathbf{Z})$ where, in addition to T-duality, we have mirror transformations. In the case of type IIA on T^4, it is because we include the R-R backgrounds and the dilaton that the modular symmetry $O(4,4;\mathbf{Z})$ is enhanced to the U-duality symmetry. In spite of the analogies, the physical meaning is different. To apreciate this, let us now consider type IIA on $K3$. The dilaton moduli can be added, but the R-R fields are not producing any new moduli. In fact, recall that $\Pi_1(K3)=0$, and $H_3=0$, so the moduli of type IIA on $K3$ is simply

$$O(4,20;\mathbf{Z})\backslash O(4,20)/O(4)\times O(20) \quad \times \mathbf{R}, \tag{3.111}$$

with $\mathbf{R}$ parametrizing the dilaton, and the modular group not acting on it.

The way to interpret the moduli (3.109) goes under the name of M-theory. Before entering a more precise definition of M-theory, the basic idea is thinking of (3.109) simply as the moduli of a toroidal compactification on T^5; however, in order to obtain a six dimensional $N=4$ theory, we need to start with some theory living in 11 dimensions. The theory satisfying this is M-theory, a theory whose low energy supergravity description is well understood: it should be such that through standard Kaluza-Klein compactification it gives the field theory limit of type IIA strings; but this a theory known as eleven dimensional type IIA supergravity.

Once we have followed the construction of the type IIA string theory moduli on T^4, let us consider the general case of compactification on T^d. The dimension of the moduli is

$$\dim \; = d^2+1+d+\frac{d(d-1)(d-2)}{3}, \tag{3.112}$$

where d^2 is the NS-NS contribution, the 1 sumand comes form the dilaton, d from the Wilson lines, and $\frac{d(d-1)(d-2)}{3}$ from the 3-form $A_{\mu\nu\rho}$. The formula (3.112) has to be completed, for $d\geq 5$, by including dual scalars. For $d=5$, the dual to the 3-form $A_{\mu\nu\rho}$ is a scalar. The result is

$$\begin{aligned} &\frac{d(d-1)(d-2)(d-3)(d-4)}{5} \quad \text{duals} \quad \text{to } A_{\mu\nu\rho}, \\ &\frac{d(d-1)\dots(d-6)}{7} \quad \text{duals} \quad \text{to } A_{\mu}. \end{aligned} \tag{3.113}$$

The moduli spaces, according to the value of the dimension of the compactification torus, are listed in the table below.

For supergravity practitioners, the appearance of E_6 and E_7 in this table should not be a surprise.

Dimension	Moduli
$d = 4$	$O(5,5;\mathbf{Z})\backslash O(5,5)/O(5) \times O(5)$
$d = 5$	$E_{6,(6)}(\mathbf{Z})\backslash E_{6,(6)}/Sp(4)$
$d = 6$	$E_{7,(7)}(\mathbf{Z})\backslash E_{7,(7)}/SU(8)$
$d = 3$	$Sl(5,\mathbf{Z})\backslash Sl(5)/SO(5)$
$d = 2$	$Sl(3,\mathbf{Z}) \times Sl(2,\mathbf{Z})\backslash Sl(3)/SO(3)\ \ Sl(2)/SO(2)$

Let us now see what happens in the type IIB case. The moduli on, for instance, T^4, is again the 16 dimensional piece coming from the NS-NS sector; now, the R-R sector is determined by the cohomology groups H^0, H^2 and H^4 (see equation (3.101)). From the Hodge diamond for T^4,

$$\begin{array}{ccccc} & & 1 & & \\ & 2 & & 2 & \\ 1 & & 4 & & 1 \\ & 2 & & 2 & \\ & & 1 & & \end{array} \tag{3.114}$$

we get 8 extra modulis, exactly the same number as in the type IIA case. This is a general result for any T^d compactification. The reason for this is that type IIA and type IIB string theories are, after toroidal compactification, related by T-duality. However, on a manifold as $K3$, with $\Pi_1 = 0$, the moduli for IIA and IIB are drastically different, as can be derived from direct inspection of the $K3$ Hodge diamond (see equation (3.39)). Therefore, for type IIB we get, from the R-R sector, 1 coming from H^0, 22 from H^2, and 1 from H^4, which sums up a total of 24 extra modulis to be added to the $58 + 22$ of the NS-NS sector. Then, including the dilaton,

$$\dim \mathrm{IIB}(K3) = 22 + 58 + 24 + 1 = 105. \tag{3.115}$$

Therefore, the natural guess for the moduli is

$$O(5,21;\mathbf{Z})\backslash O(5,21)/O(5) \times O(21). \tag{3.116}$$

Here, something quite surprising is taking place. As we can see from (3.111), when type IIA is compactified on $K3$, we do not find any appearance of U-duality or, in other words, S-duality. By contrast, in the type IIB case we find a modular group $O(5,2;\mathbf{Z})$, that contains the dilaton and, therefore, the S-duality transformation. This is what can be called the S-duality of type IIB string theory [87], which can already be observed from equation (3.101). In fact, the R-R and NS-NS sectors both contain scalar fields and the antisymmetric tensor.

3.2.2 Heterotic String.. The idea of "heterosis", one of the most beatiful and productive ideas in the recent history of string theory [88] was motivated by two basic facts. First of all, the need to find a natural way to define non abelian gauge theories in string theory, without entering the use of Chan-Paton factors, and, secondly, the sharpness of the gap in string theory between left and right moving degrees of freedom. Here, we will concentrate on some of the ideas leading to the construction of heterosis. In the toroidal compactification of the bosonic string on T^d, we have found that the momenta live in a $\Gamma^{d,d}$ lattice. This is also true for the NS sector of the superstring. The lattice $\Gamma^{d,d}$, where the momenta live, is even and self dual. Taking into account the independence between left and right sectors, we can think on the possibility to compactify the left and right components on different tori, T^{d_L} and T^{d_R}, and consider as the corresponding moduli the manifold

$$O(d_L, d_R; \mathbf{Z}) \backslash O(d_L, d_R)/O(d_L) \times O(d_R). \tag{3.117}$$

Before trying to find out the consistency of this picture, let us try to get a simple interpretation of moduli (3.117). The dimension of this moduli is $d_L \times d_R$, and we can separate it into $d_L \times d_L + d_L \times (d_R - d_L)$. Let us interpret the first part, $d_L \times d_L$, as the standard moduli for compactifications on a torus T^{d_L}; then, the second piece can be interpreted as the moduli of Wilson lines for a gauge group

$$U(1)^{d_R - d_L}. \tag{3.118}$$

With this simple interpretation, we already notice the interplay in heterosis when working with a gauge group that can be potentially non abelian, the gauge group (3.118), and differentiating left and right parts. When we were working with type II string theory, and considered toroidal compactifications, we were also adding, to the moduli space, the contribution of the Wilson lines for the RR gauge field, A_μ (in case we are in type IIA). However, in the case of type IIA on T^4, taking into account the Wilson lines did not introduce any heterosis asymmetry in the moduli of the kind (3.117). However, T^4 is not the only Ricci flat four dimensional manifold; we can also consider $K3$ surfaces. It looks like if T^4, $K3$, and its orbifold surface in between, $T^4/\mathbf{Z}_2$, saturate all compactification manifolds that can be thought in four dimensions. In the case of $K3$, the moduli of type IIA string (see equation (3.111)) really looks like the heterotic moduli, of the kind (3.117), we are looking for. Moreover, in this case, and based on the knowledge of the lattice of the second cohomology group of $K3$ (see equation (3.44)),

$$E_8 \perp E_8 \perp \mathcal{U} \perp \mathcal{U} \perp \mathcal{U}, \tag{3.119}$$

we can interpret the $16 = d_R - d_L$ units as corresponding precisely to Wilson lines of the $E_8 \times E_8$ gauge group appearing in (3.119). In other words, and following a very distant path form the historical one, what we are suggesting is interpreting moduli (3.111), of type IIA on $K3$, as some sort of heterosis, with $d_L = 4$ and $d_R = 20$. The magic of numbers is in fact playing in our team, as the numbers we get for d_L and d_R strongly suggest a left part, of critical dimension 10, and a right part, of precisely the critical dimension of the bosonic string, 26. This was, in fact, the original idea hidden under heterosis: working out a string theory looking, in its left components, as the standard superstring, and in its right components as the 26 dimensional bosonic string. However, we are still missing something in the "heterotic" interpretation of (3.111), which is the visualization, from $K3$ geometry, of the gauge group. In order to see this, some of the geometrical material introduced in subsection 3.1.5 will be needed; in terms of the concepts there introduced, we would claim that the (p_L, p_R) momentum is living in the lattice $\Gamma^{4,20}$. We can then think that p_L is in the space-like 4-plane where the holomorphic top form Ω, and the Kähler class J, are included. Recall that they define a space-like 3-plane. Now, momentum vectors, orthogonal to this 4-plane, can be considered; they are of the type

$$(0, p_R). \tag{3.120}$$

Now, whenever $p_R^2 = -2$, this vector will define a rational curve inside $K3$, with vanishing volume (in fact, the volume is given by $p_R \cdot J = 0$). The points $p_R^2 = -2$ will be at the root lattice of $E_8 \times E_8$. Now, from the mass formulas (3.26) we easily observe that $p_R^2 = -2$ is the condition for massless vector particles. In fact, if we separate, in the spirit of heterosis, the p_R of a 26 dimensional bosonic string into $(p_R^{(16)}, p_R^{(10)})$, we get, from (3.26),

$$M^2 = 4(p_R^{(16)})^2 + 8(N-1), \tag{3.121}$$

so that $M^2 = 0$, for $N=0$, if $(p_R^{(16)})^2 = 2$. The sign difference appears here because (recall subsection 3.1.5) in the $K3$ construction used for the second cohomology lattice, the E_8 lattice was defined by minus the Cartan algebra of E_8. Therefore, we observe that massless vector bosons in heterotic string are related to rational curves in $K3$ of vanishing volume, which allows to consider enhancement of symmetries when moving in moduli space [81, 89, 90]. Some of these rational curves can be blown up, which would be the geometrical analog of the Higgs mechanism, or either blown down, getting extra massless stuff. Moreover, for elliptically fibered $K3$ surfaces, the different Kodaira singularities reflect, in its Dynkin diagram, the kind of gauge symmetry to be found.

The above discussion summarizes what can be called the first quasi-theorem on string equivalence [78, 81],

Quasi-Theorem 1 Type IIA string on $K3$ is equivalent to $E_8 \times E_8$ heterotic string on T^4.

Previous arguments were so general that we can probably obtain extra equivalences by direct inspection of the different $K3$ moduli spaces that have been discussed in subsection 3.1.5. In particular, let us consider the moduli space of complex structures for an elliptically fibered $K3$ surface, a fact represented, in terms of the Picard lattice, claming that it is of $\Gamma^{1,1}$ type, generated by a section, and with the fiber satisfying relations (3.28). This moduli is

$$O(2,18;\mathbf{Z})\backslash O(2,18)/O(2)\times O(18), \tag{3.122}$$

where we have used equation (3.62), and the fact that the transcendental lattice is of type $(2,18)$. From the heterosis point of view, it would be reasonable to interpret (3.122) as heterotic $E_8 \times E_8$ string, compactified on a 2-torus, T^2. In fact, we will have 4 real moduli, corresponding to the Kähler class and complex structure of T^2, and 16 extra complex moduli associated to the Wilson lines. However, now the type II interpretation of (3.122) is far from being clear, as (3.122) is just the part of the moduli space that is preserving the elliptic fibration. Now, in order to answer how (3.122) can be understood as a type II compactification a similar problem appears as we try to work out an heterotic interpretation of the type IIB moduli on $K3$, given in (3.116). A simple way to try to interpret (3.122), as some kind of type II compactification, is of course thinking of an elliptically fibered $K3$, where the volume of the fiber is fixed to be equal zero; generically,

$$J \cdot F = 0, \tag{3.123}$$

where F indicates the class of the fiber. Now, we can think that we are compactifying a type II string on the base space of the bundle. However, this does not lead to (3.122) for the type IIA case, as the RR fields are in H^1 and H^3, which will vanish. But what about type IIB? In this case, we have the NS field ϕ, and the R field χ, and we should fix the moduli of possible configurations of these fields on the base space of the elliptic fibration. Here, type IIB S-duality, already implicit in moduli (3.116), can help enormously, mainly because we are dealing with an ellipticaly fibered $K3$ manifold [91, 92, 93]. To proceed, let us organize the fields ϕ and χ into the complex

$$\tau = \chi + ie^{-\phi}, \tag{3.124}$$

and identify this τ with the moduli of the elliptic fiber. Then, the 18 complex moduli dimension of (3.122) parametrizes the moduli of complex structures of the elliptic fibration, and therefore the moduli of τ field configurations on the base space (provided τ and $\frac{a\tau+b}{c\tau+d}$ are equivalent from the type IIB point of view). These moduli parametrize then the type IIB compactification on the base space B (it is $\mathbb{P}^1$; recall that in deriving (3.122) we have used a base space B such that $B \cdot B = -2$). There is still one moduli missing: the size of the base space B, that we can identify with the heterotic string coupling constant. Thus, we arrive to the following quasi-theorem,

Quasi-Theorem 2 Heterotic string on T^2 is equivalent to type IIB string theory on the base space of an elliptically fibered $K3$.

The previous discussion is known, in the physics literature, under the generic name of F-theory [94, 95, 96].

We have been considering, until now, type II strings on $K3$, and compared them to heteotic string on a torus. To find out what is the expected moduli for the heterotic string on $K3$, we can use the following trick: if heterotic string on T^2 is type IIB on the base space of an elliptically fibered $K3$, by quasi-theorem 2 heterotic string on an elliptically fibered $K3$ should correspond to type IIB on the base space of an elliptically fibered Calabi-Yau manifold. More precisely, type IIB string should be compactified on the basis of an elliptic fibration, which is now four dimensional, and that can be represented as a fibration of a $\mathbb{P}^1$ space over another $\mathbb{P}^1$. This type of fibrations are known in the literature as Hirzebruch spaces, $\mathbf{F}_n$. Hirzebruch spaces can simply be determined through heterotic data, given by the $E_8 \times E_8$ bundle on the $K3$ manifold. The moduli of these bundles on $K3$ will put us in contact with yet another interesting topic: small instantons.

3.2.3 Heterotic Compactifications to Four Dimensions.. Before considering some definite examples, let us simply summarize the different supersymmetries we can get when compactifying to three dimensions, depending on the holonomy of the target manifold. In order to do that, we will need the results in subsection 3.2.1, on the maximum number of supersymmetries allowed for a given spacetime dimension.

Type of String	Target Manifold	Holonomy	Supersymmetry
II	$K3 \times T^2$	$SU(2)$	$N=4$
Heterotic	T^6	Trivial	$N=4$
II	Calabi-Yau	$SU(3)$	$N=2$
Heterotic	$K3 \times T^2$	$SU(2)$	$N=2$
II	$B_{SU(4)}$	$SU(4)$	$N=1$
Heterotic	Calabi-Yau	$SU(3)$	$N=1$
II	$K3 \times T^2$	$SU(2)$	$N=4$
Heterotic	T^6	Trivial	$N=4$
II	Calabi-Yau	$SU(3)$	$N=2$
Heterotic	$K3 \times T^2$	$SU(2)$	$N=2$
II	$B_{SU(4)}$	$SU(4)$	$N=1$
Heterotic	Calabi-Yau	$SU(3)$	$N=1$

In the table above we have not differentiated between type IIA and type IIB[16]. The first two lines, corresponding to cases with $N = 4$ and $N = 2$ supersymmetry in four dimensional spacetime, will be the basic examples we will use to introduce the concept of dual pairs of string compactifications down to four dimensions.

Before entering a discussion on the ingredients of this table, we yet need to consider the holonomy of the moduli space. This holonomy will of course depend on the number of supersymmetries and the type (real, complex or quaternionic) of the representation. Hence, from subsection 3.2.1, we can complete the table below.

Spacetime Dimension	Supersymmetries	Type	Holonomy
$d = 6$	$N=2$	$\mathbf{H}^4$	$Sp(1) \oplus Sp(1)$
$d = 4$	$N=4$	$\mathbf{C}^2$	$U(4)$
$d = 4$	$N=2$	$\mathbf{C}^2$	$U(2)$

Using this results, we can now decompose the tangent vectors to the moduli according to its transformation rules with respect to the holonomy group. Let us concentrate in the $d = 4$ case. For $U(4)$, we get

$$U(4) \simeq U(1) \oplus SO(6). \tag{3.125}$$

The matter multiplets will contain 6 (real) scalars each, i. e., the number of dimensions we compactify. Then, if we have m of these matter multiplets, the part of the moduli on which the $SO(6)$ part of the holonomy group is acting should be

$$O(6, m)/O(6) \times O(m). \tag{3.126}$$

The $U(1)$ part of (3.125) will act on the supergravity multiplet so we expect, just from holonomy arguments, a moduli of type

$$O(6, m)/O(6) \times O(m) \quad \times Sl(2)/U(1). \tag{3.127}$$

Now, we need to compute m. For heterotic string, the answer is clear: $m = 22$, and the total dimension of (3.127) will be 134. Let us now consider the case of type IIA. From the table, we see that we should consider $K3 \times T^2$ as compactification manifold. Let us then first compute the dimension of the moduli space:

[16] This will be relevant when discussing the third line where, by $B_{SU(4)}$, we are thinking in the spirit of the discussion in the last part of previous section, where a Calabi-Yau fourfold of $SU(4)$ holonomy, elliptically fibered, and with a zero volume fiber, is used for compactification.

$$\begin{aligned} \text{Moduli of metrics and B fields on } K3 &= 80 \\ \text{Moduli of metrics and B fields on } T^2 &= 4 \\ b_1(K3 \times T^2) &= 2 \\ b_3(K3 \times T^2) &= 44 \\ \text{Axion-Dilaton} &= 2 \\ \text{Duals in } \mathbf{R}^4 \text{ to } 2-\text{forms} &= 2 \end{aligned} \tag{3.128}$$

which sums up to 134. Notice that the 44 in $b_3(K3 \times T^2)$ is coming from the 3-cycles obtained from one S^1 of T^2, and the 22 elements in $H^2(K3;\mathbf{Z})$. The 3-form of IIA can be compactified on the S^1 cycles of T^2 to give 2-forms in four dimensions. Now, the dual of a 2-form in $\mathbf{R}^4$ is scalar, so we get the last two extra moduli.

Now, we need to compare the two moduli spaces. If we expect S-duality in $N=4$ for the heterotic compactification, the moduli, once we have taken into account the $O(6,22;\mathbf{Z})$ T-duality, will look like

$$O(6,22;\mathbf{Z})\backslash O(6,22)/O(6) \times O(22) \;\; Sl(2,\mathbf{Z})\backslash Sl(2)/U(1). \tag{3.129}$$

Now, we have a piece in IIA looking naturally as the second term in (3.129), namely the moduli of the σ-model on T^2, where $Sl(2,\mathbf{Z})$ will simply be part of the T-duality. Thus, it is natural to relate the moduli of IIA on the torus with the part of the moduli in (3.127) coming form the supergravity multiplet.

Let us now consider dual pairs in the second line of our table. There is a simple way to visualize under what general conditions on the Calabi-Yau manifold with $SU(3)$ holonomy such dual pairs can exist. In fact, imagine that $K3$ is ellipticaly fibered in $K3 \times T^2$; then, what we get is a fibration on $\mathbb{P}^1$ of the T^4 tori. Now, heterotic on T^4 is equivalent to type IIA on $K3$, so we expect that the Calabi-Yau manifold should be a $K3$ fibration on $\mathbb{P}^1$, and that duality works fiberwise. Therefore, from general arguments, we expect to get heterotic-type II dual pairs with $N=2$ if we use Calabi-Yau manifolds which are $K3$ fibrations [97, 93]. In order to get a more precise picture, we need again to work out the holonomy, which is $U(2)$ in this case. In $N=2$ we have two types of multiplets, vector and hypermultiplets. The vector multiplet contains two real scalars, and the hypermultiplet four real scalars. Then, we decompose $U(2)$ into $U(1) \oplus Sp(1)$, and the moduli into vector and hypermultiplet part.

Let us first consider type IIA string on the Calabi-Yau manifold. The moduli will contain $h^{1,1}$ deformations of B and J, $h^{2,1}$ complex deformations and b^3 RR deformations (b^1 does not contribute, as we are working with a Calabi-Yau manifold). The total number, in real dimension, is

$$2h^{1,1} + 4(h^{2,1}+1), \tag{3.130}$$

where we have used that $b^3 = 2(h^{2,1}+1)$, in real dimension. From (3.130) we conclude that we have $h^{1,1}$ vector multiplets, and $h^{2,1}+1$ hypermultiplets.

Notice that $4(h^{2,1}+1)$ is counting the 2 coming from the dilaton and the axion so, for type II we have combined dilaton and axion into an hypermutiplet.

Now, let us consider heterotic string on $K3 \times T^2$. The moduli we must now consider, of $E_8 \times E_8$ bundles on $K3$, is much more elaborated than that of T^4, or T^6, that we have worked out. Part of the difficulty comes from anomaly conditions. However, we know, accordding to Mukai's theorem, that the moduli of holomorphic bundles on $K3$ is quaternionic, i. e., hyperkähler, and that the moduli of the σ-model on $K3$ is of dimension 80. We have yet the moduli on T^2, that will be a manifold of $O(2,m)/O(2) \times O(m)$ type, and therefore a good candidate for representing the vector multiplet. Thus, we get

$$\begin{aligned} \text{Type IIA hypermultiplets} &\leftrightarrow K3 \text{ Heterotic}, \\ \text{Vector multiplets} &\leftrightarrow T^2. \end{aligned} \tag{3.131}$$

From our previous discussion we know that vector multiplets, in type IIA are related to $h^{1,1}$. Working fiberwise on a $K3$ fibered Calabi-Yau manifold we get, for $h^{1,1}$,

$$h^{1,1} = 1 + \rho, \tag{3.132}$$

with ρ the Picard number of the $K3$ manifold. Then, in order to get a dual pair in the sense of (3.131) we need m in the heterotic to statisfy

$$m = \rho. \tag{3.133}$$

In order to control the value of m, from the heterotic point of view, we need to watch out for possible Wilson lines that can be defined on T^2 after the gauge group has been fixed from the $K3$ piece. From (3.132) (and this was the logic for the identification (3.133)), the heterotic dilaton-axion is related to the 1 term contributing in (3.132), i. e., the 2-cycle defined by the base space of the $K3$-fibration.

As can be observed from (3.133), if we do not freeze either the Kähler class or the complex structure of T^2, the minimum value for ρ is 2. This is the contribution to the Picard lattice of a Dynkin diagram of type A_2, i. e., $SU(3)$. A possible line of work opens here, in order to identify the moduli spaces of vector multiplets for type IIA theories with the quantum moduli, defined according to Seiberg and Witten, for gauge theories, with

$$\text{rank } G = \rho. \tag{3.134}$$

4. Chapter IV

4.1 M-Theory Compactifications.

Wittgenstein used to say that "meaning is use". This is the kind of philosophycal slogan able to make unhappy the platonic mathematician, but it is in essence the type of game we are going to play in order to begin the study of M-theory [98, 78, 80, 81, 86]. More precisely, we will start without saying what M-theory is from a microscopical point of view, giving instead a precise meaning to M-theory compactifications.

Recall that our first contact with the idea of M-theory was in connection with the interpretation of the moduli of type IIA string theory on T^4. In that case the moduli, after including RR fields, was of the type

$$O(5,5;\mathbf{Z})\backslash O(5,5)/O(5)\times O(5). \tag{4.1}$$

The M-theory interpretation of moduli (4.1) can be summarized according to the equivalence

$$\text{M-theory compactified on } T^5 \leftrightarrow \text{ IIA on } T^4, \tag{4.2}$$

and therefore, more generically,

$$\text{M-theory compactified on } X\times S^1 \leftrightarrow \text{ IIA on } X. \tag{4.3}$$

Let us now put rule (4.3) into work. In fact, one particular case of (4.3) will consist in considering M-theory on a manifold of type $B\times S^1\times S^1$. Then, using T-duality, we can get

$$\text{M-theory compactified on } B\times S^1\times S^1(R)\leftrightarrow$$

$$\text{IIA on } B\times S^1(R)\leftrightarrow \text{ IIB on } B\times S^1(\frac{\alpha'}{R}). \tag{4.4}$$

From (4.4) we see that in the $R\to\infty$ limit, we get type IIB string theory on B or, equivalently, M-theory on $B\times S^1$, since the second S^1 becomes uncompactified. This is in fact a very close example to the ones described in previous sections, under the generic name of F-theory compactifications. Namely, the $R\to\infty$ limit in (4.4) can be interpreted as defining a compactification of type IIB string theory on the base space B of an elliptic fibration $B\times S^1\times S^1$, in the limit where the volume of the elliptic fiber becomes zero. Following that path, we get an interesting equivalence between M-theory on $B\times S^1\times S^1$, as elliptic fibration, in the limit in which the volume of the elliptic fiber goes zero, and type IIB on B. This stands as a surprise, when compared to the result derived from the compactification rule (4.3). In fact, if B is, for instance, of dimension d, then we should expect that the compactification of an eleven dimensional theory, as M-theory, on $B\times S^1\times S^1$, should lead to $11-d-2$ dimensions. However, type IIB, which is ten dimensional, would lead, when compactified on B, to a $10-d$ dimensional theory, so that one dimension is missing. Getting rid off this contradiction requires knowledge of the microscopic nature of M-theory. The first thing to be required on M-theory is of course to have, as low energy limit, eleven dimensional supergravity. There is a connection between type IIA string theory and eleven dimensional supergravity, as the corresponding Kaluza-Klein dimensional reduction on an internal S^1, which allows an identification of the string theory spectrum with supergravity. In particular, the RR field in ten dimensions comes from the $g_{11,\mu}$ component of the metric, while the dilaton is obtained from $g_{11,11}$. The precise relation, in what is known as the string frame, is[17]

$$e^{-2\phi} = e^{-3\gamma}, \tag{4.5}$$

with ϕ the type IIA dilaton field. In terms of the radius R of the S^1,

$$R = e^{2\phi/3}. \tag{4.6}$$

Using now equation (3.21), we get a relation between the R of the internal manifold, S^1, and the string coupling constant of type IIA string theory,

$$R = g^{2/3}. \tag{4.7}$$

From (4.7) it is obvious that, as $R \to \infty$, we properly enter the M-theory region when g is large, i. e., working in the strong coupling regime of string theory. Historically, this beatiful simple argument was put forward in 1995 by Witten [81]. It is astonishing that, with all the pieces around, nobody was able before to make at least the comment relating the R of eleven dimensional supergravity with the string coupling constant, and to derive from it such a striking conjecture as it is that strongly coupled IIA strings are described by eleven dimensional supergravity. In fact, there are good reasons for such a mental obstacle in the whole community: first of all, nobody did worry about type IIA dynamics, as it was a theory with only uninteresting pure abelian gauge physics. Secondly, the Kaluza-Klein modes coming from the compactification on S^1, which have a mass of the order $\frac{1}{R}$, are charged with respect to the $U(1)$ gauge field defined by the $g_{11,\mu}$ piece of the metric. But this A_μ field in ten dimensional type IIA string is of RR type, so before the discovery of D-branes, there was no candidate in the string spectrum to be put in correspondence with these Kaluza-Klein modes, which can now be identified with D-0branes.

Witten's approach to M-theory can be the conceptual key to solve the problem concerning the missing dimension: in fact, something in the spectrum is becoming massless as the volume of the elliptic fiber, in the case of $B \times S^1 \times S^1$, is sent to zero. Moreover, the object becoming massless can be, as suggested by Sethi and Susskind, interpreted as a Kaluza-Klein mode of an opening dimension as the volume of the elliptic fiber goes zero. To understand the nature of this object we should look more carefully at M-theory. This theory is expected to contain a fundamental two dimensional membrane; if this membrane wraps the 2-torus $S^1 \times S^1$, its mass becomes zero as the volume of the fiber goes zero. Then, all what is left is to relate the area with the standard Kaluza-Klein formula for compactifications on S^1, which leads to

[17] We have identified $g_{11,11} = e^{2\gamma}$.

$$\frac{1}{R} \sim L_1 L_2 \tag{4.8}$$

solving our problem on the adequate interpretation of (4.4).

Let us now concentrate on a concrete example of (4.4): we will choose $X = B \times S^1 \times S^1_R$ as representing a Calabi-Yau fourfold of $SU(4)$ holonomy. After compactification, $SU(4)$ holonomy implies a three dimensional theory with $N = 2$ supersymmetry should be expected. Moreover, sending $R \to \infty$ leads to a four dimensional $N = 1$ theory. In order to work out the spectrum of the three dimensional theory, standard Kaluza-Klein techniques can be used. Compactification on the 2-cycles of $H^2(X; \mathbf{Z})$ of the 3-form $C_{\mu\nu\rho}$ of eleven dimensional supergravity leads to a vector in three dimensions. Moreover, the Kähler class can also be used to generate real scalars, from each 2-cycle. Thus, let us assume $r = \dim H^2(X; \mathbf{Z})$; then, the previous procedure produces r real scalars and r vector fields. In order to define r $N = 2$ vector multiplets in three dimensions, with these vector fields, another set of r scalars is yet needed, in order to build the complex fields. These extra r scalars can, as usual, be identified with the duals, in three dimensions, of the 1-form vector fields: the three dimensional dual photon.

Our next task will be reproducing, using M-theory, the well known instanton effects in $N = 2$ supersymmetric gauge theories in three dimensions.

4.2 M-Theory Instantons.

In order to define instantons in three dimensions, we will use 5-branes wrapped on the 6-cycles of a Calabi-Yau fourfold X [99]. The reason for using 6-cycles is understood as follows: the gauge bosons in three dimensions are obtained from the integration of the 3-form $C_{\mu\nu\rho}$ over 2-cycles. Thus, in order to define the dual photon, we should consider the dual, in the Calabi-Yau fourfold X, of 2-cycles, which are 6-cycles. However, not any 6-cycle can be interpreted as an instanton with topological charge equal one, and therefore no 6-cycle will contribute to the three dimensional superpotential.

If we interpret a 5-brane wrapped on a 6-cycle D of X as an instanton, we can expect a superpotential of the type

$$W = e^{-(V_D + i\phi_D)}, \tag{4.9}$$

with V_D the volume of D measured in units of the 5-brane tension, and ϕ_D the dual photon field, associated with the cycle D. In order to get, associated to D, a superpotential like (4.9), we need

i) To define a $U(1)$ transformation with respect to which three dimensional fermions are charged.
ii) To associate with the 6-cycle D a violation of $U(1)$ charge, in the adecuate amount.
iii) To prove that this $U(1)$ symmetry is not anomalous.
iv To interpret ϕ_D as the corresponding Goldstone boson.

Following these steps, we will extend to M-theory the instanton dynamics of three dimensional $N = 2$ gauge theories described in chapter I. We will start defining the $U(1)$ transformation. Let D be 6-cycle in the Calabi-Yau fourfold X, and let us denote by N the normal bundle of D in X. Since X is a Calabi-Yau manifold, its canonical bundle is trivial, and therefore we get

$$K_D \simeq N, \tag{4.10}$$

with K_D the canonical bundle of D. Locally, we can interpret X as the total space of the normal bundle. Denoting by z the coordinate in the normal direction, the $U(1)$ transformation can be defined as

$$z \to e^{i\theta} z. \tag{4.11}$$

The $U(1)$ transformation defined by (4.11) is very likely not anomalous, since it is part of the diffeomorphisms of the elevean dimensional theory; thus, it is a good candidate for the $U(1)$ symmetry we are looking for. Next, we need to get the $U(1)$ charge of the three dimensional fermions. However, before doing so, we will review some well known facts concerning fermions and Dirac operators on Kähler manifolds.

We will consider a Kähler manifold of complex dimension N. In holomorphic coordinates,

$$g_{ab} = g_{\bar{a}\bar{b}} = 0. \tag{4.12}$$

In these coordinates, the algebra of Dirac matrices becomes

$$\begin{aligned} \{\gamma^a, \gamma^b\} &= \{\gamma^{\bar{a}}, \gamma^{\bar{b}}\} = 0, \\ \{\gamma^a, \gamma^{\bar{b}}\} &= 2g^{a\bar{b}}. \end{aligned} \tag{4.13}$$

The $SO(2N)$ spinorial representations of (4.13) can be obtained in the standard Fock approach: a vacuum state is defined by condition

$$\gamma^a|\Omega>= 0, \tag{4.14}$$

and n-particle states are defined by

$$\gamma^{\bar{a}}\gamma^{\bar{b}}\ldots\gamma^{\bar{n}}|\Omega> . \tag{4.15}$$

A spinor field $\psi(z,\bar{z})$ on the Kähler manifold takes values on the spinor bundle defined by this Fock representation:

$$\psi(z,\bar{z}) = \phi(z,\bar{z})|\Omega> + \phi_{\bar{a}}(z,\bar{z})\gamma^{\bar{a}}|\Omega> + \phi_{\bar{a}\bar{b}}(z,\bar{z})\gamma^{\bar{a}}\gamma^{\bar{b}}|\Omega> + \cdots \tag{4.16}$$

The spaces $\Omega^{0,q}$ of $(0,q)$-forms, generated by the Dirac operator, define the Dolbeaut cohomology of the Kähler manifold. Using this notation, the two different chirality spinor bundles are

$$\begin{aligned} S^+ &= (K^{1/2} \otimes \Omega^{0,0}) \oplus (K^{1/2} \otimes \Omega^{0,2}) \oplus (K^{1/2} \otimes \Omega^{0,4}) \oplus \cdots, \\ S^- &= (K^{1/2} \otimes \Omega^{0,1}) \oplus (K^{1/2} \otimes \Omega^{0,3}) \oplus (K^{1/2} \otimes \Omega^{0,5}) \oplus \cdots, \end{aligned} \tag{4.17}$$

and the change of chirality (the index for the Dirac operator on the manifold X) will be given by the aritmetic genus,

$$\chi = \sum^{N} (-1)^n h_n, \tag{4.18}$$

where $h_n = \dim\, \Omega^{0,n}$.

The previous comments can be readily applied to the case of a six dimensional divisor D in a Calabi-Yau fourfold X. Now, we should take into account the normal budle N, to D, in X. Using the fact that X is Calabi-Yau, i. e., with trivial canonical bundle, we conclude that N is isomorphic to K_D, the canonical bundle on D. The spinor bundle on N will be defined by

$$K_D^{1/2} \oplus K_D^{-1/2}. \tag{4.19}$$

In fact, in this case the complex dimension of N is one, and the vacum and filled states have, respectively, $U(1)$ charges $\frac{1}{2}$ and $-\frac{1}{2}$. On the other hand, the spinor budle on D will be defined by (4.17), with $K = K_D$. Thus, spinors on D are, up to the $SO(3)$ spacetime part, taking values in the positive and negative chirality bundles

$$\begin{aligned} \hat{S}^+ &= (K^{1/2} \oplus K^{-1/2}) \otimes [(K^{1/2} \otimes \Omega^{0,0}) \oplus (K^{1/2} \otimes \Omega^{0,2})], \\ \hat{S}^- &= (K^{1/2} \oplus K^{-1/2}) \otimes [(K^{1/2} \otimes \Omega^{0,1}) \oplus (K^{1/2} \otimes \Omega^{0,3})]. \end{aligned} \tag{4.20}$$

Now, we are interested in a change of $U(1)$ charge, with the $U(1)$ charge defined by the $\frac{1}{2}$ and $-\frac{1}{2}$ charges of the spinor bundle (4.19) on N.

For spinors of a given chirality, the change of $U(1)$ charge is given by

$$\dim\, (K \otimes \Omega^{0,0}) + \dim\, (K \otimes \Omega^{0,2}) - \dim\, (\Omega^{0,0}) - \dim\, (\Omega^{0,2}). \tag{4.21}$$

Using now Serre's duality,

$$\dim\, (K \otimes \Omega^{0,3-n}) = \dim\, (\Omega^{0,n}), \tag{4.22}$$

we get that the number of holomorphic $(0,k)$-forms is equal to the number of holomorphic sections in $K \otimes \Omega^{0,3-k}$, and therefore the number of fermionic zero modes with $U(1)$ charge equal $\frac{1}{2}$ is given by $h_3 + h_1$, and the number of fermionic zero modes with $-\frac{1}{2}$ $U(1)$ charge, is given by $h_0 + h_2$ (here we have used the Dirac operator $\bar{\partial} + \bar{\partial}^*$, with $\bar{\partial}^*$ the adjoint of $\bar{\partial}$. Thus, the index for the twisted spin bundle (4.20) is given by the holomorphic Euler characteristic,

$$\chi(D) = h_0 - h_1 + h_2 - h_3. \tag{4.23}$$

Now, each of these fermionic zero modes is doubled once we tensor with spinors in $\mathbf{R}^3$. In summary, for each 6-cycle D we get an effective vertex with a net change of $U(1)$ charge equal to $\chi(D)$.

Therefore, in order to get the three dimensional in a three dimensional $N = 2$ theory, we need to look for 6-cycles D, with $\chi(D) = 1$, as the net change of $U(1)$ charge in that case is one, provided, as we did in (4.19), we normalize the $U(1)$ charge of the fermions to be $\frac{1}{2}$. More precisely, the number of fermionic zero modes for a three dimensional instanton, defined by a 6-cycle D, is $2\chi(D)$.

4.3 D-Brane Configurations in Flat Space.

We will consider a D-brane of dimension p, in flat ten dimensional Minkowski space, and with a flat $p+1$ dimensional worldvolume. The quantization of the open superstring ending on the D-brane defines a low energy field theory, which is ten dimensional $N=1$ supersymmetric Yang-Mills, with $U(1)$ gauge group. The dimensional reduction of this theory to $p+1$ dimensions describes the massless excitations propagating on the worldvolume of the p dimensional D-brane. We will use as worldvolume coordinates $x^0, x^1, \ldots, x^p$. The worldvolume lagrangian will contain a $U(1)$ massless gauge field $A_i(x_s)$, with $i, s = 0, \ldots, p$, and a set of scalar fields $\phi_j(x_s)$, $j = p+1, \ldots, 9$, transforming in the adjoint representation. We can geometrically interpret the set of fields $\phi_j(x_s)$ as representing the "location" of the flat D-brane in transverse space. The simplest generalization of the previous picture corresponds to configurations of $k > 1$ parallel D-$p$branes. In this case we have, in addition to the massless excitations, a set of k massive excitations corresponding to open strings ending on different D-branes.

The field theory interpretation of this configuration of D-branes would be that of a gauge thory with $U(k)$ gauge group, spontaneously broken to $U(1)^k$, with the strings stretching between different D-branes representing charged massive vector bosons. To get such an interpretation, we can start with $N=1$ $U(k)$ supersymmetric Yang-Mills in ten dimensions, and perform again dimensional reduction down to $p+1$ dimensions. In this case, we will get a set of scalar fields, $X^j(x_s)$, with $j = p+1, \ldots, 9$, which are now $k \times k$ matrices, transforming in the adjoint of $U(k)$. Moreover, the kinetic term in ten dimensions produces a potential of the form

$$V = \frac{g^2}{2} \sum_{i,j=p+1}^{9} \mathrm{tr}[X^i, X^j]^2 . \tag{4.24}$$

As we have already observed in many examples before, this potential possesses flat directions, correspoding to classical vacumm states. These flat directions are defined by diagonal X^i matrices,

$$X^i = \begin{pmatrix} \lambda_1^i & & \\ & \ddots & \\ & & \lambda_k^i \end{pmatrix} . \tag{4.25}$$

On each of these vacua, the $U(k)$ gauge symmetry is spontaneously broken to $U(1)^k$; thus, we can use these vacuum configurations to describe sets of k

parallel p dimensional D-branes. In fact, as we observe for the simpler case of one D-brane, the set of scalars appearing by dimensional reduction has the geometrical interpretation of the position of the D-brane. In the case (4.25), we can in fact consider λ_l^i as defining the ith-coordinate of the lth-brane. This is consistent with the idea of interpreting the strings stretching between different D-branes as massive vector bosons. In fact, the mass of this string states would be

$$M \sim g|\lambda_l - \lambda_m|, \tag{4.26}$$

for a (l, m) string. This is, in fact, the Higgs mass of the corresponding massive charged boson. In summary, merging the previous comments into a lemma: the classical moduli space of the worldvolume lagrangian of a D-brane coincides with its transversal space. It is important realizing that only the minima of the potential (4.24), i. e., the moduli space of the worldvolume lagrangian, is the one possessing this simple geometrical interpretation. In particular, the dimensional reduction, down to $p+1$ dimensions, of $N = 1$ $U(k)$ gauge theory, describes a set of k parallel branes, but its full fledged dynamics is described by the complete matrix X^i, with non vanishing off diagonal terms. A nice way to think about the meaning of (4.25) is again in terms of 't Hooft's abelian projection. In fact, we can think of (4.25) as a unitary gauge fixing, where we now allow λ_l^i to depend on the worldvolume coordinates. The case of flat parallel D-branes corresponds to a Higgs phase, with λ_l^i constant functions on the worldvolume. Moreover, we can even consider the existence of singularities, which will be points where two eigenvalues coincide,

$$\lambda_l^i = \lambda_{l+1}^i, \ \forall i. \tag{4.27}$$

It is quite obvious realizing that (4.27) imposes three constraints, so we expect, for p-dimensional D-branes, that on a $p-3$-dimensional region of the worldvolume, two consecutive D-branes can overlap. The $p-3$ region in the $p+1$ dimensional worldvolume of the D-brane will represent, from the point of view of $p+1$ dynamics, a monopole, in the very same sense as is the case in 't Hooft's abelian projection.

Next we will consider some brane configurations for type IIA and type IIB string theory (some of the widely increasing refences are those from [100] to [117]). In order to define these configurations we will first work out the allowed vertices for intersecting branes. Let us start with a vertex of type $(p, 1^F)$, corresponding to a Dirichlet p-brane, and a fundamental string ending on the D-brane worldvolume. In type IIA p should be even, and odd for type IIB. In fact, the RR fields for type IIA and type IIB string theory are

$$\begin{aligned} \text{IIA} &\rightarrow A_\mu \ A_{\mu\nu\rho}, \\ \text{IIB} &\rightarrow \chi \ B_{\mu\nu} \ A_{\mu\nu\rho\sigma}. \end{aligned} \tag{4.28}$$

The corresponding strength tensors are, respectively, two and four-forms for type IIA, and one, three and five-forms for type IIB. Thus, the sources are

D-branes of dimensions zero and two, for type IIA string theory, and one and three for type IIB. In addition, we have the (Hodge) magnetic duals, which are six and four D-branes for type IIA string theory, and five and three D-branes for type IIB (notice that the threebrane in type IIB is self dual). Besides, for the χ field in type IIB, the source is a -1 extended object, and its dual is a D-7brane.

Let us then start with a vertex of type $(p, 1^F)$ in type IIB, i. e., with p odd. We can use the $Sl(2, \mathbf{Z})$ duality symmetry of type IIB strings to transform this vertex into a $(p, 1)$ vertex, between a D-$p$brane and a D-1brane, or D-string. By performing j T-duality transformations on the spacetime directions orthogonal to the worldvolume of the D-brane and the D-string, we pass form $(p, 1^F)$ to a vertex $(p + j, 1^F + j)$ of two D-branes, sharing j common worldvolume coordinates. If j is even, we end up with a vertex in type IIB, and if j is odd with a vertex in type IIA. Namely, through a T-duality transformation we pass from type IIB string theory to type IIA. As an example, we will consider the vertex $(3, 1^F)$ in type IIB string theory. After a S-duality transformation in the $Sl(2, \mathbf{Z})$ duality group of type IIB strings, and two T-duality transformations, we get the vertex $(5, 3)$ for branes. As we are in type IIB, we can perform a duality transformation on it to generate the vertex $(5^{NS}, 3)$, between the solitonic Neveau-Schwarz fivebrane and a D-3brane.

Let us now consider some brane configurations build up using the vertices $(5, 3)$ and $(5^{NS}, 3)$ in type IIB theory [100]. In particular, we will consider solitonic fivebranes, with worldvolume coordinates x^0, x^1, x^2, x^3, x^4 and x^5, located at some definite values of x^6, x^7, x^8 and x^9. It is convenient to organize the coordinates of the fivebrane as (x^6, ω) where $\omega = (x^7, x^8, x^9)$. By construction of the vertex, the D-3brane will share two worldvolume coordinates, in addition to time, with the fivebrane. Thus, we can consider D-3branes with worldvolume coordinates x^0, x^1, x^2 and x^6. If we put a D-3brane in between two solitonic fivebranes, at x_2^6 and x_1^6 positions in the x^6 coordinate, then the worldvolume of the D-3brane will be finite in the x^6 direction (see Figure 1). Therefore, the macroscopic physics, i. e., for scales larger than $|x_2^6 - x_1^6|$, can be effectively described by a $2 + 1$ dimensional theory. In order to unravel what kind of $2 + 1$ dimensional theory, we are obtaining through this brane configuration, we must first work out the type of constraint impossed by the fivebrane boundary conditions. In fact, the worldvolume low energy lagrangian for a D-3brane is a $U(1)$ gauge theory. Once we put the D-3brane in between two solitonic fivebranes we imposse Neumann boundary conditions, in the x^6 direction, for the fields living on the D-3brane worldvolume. This means in particular that for scalar fields we imposse

$$\partial_6 \phi = 0 \tag{4.29}$$

and, for gauge fields,

$$F_{\mu 6} = 0, \qquad \mu = 0, 1, 2. \tag{4.30}$$

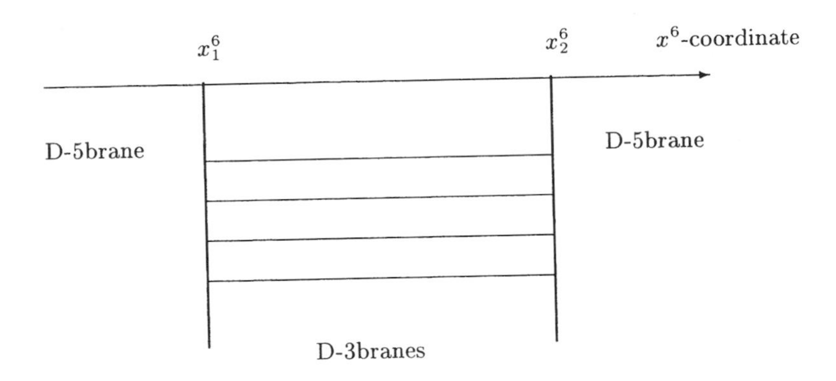

Fig. 4.1. Solitonic fivebranes with n Dirichlet threebranes stretching along them.

Thus, the three dimensional $U(1)$ gauge field, A_μ, with $\mu = 0, 1, 2$, is unconstrained which already means that we can interpret the effective three dimensional theory as a $U(1)$ gauge theory for one D-3brane, and therefore as a $U(n)$ gauge theory for n D-3branes. Next, we need to discover the amount of supersymmetry left unbroken by the brane configuration. If we consider Dirichlet threebranes, with worldvolume coordinates x^0, x^1, x^2 and x^6, then we are forcing the solitonic fivebranes to be at positions (x^6, ω_1) and (x^6, ω_2), with $\omega_1 = \omega_2$. In this particular case, the allowed motion for the D-3brane is reduced to the space $\mathbf{R}^3$, with coordinates x^3, x^4 and x^5. These are the coordinates on the fivebrane worldvolume where the D-3brane ends. Thus, we have defined on the D-3brane three scalar fields. By condition (4.29), the values of these scalar fields can be constrained to be constant on the x^6 direction. What this in practice means is that the two ends of the of the D-3brane have the same x^3, x^4 and x^5 coordinates. Now, if we combine these three scalar fields with the $U(1)$ gauge field A_μ, we get an $N = 4$ vector multiplet in three dimensions. Therefore, we can conclude that our effective three dimensional theory for n parallel D-3branes suspended between two solitonic fivebranes (Figure 1) is a gauge theory with $U(n)$ gauge group, and $N = 4$ supersymmetry. Denoting by $\mathbf{v}$ the vector (x^3, x^4, x^5), the Coulomb branch of this theory is parametrized by the v_i positions of the n D-3branes (with i labelling each brane). In addition, we have, as discussed in chapter II, the dual photons for each $U(1)$ factor. In this way, we get the hyperkähler structure of the Coulomb branch of the moduli. Hence, a direct way to get supersymmetry preserved by the brane configuration is as follows. The supersymmetry charges are defined as

$$\epsilon_L Q_L + \epsilon_R Q_R, \tag{4.31}$$

where Q_L and Q_R are the supercharges generated by the left and right-moving worldsheet degrees of freedom, and ϵ_L and ϵ_R are ten dimensional spinors.

Each solitonic $p$brane, with worldvolume extending along $x^0, x^1, \ldots, x^p$, imposses the conditions

$$\epsilon_L = \Gamma_0 \ldots \Gamma_p \epsilon_L, \ \epsilon_R = -\Gamma_0 \ldots \Gamma_p \epsilon_R, \tag{4.32}$$

in terms of the ten dimensional Dirac gamma matrices, Γ_i; on the other hand, the D-$p$branes, with worldvolumes extending along $x^0, x^1, \ldots, x^p$, imply the constraint

$$\epsilon_L = \Gamma_0 \Gamma_1 \ldots \Gamma_p \epsilon_R. \tag{4.33}$$

Thus, we see that NS solitonic fivebrane, with worldvolume located at x^0, x^1, x^2, x^3, x^4 and x^5, and equal values of $\boldsymbol{\omega}$, and Dirichlet threebranes with worldvolume along x^0, x^1, x^2 and x^6, preserve eight supersymmetries on the D-3brane worldvolume or, equivalently, $N = 4$ supersymmetry on the effective three dimensional theory.

The brane array just described allows a simple computation of the gauge coupling constant of the effective three dimensional theory: by standard Kaluza-Klein reduction on the finite x^6 direction, after integrating over the (compactified) x^6 direction to reduce the lagrangian to an effective three dimensional lagrangian, the gauge coupling constant is given by

$$\frac{1}{g_3^2} = \frac{|x_6^2 - x_6^1|}{g_4^2}, \tag{4.34}$$

in terms of the four dimensional gauge coupling constant. Naturally, (4.34) is a classical expression that is not taking into account the effect on the fivebrane position at x^6 of the D-3brane ending on its worldvolume. In fact, we can consider the dependence of x^6 on the coordinate $\mathbf{v}$, normal to the position of the D-3brane. The dynamics of the fivebranes should then be recovered when the Nambu-Goto action of the solitonic fivebrane is minimized. Far from the influence of the points where the fivebranes are located (at large values of x^3, x^4 and x^5), the equation of motion is simply three dimensional Laplace's equation,

$$\nabla^2 x^6(x^3, x^4, x^5) = 0, \tag{4.35}$$

with solution

$$x^6(r) = \frac{k}{r} + \alpha, \tag{4.36}$$

where k and α are constants depending on the threebrane tensions, and r is the spherical radius at the point (x^3, x^4, x^5). From (4.36), it is clear that there is a well defined limit as $r \to \infty$; hence, the difference $x_2^6 - x_1^6$ is a well defined constant, $\alpha_2 - \alpha_1$, in the $r \to \infty$ limit.

Part of the beauty of brane technology is that it allows to obtain very strong results by simply performing geometrical brane manipulations. We will now present one example, concerning our previous model. If we consider the brane configuration from the point of view of the fivebrane, the n suspended threebranes will look like n magnetic monopoles. This is really suggesting since, as described in chapter II, we know that the Coulomb branch moduli space of $N = 4$ supersymmetric $SU(n)$ gauge theories is isomorphic to the moduli space of BPS monopole configurations, with magnetic charge equal n. This analogy can be put more precisely: the vertex $(5^{NS}, 3)$ can, as described above, be transformed into a $(3, 1)$ vertex. In this case, from the point of view of the threebrane, we have a four dimensional gauge theory with $SU(2)$ gauge group broken down to $U(1)$, and n magnetic monopoles. Notice that by passing from the configuration build up ussing $(5^{NS}, 3)$ vertices, to that build up with the $(3, 1)$ vertex, the Coulomb moduli remains the same.

Next, we will work out the same configuration, but now with the vertex $(5, 3)$ made out of two Dirichlet branes. The main difference with the previous example comes from the boundary conditions (4.29) and (4.30), which should now be replaced by Dirichlet boundary conditions. We will choose as world-volume coordinates for the D-5branes x^0, x^1, x^2, x^7, x^8 and x^9, so that they will be located at some definite values of x^3, x^4, x^5 and x^6. As before, let us denote this positions by $(\mathbf{m}, x^6)$, where now $\mathbf{m} = (x^3, x^4, x^5)$. An equivalent configuration to the one studied above will be now a set of two D-5branes, at some points of the x^6 coordinate, that we will again call x_1^6 and x_2^6, subject to $\mathbf{m}_1 = \mathbf{m}_2$, with D-3branes stretching between them along the x^6 coordinate, with worldvolume extending again along the coordinates x^0, x^1, x^2 and x^6 (Figure 2). Our task now will

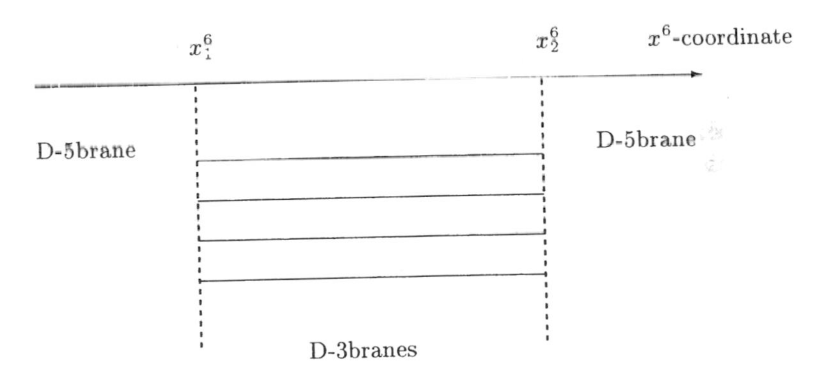

Fig. 4.2. Dirichlet threebranes extending between a pair of Dirichlet fivebranes (in dashed lines).

be the description of the effective three dimensional theory on these three-branes. The end points of the D-3branes on the fivebrane worlvolumes will now be parametrized by values of x^7, x^8 and x^9. This means that we have three scalar fields in the effective three dimensional theory. The scalar fields corresponding to the coordinates x^3, x^4, x^5 and x^6 of the threebranes are forzen to the constant values where the fivebranes are located. Next, we should consider what happens to the $U(1)$ gauge field on the D-3brane worldvolume. Impossing Dirichlet boundary conditions for this field is equivalent to

$$F_{\mu\nu} = 0, \quad \mu, \nu = 0, 1, 2, \tag{4.37}$$

i. e., there is no electromagnetic tensor in the effective three dimensional field theory. Before going on, it would be convenient summarizing the rules we have used to impose the different boundary conditions. Consider a D-$p$brane, and let M be its worldvolume manifold, and $B = \partial M$ the boundary of M. Neumann and Dirichlet boundary conditions for the gauge field on the D-$p$brane worldvolume are defined respectively by

$$\begin{aligned} N &\longrightarrow \quad F_{\mu\rho} = 0, \\ D &\longrightarrow \quad F_{\mu\nu} = 0, \end{aligned} \tag{4.38}$$

where μ and ν are directions of tangency to B, and ρ are the normal coordinates to B. If B is part of the worldvolume of a solitonic brane, we will imposse Neumann conditions, and if it is part of the worldvolume of a Dirichlet brane, we will imposse Dirichlet conditions. Returning to (4.37), we see that on the three dimensional effective theory, the only non vanishing component of the four dimensional strenght tensor is $F_{\mu 6} \equiv \partial_\mu b$. Therefore, all together we have four scalar fields in three dimensions or, equivalently, a multiplet with $N = 4$ supersymmetry. Thus, the theory defined by the n suspended D-3branes in between a pair of D-5branes, is a theory of n $N = 4$ massless hypermultiplets.

There exits a different way to interpret the theory, namely as a magnetic dual gauge theory. In fact, if we perform a duality transformation in the four dimensional $U(1)$ gauge theory, and use magnetic variables $*F$, instead of the electric field F, what we get in three dimensions, after impossing D-boundary conditions, is a dual photon, or a magnetic $U(1)$ gauge theory.

The configuration chosen for the worldvolume of the Dirichlet and solitonic fivebranes yet allows a different configuration with D-3branes suspended between a D-5brane and a NS-5brane. This is in fact consistent with the supersymmetry requirements (4.32) and (4.33). Namely, for the Dirichlet fivebrane we have

$$\epsilon_L = \Gamma_0 \Gamma_1 \Gamma_2 \Gamma_7 \Gamma_8 \Gamma_9 \epsilon_R. \tag{4.39}$$

The solitonic fivebrane imposses

$$\epsilon_L = \Gamma_0 \ldots \Gamma_5 \epsilon_L, \ \ \epsilon_R = -\Gamma_0 \ldots \Gamma_5 \epsilon_R, \tag{4.40}$$

while the suspended threebranes imply

$$\epsilon_L = \Gamma_0 \Gamma_1 \Gamma_2 \Gamma_6 \epsilon_R, \tag{4.41}$$

which are easily seen to be consistent. The problem now is that the suspended D-3brane is frozen. In fact, the position (x^3, x^4, x^5) of the end point of the NS-5brane is equal to the position $\mathbf{m}$ of the D-5brane, and the position (x^7, x^8, x^9)

of the end point on the D-5brane is forced to be equal to the position ω of the NS-5brane. The fact that the D-3brane is frozen means that the theory defined on it has no moduli, i. e., posseses a mass gap.

Using the vertices between branes we have described so far we can build quite complicated brane configurations. When Dirichlet threebranes are placed to the right and left of a fivebrane, open strings can connect the threebranes at different sides of the fivebrane. They will represent hypermultiplets transforming as $(k_1, \bar{k}_2)$, with k_1 and k_2 the number of threebranes to the left and right, respectively, of the fivebrane. In case the fivebrane is solitonic, the hypermultiplets are charged with respect to an electric group, while in case it is a D-5brane, they are magnetically charged. Another possibility is that with a pair of NS-5branes, with D-3branes extending between them, and also a D-5brane located between the two solitonic fivebranes. A massless hypermultiplet will now appear whenever the (x^3, x^4, x^5) position of the D-3brane coincides with the $\mathbf{m} = (x^3, x^4, x^5)$ position of the D-5brane.

So far we have used brane configurations for representing different gauge theories. In these brane configurations we have considered two different types of moduli. For the examples described above, these two types of moduli are as follows: the moduli of the effective three dimensional theory, corresponding to the different positions where the suspended D-3branes can be located, and the moduli corresponding to the different locations of the fivebranes, which are being used as boundaries. This second type of moduli specifies, from the point of view of the three dimensional theory, different coupling constants; hence, we can move the location of the fivebranes, and follow the changes taking place in the effective three dimensional theory. Let us then consider a case with two solitonic branes, and a Dirichlet fivebrane placed between them. Let us now move the NS-5brane on the left of the D-5brane to the right. In doing so, there is a moment when both fivebranes meet, sharing a common value of x^6. If the interpretation of the hypermultiplet we have presented above is correct, we must discover what happens to the hypermultiplet after this exchange of branes has been performed. In order to maintain the hypermultiplet, a new D-3brane should be created after the exchange, extending from the right solitonic fivebrane to the Dirichlet fivebrane. To prove this we will need D-brane dynamics at work. Let us start considering two interpenetrating closed loops, C and C', and suppose electrically charged particles are moving in C, while magnetically charged particles move in C'. The linking number $L(C, C')$ can be defined using the standard Wilson and 't Hooft loops. Namely, we can measure the electric flux passing through C' or, equivalently, compute $B(C')$, or measure the magnetic flux passing through C, i. e., the Wilson line $A(C)$. In both cases, what we are doing is integrating over C' and C the dual to the field created by the particle moving in C and C', respectively. Let us now extend this simple result to the case of fivebranes. A fivebrane is a source of 7-form tensor field, and its dual is therefore a 3-form. We will call this 3-form H_{NS} for NS-5branes, and H_D for

D-5branes. Now, let us consider the worldvolume of the two fivebranes,

$$\begin{aligned} \mathbf{R}^3 &\times Y_{NS}, \\ \mathbf{R}^3 &\times Y_D. \end{aligned} \tag{4.42}$$

We can now define the linking number as we did before, in the simpler case of a particle:

$$L(Y_{NS}, Y_D) = -\int_{Y_{NS}} H^D = \int_{Y_D} H^{NS}. \tag{4.43}$$

The 3-form H^{NS} is locally dB_{NS}. Since we have no sources for H^{NS}, we can use $H^{NS} = dB_{NS}$ globally; however, this requires B to be globally defined, or gauge invariant. In type IIB string theory, B is not gauge invariant; however, on a D-brane we can define the combination $B_{NS} - F_D$, which is invariant, with F_D the two form for the $U(1)$ gauge field on the D-brane. Now, when the D-5brane and the NS-5brane do not intersect, the linking number is obviously zero. When they intersect, this linking number changes, which means that (4.43) should, in that case, be non vanishing. Writing

$$\int_{Y_D} H^{NS} = \int_{Y_D} dB_{NS} - dF_D, \tag{4.44}$$

we observe that the only way to get linking numbers would be adding sources for F_D. These sources for F_D are point like on Y_D, and are therefore the D-3branes with worldvolume $\mathbf{R}^3 \times C$, with C ending on Y_D, which is precisely the required appearance of extra D-3branes.

4.4 D-Brane Description of Seiberg-Witten Solution.

In the previous example we have considered type IIB string theory and three and fivebranes. Now, let us consider type IIA strings, where we have four-branes that can be used to define, by analogy with the previous picture, $N = 2$ four dimensional gauge theories [103]. The idea will again be the use of solitonic fivebranes, with sets of fourbranes in between. The only difference now is that the fivebrane does not create a RR field in type IIA string theory and, therefore, the physics of the two parallel solitonic fivebranes does not have the interpretation of a gauge theory, as was the case for the type IIB configuration above described [103].

Let us consider configurations of infinite solitonic fivebranes, with worldvolume coordinates x^0, x^1, x^2, x^3, x^4 and x^5, located at $x^7 = x^8 = x^9 = 0$ and at some fixed value of the x^6 coordinate. In addition, let us introduce finite Dirichlet fourbranes, with worldvolume coordinates x^0, x^1, x^2, x^3 and x^6, which terminate on the solitonic fivebranes; thus, they are finite in the x^6 direction. On the fourbrane worldvolume, we can define a macroscopic four dimensional field theory, with $N = 2$ supersymmetry. This four dimensional theory will, as in the type IIB case considered in previous section, be defined by standard Kaluza-Klein dimensional reduction of the five dimensional theory defined on the D-4brane worldvolume. Then, the bare coupling constant of the four dimensional theory will be

$$\frac{1}{g_4^2} = \frac{|x_6^2 - x_6^1|}{g_5^2}, \tag{4.45}$$

in terms of the five dimensional coupling constant. Moreover, we can interpret as classical moduli parameters of the effective field theory on the dimensionally reduced worldvolume of the fourbrane the coordinates x^4 and x^5, which locate the points on the fivebrane worldvolume where the D-4branes terminate.

In addition to the Dirichlet fourbranes and solitonic fivebranes, we can yet include Dirichlet sixbranes, without any further break of supersymmetry on the theory in the worldvolume of the fourbranes. To prove this, we notice that each NS-5brane imposes the projections

$$\epsilon_L = \Gamma_0 \dots \Gamma_5 \epsilon_L, \quad \epsilon_R = -\Gamma_0 \dots \Gamma_5 \epsilon_R, \tag{4.46}$$

while the D-4branes, with worldvolume localized at x^0, x^1, x^2, x^3 and x^6, imply

$$\epsilon_L = \Gamma_0 \Gamma_1 \Gamma_2 \Gamma_3 \Gamma_6 \epsilon_R. \tag{4.47}$$

Conditions (4.46) and (4.47) can be recombined into

$$\epsilon_L = \Gamma_0 \Gamma_1 \Gamma_2 \Gamma_3 \Gamma_7 \Gamma_8 \Gamma_9 \epsilon_R, \tag{4.48}$$

which shows that certainly sixbranes can be added with no additional supersymmetry breaking.

The solitonic fivebranes break half of the supersymmetries, while the D-6brane breaks again half of the remaining symmetry, leaving eight real supercharges, which leads to four dimensional $N = 2$ supersymmetry.

As we will discuss later on, the sixbranes of type IIA string theory can be used to add hypermultiplets to the effective macroscopic four dimensional theory. In particular, the mass of these hypermultiplets will become zero whenever the D-4brane meets a D-6brane.

One of the main achievements of the brane representations of supersymmetric gauge theories is the ability to represent the different moduli spaces, namely the Coulomb and Higgs branches, in terms of the brane motions left free. For a configuration of k fourbranes connecting two solitonic fivebranes along the x^6 direction, as the one we have described above, the Coulomb branch of the moduli space of the four dimensional theory is parametrized by the different positions of the transversal fourbranes on the fivebranes. When N_f Dirichlet sixbranes are added to this configuration, what we are describing is the Coulomb branch of a four dimensional field theory with $SU(N_c)$ gauge group (in case N_c is the number of D-4branes we are considering), with N_f flavor hypermultiplets. In this brane representation, the Higgs branch of the theory is obtained when each fourbrane is broken into several pieces ending on different sixbranes: the locations of the D-4branes living between two D-6branes determine the Higgs branch. However, we will mostly concentrate on the study of the Coulomb branch for pure gauge theories.

As we know from the Seiberg-Witten solution of $N = 2$ supersymmetric gauge theories, the classical moduli of the theory is corrected by quantum effects. There are two types of effects that enter the game: a non vanishing beta function (determined at one loop) implies the existence, in the assymptotically free regime, of a singularity at the infinity point in moduli space, and strong coupling effects, which imply the existence of extra singularities, where some magnetically charged particles become massless. The problem we are facing now is how to derive such a complete characterization of the quantum moduli space of four dimensional $N = 2$ supersymmetric field theory directly from the dynamics governing the brane configuration. The approach to be used is completely different from a brane construction in type IIA string theory to a type IIB brane configuration. In fact, in the type IIB case, employed in the description of the preceding section of three dimensional $N = 4$ supersymmetric field theories, we can pass from weak to strong coupling through the standard $Sl(2, \mathbf{Z})$ duality of type IIB strings; hence, the essential ingredient we need is to know how brane configurations transform under this duality symmetry. In the case of type IIA string theory, the situation is more complicated, as the theory is not $Sl(2, \mathbf{Z})$ self dual. However, we know that the strong coupling limit of type IIA dynamics is described by the eleven dimensional M-theory; therefore, we should expect to recover the strong coupling dynamics of four dimensional $N = 2$ supersymmetric gauge theories using the M-theory description of strongly coupled type IIA strings.

Let us first start by considering weak coupling effects. The first thing to be noticed, concerning the above described configuration of N_c Dirichlet fourbranes extending along the x^6 direction between two solitonic fivebranes, where only a rigid motion of the transversal fourbranes is allowed, is that this simple image is missing the classical dynamics of the fivebranes. In fact, in this picture we are assuming that the x^6 coordinate on the fivebrane worldvolume is constant, which is in fact a very bad approximation. Of course, one physical requirement we should impose to a brane configuration, as we did in the case of the type IIB configurations of the previous section, is that of minimizing the total worldvolume action. More precisely, what we have interpreted as Coulomb or Higgs branches in term of free motions of some branes entering the configuration, should correspond to zero modes of the brane configuration, i. e., to changes in the configuration preserving the condition of minimum worldvolume action (in other words, changes in the brane configuration that do not constitute an energy expense). The coordinate x^6 can be assumed to only depend on the "normal" coordinates x^4 and x^5, which can be combined into the complex coordinate

$$v \equiv x^4 + ix^5, \tag{4.49}$$

representing the normal to the position of the transversal fourbranes. Far away from the position of the fourbranes, the equation for x^6 reduces now to a two dimensional laplacian,

$$\nabla^2 x^6(v) = 0, \tag{4.50}$$

with solution

$$x^6(v) = k \ln |v| + \alpha, \tag{4.51}$$

for some constants k and α, that will depend on the solitonic and Dirichlet brane tensions. As we can see from (4.51), the value of x^6 will diverge at infinity. This constitutes, as a difference with the type IIB case, a first problem for the interpretation of equation (4.45). In fact, in deriving (4.45) we have used a standard Kaluza-Klein argument, where the four dimensional coupling constant is defined by the volume of the internal space (in this ocasion, the x^6 interval between the two solitonic fivebranes). Since the Dirichlet four branes will deform the solitonic fivebrane, the natural way to define the internal space would be as the interval defined by the values of the coordinate x^6 at v equal to infinity, which is the region where the disturbing effect of the four brane is very likely vanishing, as was the case in the definition of the effective three dimensional coupling in the type IIB case. However, equations (4.50) and (4.51) already indicate us that this can not be the right picture, since these values of the x^6 coordinate are divergent. Let us then consider a configuration with N_c transversal fourbranes. From equations (4.45) and (4.51), we get, for large v,

$$\frac{1}{g_4^2} = -\frac{2kN_c \ln(v)}{g_5^2}, \tag{4.52}$$

where we have differentiated the direction in which the fourbranes pull the fivebrane. Equation (4.52) can have a very nice meaning if we interpret it as the one loop renormalization group equation for the effective coupling constant. In order to justify this interpretation, let us first analyze the physical meaning of the parameter k. From equation (4.51), we notice that if we move in v around a value where a fourbrane is located (that we are assuming is $v = 0$), we get the monodromy transformation

$$x^6 \to x^6 + 2\pi i k. \tag{4.53}$$

This equation can be easily understood in M-theory, where we add an extra eleventh dimension, x^{10}, that we use to define the complex coordinate

$$x^6 + i x^{10}. \tag{4.54}$$

Now, using the fact that the extra coordinate is compactified on a circle of radius R we can, from (4.53), identify k with R. From a field theory point of view, we have a similar interpretation of the monodromy of (4.52), but now in terms of a change in the theta parameter. Let us then consider the one loop renormalization group equation for $SU(N_c)$ $N=2$ supersymmetric gauge theories without hypermultiplets,

$$\frac{4\pi}{g_4^2(u)} = \frac{4\pi}{g_0^2} - \frac{2N_c}{4\pi} \ln\left(\frac{u}{\Lambda}\right), \tag{4.55}$$

with Λ the dynamically generated scale, and g_0 the bare coupling constant. The bare coupling constant can be absorbed through a change in Λ; in fact, when going from Λ to a new scale Λ', we get

$$\frac{4\pi}{g_4^2(u)} = \frac{4\pi}{g_0^2} - \frac{2N_c}{4\pi} \ln\left(\frac{u}{\Lambda'}\right) - \frac{2N_c}{4\pi} \ln\left(\frac{\Lambda'}{\Lambda}\right). \tag{4.56}$$

Thus, once we fix a reference scale Λ_0, the dependence on the scale Λ of the bare coupling constant is given by

$$-\frac{2N_c}{4\pi} \ln\left(\frac{\Lambda}{\Lambda_0}\right). \tag{4.57}$$

It is important to distinguish the dependence on Λ of the bare coupling constant, and the dependence on u of the effective coupling. In the brane configuration approach, the coupling constant defined by (4.52) is the bare coupling constant of the theory, as determined by the definite brane configuration. Hence, it is (4.57) that we should compare with (4.52); naturally, some care is needed concerning units and scales. Once we interpret k as the radius of the internal S^1 of M-theory we can, in order to make contact with (4.57), identify g_5^2 with the radius of S^1, which in M-theory units is given by

$$R = gl_s, \tag{4.58}$$

with g the string coupling constant, and l_s the string length, $\frac{1}{\alpha'}$. Therefore, (4.45) should be modified to

$$\frac{1}{g_4^2} = \frac{x_2^6 - x_1^6}{gl_s} = -2N_c \ln(v), \tag{4.59}$$

which should be dimensionless. Then, we should interpret v in (4.59) as a dimensionless variable or, equivalently, as $\frac{v}{R}$, with R playing the role of natural unit of the theory. Then, comparing (4.57) and (4.59), $\frac{v}{R}$ becomes the scale $\frac{\Lambda}{\Lambda_0}$ in the formula for the bare coupling constant. In summary, v fixes the scale of the theory. From the previous discussion, an equivalent interpretation follows, where R fixes Λ_0, and therefore changes in the scale are equivalent to changes in the radius of the internal S^1.

Defining now an adimensional complex variable,

$$s \equiv (x^6 + ix^{10})/R, \tag{4.60}$$

and a complexified coupling constant,

$$\tau = \frac{4\pi i}{g^2} + \frac{\theta}{2\pi}, \tag{4.61}$$

we can generalize (4.59) to

$$-i\tau_\alpha(v) = s_2(v) - s_1(v), \tag{4.62}$$

for the simple configuration of branes defining a pure gauge theory. Now, we can clearly notice how the monodromy, as we move around $v = 0$, means a change $\theta \to +2\pi N_c$.

Let us now come back, for a moment, to the bad behaviour of $x^6(v)$ at large values of v. A possible way to solve this problem is modifying the configuration of a single pair of fivebranes, with N_c fourbranes extending between them, to consider a larger set of solitonic fivebranes. Labelling this fivebranes by α, with $\alpha = 0, \ldots, n$, the corresponding x^6_α coordinate will depend on v as follows:

$$x^6(v)_\alpha = R\sum_{i=1}^{q_L} \ln|v - a_i| - R\sum_{j=1}^{q_R} \ln|v - b_j|, \tag{4.63}$$

where q_L and q_R represent, respectively, the number of D-4branes to the left and right of the αth fivebrane. As is clear from (4.63), a good behaviour at large v will only be possible if the numbers of fourbranes to the right and left of a fivebrane are equal, $q_L = q_R$, which somehow amounts to compensating the perturbation created by the fourbranes at the sides of a fivebrane. The four dimensional field theory represented now by this brane array will have a gauge group $\prod_\alpha U(k_\alpha)$, where k_α is the number of transversal fourbranes between the $\alpha - 1$ and α^{th} solitonic fivebranes. Now, minimization of the worldvolume action will require not only taking into account the dependence of x^6 on v, but also the fourbrane positions on the NS-5brane, represented by a_i and b_j in (4.63), on the four dimensional worldvolume coordinates x^0, x^1, x^2 and x^3. Using (4.63), and the Nambu-Goto action for the solitonic fivebrane, we get, for the kinetic energy,

$$\int d^4x d^2v \sum_{\mu=0}^{3} \partial_\mu x^6(v, a_i(x^\mu), b_j(x^\mu))\partial^\mu x^6(v, a_i(x^\mu), b_j(x^\mu)). \tag{4.64}$$

Convergence of the v integration implies

$$\partial_\mu(\sum_i a_i - \sum_j b_j) = 0 \tag{4.65}$$

or, equivalently,

$$\sum_i a_i - \sum_j b_j = \text{constant}. \tag{4.66}$$

This "constant of motion" is showing how the average of the relative position between left and right fourbranes must be hold constant. Since the

Coulomb branch of the $\prod_\alpha U(k_\alpha)$ gauge theory will be associated with different configurations of the transversal fourbranes, constraint (4.66) will reduce the dimension of this space. As we know from our general discussion on D-branes, the $U(1)$ part of the $U(k_\alpha)$ gauge group can be associated to the motion of the center of mass. Constraint (4.66) implies that the center of mass is frozen in each sector. With no semi-infinite fourbranes to the right, we have that $\sum_i a_i = 0$; now, this constraint will force the center of mass of all sectors to vanish, which means that the field theory we are describing is $\prod_\alpha SU(k_\alpha)$, instead of $\prod_\alpha U(k_\alpha)$. The same result can be derived if we include semi-infinite fourbranes to the left and right of the first and last solitonic fivebranes: as they are infinitely massive, we can assume that they do not move in the x^4 and x^5 directions. An important difference will appear if we consider periodic configurations of fivebranes, upon compactification of the x^6 direction to a circle: in this case, constraint (4.66) is now only able to reduce the group to $\prod_\alpha SU(k_\alpha) \times U(1)$, leaving alive a $U(1)$ factor.

Hypermultiplets in this gauge theory are understood as strings connecting fourbranes on different sides of a fivebrane; therefore, whenever the positions of the fourbranes to the left and right of a solitonic brane become coincident, a massless hypermultiplet arises. As the hypermultiplets are charged under the gauge groups at both sides of a certain $\alpha+1$ fivebrane, they will transform as $(k_\alpha, \bar{k}_{\alpha+1})$.

However, as the position of the fourbranes on both sides of a fivebrane varies as a function of x^0, x^1, x^2 and x^3, the existence of a well defined hypermultiplet can only be accomplished thanks to the fact that its variation rates on both sides are the same, as follows again from (4.65): $\partial_\mu(\sum_i a_{i,\alpha}) = \partial_\mu(\sum_j a_{j,\alpha+1})$. The definition of the bare massses comes then naturally from constraint (4.66):

$$m_\alpha = \frac{1}{k_\alpha}\sum_i a_{i,\alpha} - \frac{1}{k_{\alpha+1}}\sum_j a_{j,\alpha+1}. \tag{4.67}$$

With this interpretation, the constraint (4.66) becomes very natural from a physical point of view: it states that the masses of the hypermultiplets do not depend on the spacetime position.

The consistency of the previous definition of hypermultiplets can be checked using the previous construction of the one-loop beta function. In fact, from equation (4.62), we get, for large values of v,

$$-i\tau_\alpha(v) = (2k_\alpha - k_{\alpha-1} - k_{\alpha+1}) \ln v. \tag{4.68}$$

The number k_α of branes in the αth is, as we know, the number of colours, N_c. Comparing with the beta function for $N=2$ supersymmetric $SU(N_c)$ gauge theory with N_f flavors, we conclude that

$$N_f = k_{\alpha-1} + k_{\alpha+1}, \tag{4.69}$$

so that the number of fourbranes (hypemultiplets) at both sides of a certain pair of fivebranes, $k_{\alpha+1} + k_{\alpha-1} \equiv N_f$, becomes the number of flavors.

Notice, from (4.67), that the mass of all the hypermultiplets associated with fourbranes at both sides of a solitonic fivebrane are the same. This implies a global flavor symmetry. This global flavor symmetry is the gauge symmetry of the adjacent sector. This explains the physical meaning of (4.67).

Let us now come back to equation (4.59). What we need in order to unravel the strong coupling dynamics of our effective four dimensional gauge theory is the u dependence of the effective coupling constant, dependence that will contain non perturbative effects due to instantons. It is from this dependence that we read the Seiberg-Witten geometry of the quantum moduli space. Strong coupling effects correspond to u in the infrared region, i. e., small u or, equivalently, large Λ. From our previous discussion of (4.59), we conclude that the weak coupling regime corresponds to the type IIA string limit, $R \to 0$, and the strong coupling regime to the M-theory reime, at large values of R (recall that changes of scale in the four dimensional theory correspond to changes of the radius of the internal S^1). This explains our hopes that M-theory could describe the strong coupling regime of the four dimensional theory). We will then see now how M-theory is effectively working.

4.4.1 M-Theory and Strong Coupling.. From the M-theory point of view, the brane configuration we are considering can be interpreted in a different way. In particular, the D-4branes we are using to define the four dimensional macroscopic gauge theory can be considered as fivebranes wrapping the eleven dimensional S^1. Moreover, the trick we have used to make finite these fourbranes in the x^6 direction can be directly obtained if we consider fivebranes with worldvolume $\mathbf{R}^4 \times \Sigma$, where $\mathbf{R}^4$ is parametrized by the coordinates x^0, x^1, x^2 and x^4, and Σ is two dimensional, and embedded in the four dimesional space of coordinates x^4, x^5, x^6 and x^{10}. If we think in purely classical terms, the natural guess for Σ would be a cylinder with the topology $S^1 \times [x_2^6, x_1^6]$, for a configuration of k D-4branes extending along the x^6 direction between two solitonic fivebranes. This is however a very naive compactification, because there is no reason to believe that a fivebrane wrapped around this surface will produce, on the four dimensional worldvolume $\mathbf{R}^4$, any form of non abelian gauge group. In fact, any gauge field on $\mathbf{R}^4$ should come from integrating the chiral antisymmetric tensor field β of the M-theory fivebrane worldvolume, on some one-cycle of Σ. If we wnat to reproduce, in four dimensions, some kind of $U(k)$ or $SU(k)$ gauge theory, we should better consider a surface Σ with a richer first homology group. However, we can try to do something better when including the explicit dependence of the x^6 coordinate on v. In this case, we will get a picture that is closer to the right answer, but still far away from the true solution. Including the v dependence of the x^6 coordinate leads to a family of surfaces, parametrized by v, Σ_v, defined by $S^1 \times [x_2^6, x_1^6](v)$. The nice feature about this picture is that v, which is the transverse coordinate of Σ in the space Q, defined by the coordinates

x^4, x^5, x^6 and x^{10}, becomes now similar to the moduli of Σ_v; however, we have yet the problem of the of the genus or, in more general terms, the first homology group of Σ. The reason for following the previous line of thought, is that we are trying to keep alive the interpretation of the v coordinate as moduli, or coordinate of the Coulomb branch. This is, in fact, the reason giving rise to the difficulties with the genus, as we are using just one complex coordinate, independently of the rank of the gauge group, something we are forced to do because of the divergences in equation (4.51).

The right M-theory approach is quite different. In fact, we must try to get Σ directly from the particular brane configuration we are working with, and define the Colomb branch of the theory by the moduli space of brane configurations. Let us then define the single valued coordinate t,

$$t \equiv \exp -s, \tag{4.70}$$

and define the surface Σ we are looking for through

$$F(t, v) = 0. \tag{4.71}$$

From the classical equations of motion of the fivebrane we know the assymptotic behaviour for very large t,

$$t \sim v^k, \tag{4.72}$$

and for very small t,

$$t \sim v^{-k}. \tag{4.73}$$

Conditions (4.72) and (4.73) imply that $F(t, v)$ will have, for fixed values of t, k roots, while two different roots for fixed v. It must be stressed that the assymptotic behaviour (4.72) and (4.73) corresponds to the one loop beta function for a field theory with gauge group $SU(k)$, and without hypermultiplets. A function satisfying the previous conditions will be of the generic type

$$F(t, v) = A(v)t^2 + B(v)t + C(v), \tag{4.74}$$

with A, B and C polynomials in v of degree k. From (4.72) and (4.73), the function (4.74) becomes

$$F(t, v) = t^2 + B(v)t + \text{ constant}, \tag{4.75}$$

with one undetermined constant. In order to kill this constant, we can rescale t to $t/\text{constant}$. The meaning of this rescaling can be easily understood in terms of of the one loop beta function, written as (4.72) and (4.73). In fact, these equations can be read as

$$s = -k \ln \left(\frac{v}{R} \right), \tag{4.76}$$

and therefore the rescaling of R goes like

$$s \to -k \ln \left(\frac{v}{R'} \frac{R'}{R} \right) \tag{4.77}$$

or, equivalently,

$$t \to t \left(\frac{R'}{R} \right)^k . \tag{4.78}$$

Thus, and based on the above discussion on the definition of the scale, we observe that the constant in (4.75) defines the scale of the theory. With this interpretation of the constant, we can get the Seiberg-Witten solution for $N = 2$ pure gauge theories, with gauge group $SU(k)$. If $B(v)$ is chosen to be

$$B(v) = v^k + u_2 v^{k-2} + u_3 v^{k-3} + \cdots + u_k, \tag{4.79}$$

we finally get the Riemann surface

$$t^2 + B(v)t + 1 = 0, \tag{4.80}$$

a Riemann surface of genus $k - 1$, which is in fact the rank of the gauge group. Moreover, we can now try to visualize this Riemann surface as the worldvolume of the fivebrane describing our original brane configuration: each v-plane can be compactified to $\mathbf{P}^1$, and the transversal fourbranes cna be interpreted as gluing tubes, which clearly represents a surface with $k - 1$ handles. This image corresponds to gluing two copies of $\mathbf{P}^1$, with k disjoint cuts on each copy or, equivalently, $2k$ branch points. Thus, as can be observed from (4.80), to each transversal D-4brane there correspond two branch points and one cut on $\mathbf{P}^1$.

If we are interested in $SU(k)$ gauge theories with hypermultiplets, then we should first replace (4.72) and (4.73) by the corresponding relations,

$$t \sim v^{k-k_{\alpha-1}}, \tag{4.81}$$

and

$$t \sim v^{-k-k_{\alpha+1}}, \tag{4.82}$$

for t large and small, respectively. These are, in fact, the relations we get from the beta functions for these theories. If we take $k_{\alpha_1} = 0$, and $N_f = k_{\alpha+1}$, the curve becomes

$$t^2 + B(v)t + C(v) = 0, \tag{4.83}$$

with $C(v)$ a polynomial in v, of degree N_f, parametrized by the masses of the hypermultiplets,

$$C(v) = f \prod_{j=1}^{N_f} (v - m_j), \tag{4.84}$$

with f a complex constant.

Summarizing, we have been able to find a moduli of brane configurations reproducing four dimensional $N = 2$ supersymmetric $SU(k)$ gauge theories.

The exact Seiberg-Witten solution is obtained by reduction of the worldvolume fivebrane dynamics on the surface $\Sigma_{\mathbf{u}}$ defined at (4.80) and (4.82). Obviously, reducing the fivebrane dynamics to $\mathbf{R}^4$ on $\Sigma_{\mathbf{u}}$ leads to an effective coupling constant in $\mathbf{R}^4$, the $k-1 \times k-1$ Riemann matrix $\tau(\mathbf{u})$ of $\Sigma_{\mathbf{u}}$.

Before finishing this section, it is important to stress some peculiarities of the brane construction. First of all, it should be noticed that the definition of the curve Σ, in terms of the brane configuration, requires working with uncompactified x^4 and x^5 directions. This is part of the brane philosophy, where we must start with a particular configuration in flat spacetime. A different approach will consist in directly working with a spacetime $\mathcal{Q} \times R^7$, with $\mathcal{Q}$ some Calabi-Yau manifold, and consider a fivebrane worldvolume $\Sigma \times \mathbf{R}^4$, with $\mathbf{R}^4 \subset \mathbf{R}^7$, and Σ a lagrangian submanifold of $\mathcal{Q}$. Again, by Mc Lean's theorem, the $N=2$ theory defined on $\mathbf{R}^4$ will have a Coulomb branch with dimension equal to the first Betti number of Σ, and these deformations of Σ in $\mathcal{Q}$ will represent scalar fields in the four dimensional theory. Moreover, the holomorphic top form Ω of $\mathcal{Q}$ will define the meromorphic λ of the Seiberg-Witten solution. If we start with some Calabi-Yau manifold $\mathcal{Q}$, we should provide some data to determine Σ (this is what we did in the brane case, with $\mathcal{Q}$ non compact and flat. If, on the contrary, we want to select Σ directly from $\mathcal{Q}$, we can only do it in some definite cases, which are those related to the *geometric mirror construction* [118, 119]. Let us then recall some facts about the geometric mirror. The data are

- The Calabi-Yau manifold $\mathcal{Q}$.
- A lagrangian submanifold $\Sigma \to \mathcal{Q}$.
- A $U(1)$ flat bundle on Σ.

The third requirement is equivalent to interpreting Σ as a D-brane in $\mathcal{Q}$. This is a crucial data, in order to get from the above points the structure of abelian manifold of the Seiberg-Witten solution. Namely, we frist use Mc Lean's theorem to get the moduli of deformations of $\Sigma \to \mathcal{Q}$, preserving the condition of lagrangian submanifold. This space is of dimension $b_1(\Sigma)$. Secondly, on each of these points we fiber the jacobian of Σ, which is of dimension g. This family of abelian varieties defines the quantum moduli of a gauge theory, with $N=2$ supersymmetry, with a gauge group of rank equal $b_1(\Sigma)$. Moreover, this family of abelian varieties is the moduli of the set of data of the second and third points above, i. e., the moduli of Σ as a D-2brane. In some particular cases, this moduli is $\mathcal{Q}$ itself or, more properly, the geometric mirror of $\mathcal{Q}$. This will be the case for Σ of genus equal one, i. e., for the simple $SU(2)$ case. In this cases, the characterization of Σ in $\mathcal{Q}$ is equivalent to describing $\mathcal{Q}$ as an elliptic fibration. The relation between geometric mirror and T-duality produces a completely different physical picture. In fact, we can, when Σ is a torus, consider in type IIB a threebrane with classical moduli given by $\mathcal{Q}$. After T-duality or mirror, we get the type IIA description in terms of a fivebrane. In summary, it is an important problem to understand the relation of quantum mirror between type IIA and type IIB string theory, and the M-theory strong coupling description of type IIA strings.

4.5 Brane Description of $N = 1$ Four Dimensional Field Theories.

In order to consider field theories with $N = 1$ supersymmetry, the first thing we will study will be R-symmetry. Let us then recall the way R-symmetries were defined in the case of four dimensional $N = 2$ supersymmetry, and three dimensional $N = 4$ supersymmetry, through compactification of six dimensional $N = 1$ supersymmetric gauge field theories. The $U(1)_R$ in four dimensions, or $SO(3)_R$ in three dimensions, are simply the euclidean group of rotations in two and three dimensions, respectively. Now, we have a four dimensional space $\mathcal{Q}$, parametrized by coordinates t and v, and a Riemann surface Σ, embedded in $\mathcal{Q}$ by equations of the type (4.74). To characterize R-symmetries, we can consider transformations on $\mathcal{Q}$ which transform non trivially its holomorphic top form Ω. The unbroken R-symmetries will then be rotations in $\mathcal{Q}$ preserving the Riemann surface defined by the brane configuration. If we consider only the assymptotic behaviour of type (4.72), or (4.81), we get $U(1)_R$ symmetries of type

$$\begin{aligned} t &\rightarrow \lambda^k t, \\ v &\rightarrow \lambda v. \end{aligned} \tag{4.85}$$

This $U(1)$ symmetry is clearly broken by the curve (4.80). This spontaneous breakdown of the $U(1)_R$ symmetry is well understood in field theory as an instanton induced effect. If instead of considering $\mathcal{Q}$, we take the larger space $\hat{\mathcal{Q}}$, containing the x^7, x^8 and x^9 coordinates, we see that the $N = 2$ curve is invariant under rotations in the (x^7, x^8, x^9) space.

Let us now consider a brane configuration which reproduces $N = 1$ four dimensional theories [117]. We will again start in type IIA string theory, and locate a solitonic fivebrane at $x^6 = x^7 = x^8 = x^9 = 0$ with, as usual, worldvolume coordinates x^0, x^1, x^2, x^3, x^4 and x^5. At some definite value of x^6, say x_0^6, we locate another solitonic fivebrane, but this time with worldvolume coordinates x^0, x^1, x^2, x^3, x^7 and x^8, and $x^4 = x^5 = x^9 = 0$. As before, we now suspend a set of k D-4branes in between. They will be parametrized by the positions $v = x^4 + ix^5$, and $w = x^7 + ix^8$, on the two solitonic fivebranes. The worldvolume coordinates on this D-4branes are, as in previous cases, x^0, x^1, x^2 and x^3. The effective field theory defined by the set of fourbranes is macroscopically a four dimensional gauge theory, with coupling constant

$$\frac{1}{g^2} = \frac{x_0^6}{gl_s}. \tag{4.86}$$

Moreover, now we have only $N = 1$ supersymmetry, as no massless bosons can be defined on the four dimensional worldvolume (x^0, x^1, x^2, x^3). In fact, at the line $x^6 = 0$ the only possible massless scalar would be v, since $w = 0$ and $x^9 = 0$, so that we project out x^9 and w. On the other hand, at x_0^6 we have $v = 0$ and $x^9 = 0$ and, therefore, we have projected out all massless scalars. Notice that by the same argument, in the case of two solitonic fivebranes located at different values of x^6 but at $x^7 = x^8 = x^9 = 0$, we have one complex massless scalar that is not projected out, which leads to $N = 2$ supersymmetry in four dimensions. The previous discussion means that v, w and x^9 are projected out as four dimensional scalar fields; however, w and v are still classical moduli parameters of the brane configuration.

Now, we return to a comment already done in previous section: each of the fourbranes we are suspending in between the solitonic fivebranes can be interpreted as a fivebrane wrapped around a surface defined by the eleven dimensional S^1 of M-theory, multiplied by the segment $[0, x_0^6]$. Classically, the four dimensional theory can be defined through dimensional reduction of the fivebrane worldvolume on the surface Σ. The coupling constant will be given by the moduli τ of this surface,

$$\frac{1}{g^2} = \frac{2\pi R}{S}, \tag{4.87}$$

with S the length of the interval $[0, x_0^6]$, in M-theory units. In $N = 1$ supersymmetric field theories, on the contrary of what takes place in the $N = 2$ case, we have not a classical moduli and, therefore, we can not define a wilsonian coupling constant depending on some mass scale fixed by a vacuum expectation value. This fact can produce some problems, once we take into account the classical dependence of x^6 on v and w. In principle, this dependence should be the same as that in the case studied in previous section,

$$\begin{aligned} x^6 &\sim \tilde{k} \ln v, \\ x^6 &\sim \tilde{k} \ln w. \end{aligned} \tag{4.88}$$

Using the t coordinate defined in (4.70), equations (4.88) become

$$\begin{aligned} t &\sim v^k, \\ t &\sim w^k, \end{aligned} \tag{4.89}$$

for large and small t, respectively, or, equivalently, $t \sim v^k$, $t^{-1} \sim w^k$. Now, we can use these relations in (4.86). Taking into account the units, we can write

$$\frac{1}{g^2} \sim N_c[\ln \frac{v}{R} + \ln \frac{w}{R}], \tag{4.90}$$

with $k \equiv N_c$. As we did in the $N = 2$ case, we can try to compare (4.90) with the one loop beta function for $N = 1$ supersymmetric $SU(N_c)$ pure Yang-Mills theory,

$$\Lambda = \mu \exp - \frac{8\pi^2}{3N_c g(\mu)^2}. \tag{4.91}$$

In order to get the scale from (4.90) we impose

$$v = \zeta w^{-1}, \tag{4.92}$$

with ζ some constant with units of (length)^2. Using (4.92) and (4.90) we get

$$\frac{1}{g^2} \sim N_c \ln\left(\frac{\zeta}{R^2}\right). \tag{4.93}$$

In order to make contact with (4.91) we must impose

$$\frac{\zeta}{R^2} = (\Lambda R)^3, \tag{4.94}$$

where we have used $\frac{1}{R}$ in order to measure Λ. Using (4.92), we get the curve associated to four dimensional $N = 1$ field theory,

$$\begin{aligned} t &= v^k, \\ \zeta^k t^{-1} &= w^k \\ v &= \zeta w^{-1}. \end{aligned} \tag{4.95}$$

The curve defined by (4.95) will only depend on ζ^k. The different set of brane configurations compatible with (4.95) are given by values of ζ, with fixed ζ^k. These N_c roots parametrize the N_c different vacua predicted by $\mathrm{tr}(-1)^F$ arguments. It is important to observe that the coupling constant $\frac{1}{g^2}$ we are defining is the so called wilsonian coupling. We can interpret it as a complex number with $\mathrm{Im}\,\frac{1}{g^2} \equiv \frac{\theta}{8\pi^2}$. Hence, the value of $\mathrm{Im}\,\zeta^k$ fixes the θ parameter of the four dimensional theory.

For a given value of ζ, (4.95) defines a Riemann surface of genus zero, i. e., a rational curve. This curve is now embedded in the space of (t, v, w) coordinates. We will next observe that these curves, (4.95), are the result of "rotating" [106] the rational curves in the Seiberg-Witten solution, corresponding to the singular points. However, before doing that let us comment on $U(1)_R$ symmetries. As mentioned above, in order to define an R-symmetry we need a transformation on variables (t, v, w) not preserving the holomorphic top form,

$$\Omega = dv \wedge dw \wedge \frac{dt}{t} R. \tag{4.96}$$

A rotation in the w-plane, compatible with the assymptotic conditions (4.89), and defining an R-symmetry, is

$$\begin{aligned} v &\to v, \\ t &\to t, \\ w &\to e^{2\pi i/k} w. \end{aligned} \tag{4.97}$$

Now, it is clear that this symmetry is broken spontaneously by the curve (4.95). More interesting is an exact $U(1)$ symmetry, that can be defined for the curve (4.95):

$$
\begin{aligned}
v &\rightarrow e^{i\delta} v, \\
t &\rightarrow e^{i\delta k} t, \\
w &\rightarrow e^{-i\delta} w.
\end{aligned}
\tag{4.98}
$$

As can be seen from (4.96), this is not an R-symmetry, since Ω is invariant. Fields charged with respect to this $U(1)$ symmetry should carry angular momentum in the v or w plane, or linear momentum in the eleventh dimension interval (i. e., zero branes) The fields of $N = 1$ SQCD do not carry any of these charges, so all fields with $U(1)$ charge should be decoupled from the $N = 1$ SQCD degrees of freedom. This is equivalent to the way we have projected out fields in the previous discussion on the definition of the effective $N = 1$ four dimensional field theory.

4.5.1 Rotation of Branes.. A different way to present the above construction is by performing a rotation of branes. We will now concentrate on this procedure. The classical configuration of NS-5branes with worldvolumes extending along x^0, x^1, x^2, x^3, x^4 and x^5, can be modified to a configuration where one of the solitonic fivebranes has been rotated, from the $v = x^4 + ix^5$ direction, to be also contained in the (x^7, x^8)-plane, so that, by moving it a finite angle μ, it is localized in the (x^4, x^5, x^7, x^8) space. Using the same notation as in previous section, the brane configuration, where a fivebrane has been moved to give rise to an angle μ in the (v, w)-plane, the rotation is equivalent to imposing

$$
w = \mu v. \tag{4.99}
$$

In the brane configuration we obtain, points on the rotated fivebrane are parametrized by the (v, w) coordinates in the (x^4, x^5, x^7, x^8) space. We can therefore imposse the following assymptotic conditions [116]:

$$
\begin{aligned}
t &= v^k, \quad w = \mu v, \\
t &= v^{-k}, \quad w = 0,
\end{aligned}
\tag{4.100}
$$

respectively for large and small t. Let us now assume that this brane configuration describes a Riemann surface, $\hat{\Sigma}$, embedded in the space $(x^6, x^{10}, x^4, x^5, x^7, x^8)$, and let us denote by Σ the surface in the $N = 2$ case, i. e., for $\mu = 0$. In these conditions, $\hat{\Sigma}$ is simply the graph of the function w on Σ. We can interpret (4.99) as telling us that w on Σ posseses a simple pole at infinity, extending holomorphically over the rest of the Riemann surface. If we imposse this condition, we get that the projected surface Σ, i. e., the one describing the $N = 2$ theory, is of genus zero. In fact, it is a well known result in the theory of Riemann surfaces that the order of the pole at infinity depends on the genus of the surface in such a way that for genus larger than zero, we will be forced to replace (4.99) by $w = \mu v^a$ for some power a depending on the genus. A priori, there is no problem in trying to rotate using, instead of $w = \mu v$, some higher pole modification of the type $w = \mu v^a$, for $a > 1$. This would provide Σ surfaces with genus different from zero; however, we would immediately find problems with equation (4.90), and we will be unable to kill all dependence of the coupling constant on v and w. Therefore, we conclude that the only curves that can be rotated to produce a four dimensional $N = 1$ theory are those with zero genus. This is in perfect agreement with the physical picture we get from the Seiberg-Witten solution. Namely, once we add a soft breaking term of the type $\mu \mathrm{tr} \phi^2$, the only points remaining in the moduli space as real vacua of the theory are the singular points, where the Seiberg-Witten curve degenerates.

4.5.2 QCD Strings and Scales.. In all our previous discussion we have not been careful enough in separating arguments related to complex or *holomorphic structure*, and those related to *Kähler structure*. The M-theory description contains however relevant information on both aspects. For instance, in our previous derivation of curves, we were mostly interested in reproducing the complex structure of the Seiberg-Witten solution, as is, for instance, the moduli dependence on vacuum expectation values, i. e., the effective wilsonian coupling constant. However, we can also ask ourselves on BPS masses and, in that case, we will need the definite embedding of Σ in the ambient space $\mathcal{Q}$, and the holomorphic top form defined on $\mathcal{Q}$. As is clear from the fact that we are working in M-theory, the holomorphic top form on $\mathcal{Q}$ will depend explicitely on R, i. e., on the string coupling constant, and we will therefore find BPS mass formulas that will depend explicitely on R. We will discuss this type of dependence on R first in the case of $N = 1$ supersymmetry. The $N = 1$ four dimensional field theory we have described contains, in principle, two parameters. One is the constant ζ introduced in equation (4.92) which, as we have already mentioned, is, because of (4.90), intimately connected with Λ, and the radius R of the eleven dimensional S^1. Our first task would be to see what kind of four dimensional dynamics is dependent on the particular value of R, and in what way. The best example we can of course use is the computation of gaugino-gaugino condensates. In order to do that, we should try to minimize a four dimensional suerpotential for the $N = 1$ theory. Following Witten, we will define this superpotential W as an holomorphic function of Σ, and with critical points precisely when the surface Σ is a holomorphic curve in $\mathcal{Q}$. The space $\mathcal{Q}$ now is the one with coordinates x^4, x^5, x^6, x^7, x^8 and x^{10} (notice that this second condition was the one used to prove that rotated curves are necesarily of genus equal zero). Moreover, we need to work with a holomorphic curve because of $N = 1$ supersymmetry. A priori, there are two different ways we can think about this superpotential: maybe the simplest one, from a physical point of view, is as a functional defined on the volume of Σ, where this volume is given by

$$\mathrm{Vol}(\Sigma) = J.\Sigma, \tag{4.101}$$

with J the Kähler class of $\mathcal{Q}$. The other posibility is defining

$$W(\Sigma) = \int_B \Omega, \tag{4.102}$$

with B a 3-surface such that $\Sigma = \partial B$, and Ω the holomorphic top form in $\mathcal{Q}$. Definition (4.102) automatically satisfies the condition of being stationary, when Σ is a holomorphic curve in $\mathcal{Q}$. Notice that the holomorphy condition on Σ means, in mathematical terms, that Σ is an element of the Picard lattice of $\mathcal{Q}$, i. e., an element in $H_{1,1}(\mathcal{Q}) \cap H_2(\mathcal{Q})$. This is what allows us to use (4.101), however, and this is the reason for temporarily abandoning the approach based on (4.101). What we require to W is being stationary for holomorphic curves, but it should, in principle, be defined for arbitrary surfaces Σ, even those which are not part of the Picard group. Equation (4.102) is only well defined if Σ is contractible, i. e., if the homology class of Σ in $H_2(\mathcal{Q};\mathbf{Z})$ is trivial. If that is not the case, a reference surface Σ_0 needs to be defined, and (4.102) is modified to

$$W(\Sigma) - W(\Sigma_0) = \int_B \Omega, \tag{4.103}$$

where now $\partial B = \Sigma \cup \Sigma_0$. For simplicity, we will assume $H_3(\mathcal{Q};\mathbf{Z}) = 0$. From physical arguments we know that the set of zeroes of the superpotential should be related by $\mathbf{Z}_k$ symmetry, with k the number of transversal fourbranes. Therefore, if we choose Σ_0 to be $\mathbf{Z}_k$ invariant, we can write $W(\Sigma_0) = 0$, and $W(\Sigma) = \int_B \Omega$. Let us then take B as the complex plane multiplied by an interval $I = [0, 1]$, and let us first map the complex plane into Σ. Denoting r the coordinate on this complex plane, Σ, as given by (4.95), is defined by

$$\begin{aligned} t &= r^k, \\ v &= r, \\ w &= \zeta r^{-1}. \end{aligned} \tag{4.104}$$

Writing $r = e^{\rho} e^{i\theta}$, we can define Σ_0 as

$$\begin{aligned} t &= r^k, \\ v &= f(\rho) r, \\ w &= \zeta f(-\rho) r^{-1}, \end{aligned} \tag{4.105}$$

with $f(\rho) = 1$ for $\rho > 2$, and $f(\rho) = 0$ for $\rho < 1$. The $\mathbf{Z}_k$ transformation $t \to t$, $w \to e^{2\pi i/k} w$ and $v \to v$, is a symmetry of (4.105) if, at the same time, we perform the reparametrization of the r-plane

$$\begin{aligned} \rho &\to \rho, \\ \theta &\to \theta + b(\rho), \end{aligned} \tag{4.106}$$

with $b(\rho) = 0$ for $\rho \geq 1$, and $b(\rho) = -\frac{2\pi}{k}$ for $\rho \leq -1$. Thus, the 3-manifold entering the definition of B, is given by

$$\begin{aligned} t &= r^k, \\ v &= g(\rho,\sigma)r, \\ w &= \zeta g(-\rho,\sigma)r^{-1}, \end{aligned} \tag{4.107}$$

such that for $\sigma = 0$ we have $g = 1$, and for $\sigma = 1$, we get $g(\rho) = f(\rho)$. Now, with

$$\Omega = R dv \wedge dw \wedge \frac{dt}{t}, \tag{4.108}$$

we get

$$W(\Sigma) = kR \int_B dv \wedge dw \wedge \frac{dr}{r}. \tag{4.109}$$

The dependence on R is already clear from (4.109). In order to get the dependence on ζ we need to use (4.107),

$$W(\Sigma) = kR\zeta \int d\sigma d\theta d\rho \left(\frac{\partial g_+}{\partial \sigma}\frac{\partial g_-}{\partial \rho} - \frac{\partial g_+}{\partial \rho}\frac{\partial g_-}{\partial \sigma} \right), \tag{4.110}$$

for $g_\pm = g(\pm\rho,\sigma)$. Thus we get

$$W(\Sigma) \sim kR\zeta, \tag{4.111}$$

Notice that the superpotential (4.111) is given in units $(\text{length})^3$, as corresponds to the volume of a 3-manifold. In order to make contact with the gaugino-gaugino condensate, we need to obtain $(\text{length})^{-3}$ units. We can do this multiplying by $\frac{1}{R^6}$; thus, we get

$$< \lambda\lambda > \sim kR\zeta \frac{1}{R^6} \sim \Lambda^3, \tag{4.112}$$

where we have used equation (4.108). A different way to connect ζ with Λ is defining, in the M-theory context, the QCD string and computing its tension. Following Witten, we will then try an interpretation of ζ independent of (4.90), by computing in terms of ζ the tension of the QCD string. We will then, to define the tension, consider the QCD string as a membrane, product of a string in $\mathbf{R}^4$, and a string living in Q. Let us then denote by C a curve in Q, and assume that C ends on Σ in such a way that a membrane wrapped on C defines a string in $\mathbf{R}^4$ [18]. Moreover, we can simply think of C as a closed curve in Q, going around the eleven dimensional S^1,

$$\begin{aligned} t &= t_0 \exp(-2\pi i \sigma), \\ v &= t_0^{1/k}, \\ w &= \zeta v^{-1}. \end{aligned} \tag{4.113}$$

This curve is a non trivial element in $H_1(Q;\mathbf{Z})$, and a membrane wrapped on it will produce an ordinary type IIA string; however, we can not think that

[18] Notice that if we were working in type IIB string theory, we would have the option to wrap a threebrane around Σ, in order to define a string on $\mathbf{R}^4$.

the QCD string is a type IIA string. If $\mathcal{Q} = \mathbf{R}^3 \times S^1$, then $H_1(\mathcal{Q};\mathbf{Z}) = \mathbf{Z}$, and curves of type (4.113) will be the only candidates for non trivial 1-cycles in $\mathcal{Q}$. However, we can define QCD strings using cycles in the relative homology, $H_1(\mathcal{Q}/\Sigma;\mathbf{Z})$, i. e., considering non trivial cycles ending on the surface Σ. To compute $H_1(\mathcal{Q}/\Sigma;\mathbf{Z})$, we can use the exact sequence

$$H_1(\Sigma;\mathbf{Z}) \to H_1(\mathcal{Q};\mathbf{Z}) \xrightarrow{\imath} H_1(\mathcal{Q}/\Sigma;\mathbf{Z}), \tag{4.114}$$

which implies

$$H_1(\mathcal{Q}/\Sigma;\mathbf{Z}) = H_1(\mathcal{Q};\mathbf{Z})/\imath H_1(\Sigma;\mathbf{Z}). \tag{4.115}$$

The map $\imath$ is determined by the map defining Σ $(t = v^k)$, and thus we can conclude that, very likely,

$$H_1(\mathcal{Q}/\Sigma;\mathbf{Z}) = \mathbf{Z}_k. \tag{4.116}$$

A curve in $H_1(\mathcal{Q}/\Sigma;\mathbf{Z})$ can be defined as follows:

$$\begin{aligned} t &= t_0, \\ v &= t_0^{1/k} e^{2\pi i \sigma/k}, \\ w &= \zeta v^{-1}, \end{aligned} \tag{4.117}$$

with $t_0^{1/k}$ one of the k roots. The tension of (4.117), by construction, is independent of R, because t is fixed. Using the metric on $\mathcal{Q}$, the length of (4.117) is given by

$$\left(\frac{\zeta^2 t^{-2/n}}{n^2} + \frac{t^{2/n}}{n^2} \right)^{1/2}, \tag{4.118}$$

and its minimum is obtained when $t^{2/n} = \zeta$. Thus, the length of the QCD string should be

$$\frac{|\zeta|^{1/2}}{n}, \tag{4.119}$$

which has the right length units, as ζ behaves as $(\text{length})^2$. In order to define the tension we need to go to $(\text{length})^{-1}$ units, again using $\frac{1}{R^2}$. Then, if we identify this tension with Λ, we get

$$\Lambda \sim \frac{|\zeta|^{1/2}}{n} \frac{1}{R^2} \tag{4.120}$$

or, equivalently,

$$\Lambda^2 \sim \Lambda^3 R. \tag{4.121}$$

Thus, consistency with QCD results requires $\Lambda \sim \frac{1}{R}$. These are not good news, as they imply that the theory we are working with, in order to match QCD, posseses 0-brane modes, with masses of the order of Λ, and therefore we have not decoupled the M-theory modes.

Next, we would like to compare the superpotential described above with the ones obtained using standard instanton techniques in M-theory. However, before doing that we will conclude this brief review on brane configurations with the description of models with $N = 4$ supersymmetry.

4.5.3 $N = 2$ Models with Vanishing Beta Function.. Let us come back to brane configurations with $n+1$ solitonic fivebranes, with k_α Dirichlet fourbranes extending between the αth pair of NS-5branes. The beta function, derived in (4.68), is

$$-2k_\alpha + k_{\alpha+1} + k_{\alpha-1}, \tag{4.122}$$

for each $SU(k_\alpha)$ factor in the gauge group. In this section, we will compactify the x^6 direction to a circle of radius L. Impossing the beta function to vanish in all sectors immediately implies that all k_α are the same. Now, the compactification of the x^6 direction does not allow to eliminate all $U(1)$ factors in the gauge group: one of them can not be removed, so that the gauge group is reduced from $\prod_{\alpha=1}^{n} U(k_\alpha)$ to $U(1) \times SU(k)^n$. Moreover, using the definition (4.67) of the mass of the hypermultiplets we get, for periodic configurations,

$$\sum_\alpha m_\alpha = 0. \tag{4.123}$$

The hypermultiplets are now in representations of type $k \otimes \bar{k}$, and therefore consists of a copy of the adjoint representation, and a neutral singlet.

Let us consider the simplest case, of $N = 2$ $SU(2) \times U(1)$ four dimensional theory, with one hypermultiplet in the adjoint representation [103]. The corresponding brane configuration contains a single solitonic fivebrane, and two Dirichlet fourbranes. The mass of the hypermultiplet is clearly zero, and the corresponding four dimensional theory has vanishing beta function. A geometric procedure to define masses for the hypermultiplets is a fibering of the v-plane on the x^6 S^1 direction, in a non trivial way, so that the fourbrane positions are identified modulo a shift in v,

$$\begin{aligned} x^6 &\rightarrow x^6 + 2\pi L, \\ v &\rightarrow v + m, \end{aligned} \tag{4.124}$$

so that now, the mass of the hypermultiplet, is the constant m appearing in (4.124), as $\sum_\alpha m_\alpha = m$.

From the point of view of M-theory, the x^{10} coordinate has also been compactified on a circle, now of radius R. The (x^6, x^{10}) space has the topology of $S^1 \times S^1$. This space can be made non trivial if, when going around x^6, the value of x^{10} is changed as follows:

$$\begin{aligned} x^6 &\rightarrow x^6 + 2\pi L, \\ x^{10} &\rightarrow x^{10} + \theta R, \end{aligned} \tag{4.125}$$

and, in addition, $x^{10} \rightarrow x^{10} + 2\pi R$. Relations (4.125) define a Riemann surface of genus one, and moduli depending on L and θ for fixed values of R. θ in (4.125) can be understood as the θ-angle of the four dimensional field theory: the θ-angle can be defined as

$$\frac{x_1^{10} - x_2^{10}}{R}, \tag{4.126}$$

with $x_2^{10} = x^{10}(2\pi L)$, and $x_1^{10} = x^{10}(0)$. Using (4.125), we get θ as the value of (4.126). This is the bare θ-angle of the four dimensional theory.

A question inmediately appears concerning the value of the bare coupling constant: the right answer should be

$$\frac{1}{g^2} = \frac{2\pi L}{R}. \tag{4.127}$$

It is therefore clear that we can move the bare coupling constant of the theory keeping fixed the value of R, and changing L and θ. Let us now try to solve this model for the massless case. The solution will be given by a Riemann surface Σ, living in the space $E \times C$, where E is the Riemann surface defined by (4.125), and C is the v-plane. Thus, all what we need is defining Σ through an equation of the type

$$F(x, y, z) = 0, \tag{4.128}$$

with x and y restricted by the equation of E,

$$y^2 = (x - e_1(\tau))(x - e_2(\tau))(x - e_3(\tau)), \tag{4.129}$$

with τ the bare coupling constant defined by (4.126) and (4.127) [120]. In case we have a collection of k fourbranes, we will require F to be a polynomial of degree k in v,

$$F(x, y, z) = v^k - f_1(x, y)v^{k-1} + \cdots \tag{4.130}$$

The moduli parameters of Σ are, at this point, hidden in the functions $f_i(x, y)$ in (4.130). Let us denote $v_i(x, y)$ the roots of (4.130) at the point (x, y) in E. Notice that (4.130) is a spectral curve defining a branched covering of E, i. e., (4.130) can be interpreted as a spectral curve in the sense of Hitchin's integrable system [121]. If f_i has a pole at some point (x, y), then the same root $v_i(x, y)$ should go to infinity. These poles have the interpretation of locating the position of the solitonic fivebranes. In the simple case we are considering, with a single fivebrane, the Coulomb branch of the theory will be parametrized by meromorphic functions on E with a simple pole at one point, which is the position of the fivebrane. As we have k functions entering (4.130), the dimension of the Coulomb branch will be k, which is the right one for a theory with $U(1) \times SU(k)$ gauge group.

Now, after this discussion of the model with massless hypermultiplets, we will introduce the mass. The space where now we need to define Σ is not $E \times C$, but the non trivial fibration defined through

$$\begin{aligned} x^6 &\rightarrow x^6 + 2\pi L, \\ x^{10} &\rightarrow x^{10} + \theta R, \\ v &\rightarrow v + m \end{aligned} \tag{4.131}$$

or, equivalently, the space obtained by fibering C non trivially on E. We can flat this bundle over all E, with the exception of one point p_0. Away from this point, the solution is given by (4.130). If we write (4.130) in a factorized form,

$$F(x, y, z) = \prod_{i=1}^{k} (v - v_i(x, y)), \tag{4.132}$$

we can write f_1 in (4.130) as the sum

$$f_1 = \sum_{i=1}^{k} v_i(x, y); \tag{4.133}$$

therefore, f_1 will have poles at the positions of the fivebrane. The mass of the hypermultiplet will be identified with the residue of the differential $f_1\omega$, with ω the abelian differential, $\omega = \frac{dx}{y}$. As the sum of the residues is zero, this means that at the point at infinity, that we identify with p_0, we have a pole with residue m.

4.6 M-Theory and String Theory.

In this section we will compare the M-theory description of $N = 2$ and $N = 1$ four dimensional gauge theories, with that obtained in string theory upon performing the point particle limit [122, 123, 124, 125]. Let us then return for a moment to the brane representation of $N = 2$ four dimensional gauge theories. In the M-theory approach, we will consider M-theory on flat spacetime, $\mathbf{R}^7 \times Q$, with $Q = \mathbf{R}^5 \times S^1$. The S^1 stands for the (compactified) eleventh dimension, with the radius R proportional to the string coupling constant. The brane configuration in $\mathbf{R}^7 \times Q$ turns out to be equivalent to a solitonic fivebrane, with worldvolume $\Sigma \times \mathbf{R}^4$, where Σ is a complex curve in Q, defined by

$$F(t, v) = 0. \tag{4.134}$$

This is equivalent to defining an embedding

$$\Phi : \Sigma \hookrightarrow Q. \tag{4.135}$$

If Σ is a lagrangian manifold of Q, then we can interpret the moduli space of the effective four dimensional $N = 2$ theory as the space of deformations of Φ in (4.135) preserving the condition of lagrangian submanifold[19]. By Mc Lean's theorem, we know that the dimension of this space of deformations is $b_1(\Sigma)$, in agreement with the existing relation between the genus of Σ and

[19] Recall that a lagrangian manifold is defined by the condition that

$$\int_\Sigma \Phi^*(\Omega) = \text{Vol }(\Sigma),$$

with Φ^* such that $\Phi^*(\omega) = 0$ (where ω is the Kähler class of Q), and Ω the holomorphic top form of Q.

the rank of the gauge group in the effective four dimensional theory. It is important keeping in mind that in the M-theory approach two ingredients are being used: the curve defined by (4.134), and the holomorphic top form Ω of $\mathcal{Q}$, which explicitely depends on the radius R of the eleventh dimension. This will be very important, as already noticed in the discussion of the $N=1$ superpotentials, because an explicit dependence on the string coupling constant will be induced in the BPS mass formulas.

A different approach to (4.134) and (4.135) is that based on geometric engineering [127]. In this case, the procedure is based on the following set of steps:

1. String theory is compactified on a Calabi-Yau threefold X, with the apropiate number of vector multiplets in four dimensions.
2. A point corresponding to classical enhancement of gauge symmetry in the moduli space of the Calabi-Yau threefold must be localized.
3. A rigid Calabi-Yau threefold is defined by performing a point particle limit.
4. The rigid Calabi-Yau manifold is used to define the Seiberg-Witten surface Σ.
5. Going form type IIB to type IIA string theory represents a brane configuration corresponding to an ALE space with singularity of some Dynkin type into a set of fivebranes that can be interpreted as a fivebrane with worldvolume $\Sigma \times \mathbf{R}^4$.
6. The BPS states are defined through the meromorphic one-form λ, derived from the the Calabi-Yau holomorphic top form, in the rigid point particle limit.

As we can see from the previous set of steps, that we will explicitely show at work in one definite example, the main difference between both approaches is at the level of the meromorphic form in Seiberg-Witten theory. There is also an important difference in the underlying philosophy, related to the implicit use in the string approach, described in the above steps, of the heterotic-type II dual pairs, driving us to the choice of a particular Calabi-Yau manifold. The most elaborated geometric engineering approach uses, instead of a certain heterotic-type II dual pair, a set of local geometrical data, determined by the type of gauge symmetry we are interested on, and generalizes mirror maps to this set of local data. In all these cases, the four dimensional field theory we are going to obtain will not depend on extra parameters, as the string coupling constant. On the other hand, the M-theory approach, where field theories are obtained depending explicitely on the string coupling constant, might be dynamically rich enough as to provide a direct explanation of phenomena that can not be easily understood in the more restricted context of the point particle limit of string theory.

Next, we will follow steps 1 to 6 through an explicit example [124]. In order to obtain a field theory with gauge group $SU(n)$ we should start with a Calabi-Yau manifold with $h_{2,1}=n$, and admiting the structure of a $K3$-fibered threefold (see chapter II for definitions, and additional details). We will consider the $SU(3)$ case, corresponding to a Calabi-Yau manifold whose mirror is the weighted projective space $\mathbb{P}^{24}_{1,1,2,8,12}$,

$$\frac{1}{24}(x_1^{24}+x_2^{24})+\frac{1}{12}x_3^{12}+\frac{1}{2}x_5^2-\psi_0 x_1x_2x_3x_4x_5-\frac{1}{6}(x_1x_2x_3)^6-\frac{1}{12}(x_1x_2)^{12}=0. \tag{4.136}$$

In order to clearly visualize (4.136) as a $K3$-fibration we will perform the change of variables

$$x_1/x_2 \equiv \hat{z}^{1/12}b^{-1/24}, \quad x_1^2 \equiv x_0\hat{z}^{1/12}, \tag{4.137}$$

so that (4.136) can be rewriten in the form

$$\frac{1}{24}(\hat{z}+\frac{b}{\hat{z}}+2)x_0^{12}+\frac{1}{12}x_3^{12}+\frac{1}{3}x_4^3+\frac{1}{2}x_5^2+\frac{1}{6\sqrt{c}}(x_0x_3)^6+\left(\frac{a}{\sqrt{c}}\right)^{1/6}x_0x_3x_4x_5=0, \tag{4.138}$$

which represents a $K3$ surface, fibered over a $\mathbb{P}^1$ space parametrized by the coordinate z. Parameters in (4.138) are related to those in (4.136) through

$$a=-\psi_0^6/\psi_1, \quad b=\psi_2^{-2}, \quad c=\psi_2/\psi_1^2. \tag{4.139}$$

The parameter b can be interpreted as the volume of $\mathbb{P}^1$:

$$-\log b = \text{Vol}\ (\mathbb{P}^1). \tag{4.140}$$

Next, we should look for the points $\hat{z}$ in $\mathbb{P}^1$ over which the $K3$ surface is singular. The discriminant can be written as

$$\Delta_{K3}=\prod_{i=0}^{2}(\hat{z}-e_i^+(a,b,c))(\hat{z}-e_i^-(a,b,c)), \tag{4.141}$$

where

$$\begin{aligned} e_0^\pm &= -1\pm\sqrt{1-b}, \\ e_1^\pm &= \frac{1-c\pm\sqrt{(1-c)^2-bc^2}}{c}, \\ e_2^\pm &= \frac{(1-a)^2-c\pm\sqrt{((1-a)^2-c)^2-bc^2}}{c}. \end{aligned} \tag{4.142}$$

The Calabi-Yau manifold will be singular whenever two roots e_i coalesce, as

$$\Delta_{\text{Calabi-Yau}}=\prod_{i<j}(e_i-e_j)^2. \tag{4.143}$$

We will consider the singular point in the moduli space corresponding to $SU(3)$ symmetry. Around this point we will introduce new coordinates, through

$$\begin{aligned} a &= -2(\alpha' u)^{3/2}, \\ b &= \alpha' \Lambda^6, \\ c &= 1 - \alpha'^{3/2}(-2u^{3/2} + 3\sqrt{3}v). \end{aligned} \tag{4.144}$$

Going now to the $\alpha' \to 0$ limit in (4.143), we get a set of roots $e_i(u, v; \Lambda^6)$ on a z-plane, with z defined in $\alpha'^{3/2} z \equiv \hat{z}$:

$$\begin{aligned} e_0 &= 0, \; e_\infty = \infty, \\ e_1^\pm &= 2u^{3/2} + 3\sqrt{3}v \pm \sqrt{(2u^{3/2} + 3\sqrt{3}v)^2 - \Lambda^6}, \\ e_2^\pm &= -2u^{3/2} + 3\sqrt{3}v \pm \sqrt{(2u^{3/2} - 3\sqrt{3}v)^2 - \Lambda^6}. \end{aligned} \tag{4.145}$$

Now, we can use (4.145) as the definition of a Riemann surface Σ, defined by the Calabi-Yau data at the singular $SU(3)$ point, and in the point particle limit. There exits a natural geometrical picture for understanding the parameters u and v in (4.144), which is the definition of the blow up, in the moduli space of complex structures of $\mathbb{P}^{24}_{1,1,2,8,12}$, of the $SU(3)$ singular point. From this point of view, the parameters u and v in (4.144) will be related to the volume of the set of vanishing two-cycles associated with a *rational* singularity, i. e., an orbifold singularity of type A_{n-1} (in the case we are considering, $n = 3$). These vanishing cycles, as is the case with rational singularities, are associated with Dynkin diagrams of non affine type. The branch points (4.145) on the z-plane define the curve

$$y^2 = \prod_i (x - e_i(u, v; \Lambda^6)), \tag{4.146}$$

which can also be represented as the vanishing locus of a polynomial $F(x, z) = 0$, with F given by [129, 130]

$$F(x, z) = z + \frac{\Lambda^6}{z} + B(x), \tag{4.147}$$

where $B(x)$ is a polynomial in x of degree three; in the general case of $SU(n)$ theories, the polynomial will be of degree n.

This has exactly the same look as what we have obtained using brane configurations, with the space Q replaced by the (x, z) space. The difference is that now we are not considering the (x, z) space as a part of spacetime, and Σ as embedded in it, but we use Σ as defined in (4.147) to define a Calabi-Yau space in a rigid limit by the equation

$$F(x, z) + y^2 + w^2 = 0, \tag{4.148}$$

which defines a threefold in the (x, y, z, w) space. And, in addition, we think of (4.148) as a Calabi-Yau representation of the point particle limit. In order to get the meromorphic one-form λ, and the BPS states, we need to define a map from the third homology group, $H_3(CY)$, of the Calabi-Yau manifold, into $H_1(\Sigma)$. This can be done as follows. The three-cycles in $H_3(CY)$ of the general type $S^2 \times S^1$, with S^2 a vanishing cycle of $K3$, correspond to S^1 circles in the z-plane. The three-cycles with the topology of S^3 can be interpreted as a path from the north to the south pole of S^3, starting with a vanishing two-cycle, and ending at another vanishing two-cycle of $K3$. This corresponds, in the z-plane, to paths going from e_i^+ to e_i^-. Once we have defined this map,

$$f : H_3(CY) \longrightarrow H_1(\Sigma), \tag{4.149}$$

we define

$$\lambda(f(C)) = \Omega(C), \tag{4.150}$$

with Ω the holomorphic top form.

A similar analysis can be done for computing the mass of BPS states, and the meromorphic one-form λ in the brane framework. In fact, we can consider a two-cycle C in $\mathcal{Q}$ such that

$$\partial C \subset \Sigma, \tag{4.151}$$

or, in other words, $C \in H_2(\mathcal{Q}/\Sigma; \mathbf{Z})$. The holomorphic top form on $\mathcal{Q}$ is given by

$$\Omega = R\frac{dt}{t} \wedge dv, \tag{4.152}$$

and thus the BPS mass will be given by

$$M \sim R \int_C \frac{dt}{t} \wedge dv = R \int_{\partial C} \frac{dt}{t} v(t), \tag{4.153}$$

with $v(t)$ given by

$$F(t, v) = 0, \tag{4.154}$$

for the corresponding Seiberg-Witten curve, Σ. Notice that the same analysis, using (4.150) and the holomorphic top form for (4.148) will give, by contrast to the brane case, a BPS mass formula independent of R.

Next, we will compare the brane construction and geometric engineering in the more complicated case of $N = 1$ [126, 128].

4.7 Local Models for Elliptic Fibrations.

Let V be an elliptic fibration,

$$\Phi : V \to \Delta, \tag{4.155}$$

with Δ an algebraic curve, and $\Phi^{-1}(a)$, with a any point in Δ, an elliptic curve. Let us denote $\{a_\rho\}$ the finite set of points in Δ such that $\Phi^{-1}(a_\rho) = \mathcal{C}_\rho$ is a singular fiber. Each singular fiber $\mathcal{C}_\rho$ can be written as

$$\mathcal{C}_\rho = \sum_i n_{i\rho} \Theta_{i\rho}, \tag{4.156}$$

where $\Theta_{i\rho}$ are non singular rational curves, with $\Theta_{i\rho}^2 = -2$, and $n_{i\rho}$ are integer numbers. Different types of singularities are characterized by (4.156) and the intersection matrix $(\Theta_{i\rho}.\Theta_{j\rho})$. All different types of Kodaira singularities satisfy the relation

$$\mathcal{C}_\rho^2 = 0. \tag{4.157}$$

Let $\tau(u)$ be the elliptic modulus of the elliptic fiber at the point $u \in \Delta$. For each path α in $\Pi_1(\Delta')$, with $\Delta' = \Delta - \{a_\rho\}$, we can define a monodromy transformation S_α, in $Sl(2,\mathbf{Z})$, acting on $\tau(u)$ as follows:

$$S_\alpha \tau(u) = \frac{a_\alpha \tau(u) + b_\alpha}{c_\alpha \tau(u) + d_\alpha}. \tag{4.158}$$

Each type of Kodaira singularity is characterized by a particular monodromy matrix.

In order to define an elliptic fibration [45], the starting point will be an algebraic curve Δ, that we will take, for simplicity, to be of genus zero, and a meromorphic function $\mathcal{J}(u)$ on Δ. Let us assume $\mathcal{J}(u) \neq 0, 1, \infty$ on $\Delta' = \Delta - \{a_\rho\}$. Then, there exists multivalued holomorphic function $\tau(u)$, with Im $\tau(u) > 0$, satisfying $\mathcal{J}(u) = j(\tau(u))$, with j the elliptic modular j-function on the upper half plane. As above, for each $\alpha \in \Pi_1(\Delta')$ we define a monodromy matrix S_α, acting on $\tau(u)$ in the form defined by (4.158). Associated to these data we will define an elliptic fibration, (4.155). In order to do that, let us first define the universal covering $\tilde{\Delta}'$, of Δ', and let us identify the covering transformations of $\tilde{\Delta}'$ over Δ', with the elements in $\Pi_1(\Delta')$. Denoting by $\tilde{u}$ a point in $\tilde{\Delta}'$, we define, for each $\alpha \in \Pi_1(\Delta')$, the covering transformation $\tilde{u} \to \alpha\tilde{u}$, by

$$\tau(\alpha\tilde{u}) = S_\alpha \tau(\tilde{u}); \tag{4.159}$$

in other words, we consider τ as a single valued holomorphic function on $\tilde{\Delta}'$. Using (4.158), we define

$$f_\alpha(\tilde{u}) = (c_\alpha \tau(\tilde{u}) + d_\alpha)^{-1}. \tag{4.160}$$

Next, we define the product $\tilde{\Delta}' \times \mathbf{C}$ and, for each (α, n_1, n_2), with $\alpha \in \Pi_1(\Delta')$, and n_1, n_2 integers, the automorphism

$$g(\alpha, n_1, n_2) : (\tilde{u}, \lambda) \to (\alpha\tilde{u}, f_\alpha(\tilde{u})(\lambda + n_1\tau(\tilde{u}) + n_2)). \tag{4.161}$$

Denoting by $\mathcal{G}$ the group of automorphisms (4.161), we define the quotient space

$$B' \equiv (\tilde{\Delta}' \times \mathcal{C})/\mathcal{G}. \tag{4.162}$$

This is a non singular surface, since g, as defined by (4.161), has no fixed points in $\tilde{\Delta}'$. From (4.161) and (4.162), it is clear that B' is an elliptic fibration on Δ', with fiber elliptic curves of elliptic modulus $\tau(u)$. Thus, by the previous construction, we have defined the elliptic fibration away from the singular points a_ρ.

Let us denote E_ρ a local neighbourhood of the point a_ρ, with local coordinate t, and such that $t(a_\rho) = 0$. Let S_ρ be the monodromy associated with a small circle around a_ρ. By $\mathcal{U}_\rho$ we will denote the universal covering of $E'_\rho = E_\rho - a_\rho$, with coordinate ρ defined by

$$\rho = \frac{1}{2\pi i} \log t. \tag{4.163}$$

The analog of (4.159) will be

$$\tau(\rho + 1) = S_\rho \tau(\rho). \tag{4.164}$$

If we go around the points a_ρ, k times, we should act with S_ρ^k; hence, we parametrize each path by the winding number k. The group of automorphisms (4.161), reduced to small closed paths around a_ρ, becomes

$$g(k, n_1, n_2)(\rho, \lambda) = (\rho + k, f_k(\rho)[\lambda + n_1\tau(\rho) + n_2]). \tag{4.165}$$

Denoting by $\mathcal{G}_\rho$ the group (4.165), we define the elliptic fibration around a_ρ as

$$(\mathcal{U}_\rho \times \mathbf{C})/\mathcal{G}_\rho. \tag{4.166}$$

Next, we will extend the elliptic fibration to the singular point a_ρ. We can consider two different cases, depending on the finite or infinite order of S_ρ.

4.8 Singularities of Type $\hat{D}_4$: $\mathbf{Z}_2$ Orbifolds.

Let us assume S_ρ is of finite order,

$$(S_\rho)^m = \mathbf{1}_d. \tag{4.167}$$

In this case, we can extend (4.166) to the singular points, simply defining a new variable σ as

$$\sigma^m = t. \tag{4.168}$$

Let us denote D a local neighbourhood in the σ-plane of the point $\sigma = 0$, and define the group G_D of automorphisms

$$g(n_1, n_2) : (\sigma, \lambda) = (\sigma, \lambda + n_1\tau(\sigma) + n_2), \tag{4.169}$$

and the space

$$F = (D \times \mathbf{C})/G_D. \tag{4.170}$$

Obviously, F defines an elliptic fibration over D, with fiber F_σ at each point $\sigma \in D$, an elliptic curve of modulus $\tau(\sigma)$. From (4.167) and (4.160), it follows that

$$f_k(\sigma) = 1, \tag{4.171}$$

with $k = O(m)$. Thus, we can define a normal subgroup $\mathcal{N}$ of $\mathcal{G}_\rho$ as the set of transformations (4.165):

$$g(k, n_1, n_2) : (\rho, \lambda) \rightarrow (\rho + k, \lambda + n_1 \tau(\rho) + n_2). \tag{4.172}$$

Comparing now (4.169) and (4.172), we get

$$(\mathcal{U}_\rho \times \mathbf{C})/\mathcal{N} = (D' \times \mathbf{C})/G_D \equiv F - F_0. \tag{4.173}$$

Using (4.172) and (4.165) we get

$$\mathcal{C} = \mathcal{G}/\mathcal{N}, \tag{4.174}$$

with $\mathcal{C}$ the cyclic group of order m, defined by

$$g_k : (\sigma, \lambda) \rightarrow (e^{2\pi i k/m}\sigma, f_k(\sigma)\lambda). \tag{4.175}$$

From (4.174) and (4.173), we get the desired extension to a_ρ, namely

$$F/\mathcal{C} = (\mathcal{U}_\rho \times \mathbf{C})/\mathcal{G}_\rho \cup F_0/\mathcal{C}. \tag{4.176}$$

Thus, the elliptic fibration extended to a_ρ, in case S_ρ is of finite order, is defined by $F/\mathcal{C}$. Now, $F/\mathcal{C}$ can have singular points that we can regularize. The simplest example corresponds to

$$S_\rho = \begin{pmatrix} -1 & 0 \\ 0 & -1 \end{pmatrix}, \tag{4.177}$$

i. e., a parity transformation. In this case, the order is $m = 2$, and we define σ by $\sigma^2 \equiv t$. The cyclic group (4.175) in this case simply becomes

$$(\sigma, \lambda) \rightarrow (-\sigma, -\lambda), \tag{4.178}$$

since from (4.177) and (4.160) we get $f_1 = -1$. At the point $\sigma = 0$ we have four fixed points, the standard $\mathbf{Z}_2$ orbifold points,

$$(0, \frac{a}{2}\tau(0) + \frac{b}{2}), \tag{4.179}$$

with $a, b = 0, 1$. The resolution of these four singular points will produce four irreducible components, $\Theta^1, \ldots, \Theta^4$. In addition, we have the irreducible component Θ_0, defined by the curve itself at $\sigma = 0$. Using the relation $\sigma^2 = t$, we get the $\hat{D}_4$ cycle,

$$\mathcal{C} = 2\Theta_0 + \Theta^1 + \Theta^2 + \Theta^3 + \Theta^4, \tag{4.180}$$

with $(\Theta_0, \Theta^1) = (\Theta_0, \Theta^2) = (\Theta_0, \Theta^3) = (\Theta_0, \Theta^4) = 1$. In general, the four external points of D-diagrams can be associated with the four $\mathbf{Z}_2$ orbifold points of the torus.

4.9 Singularities of Type $\hat{A}_{n-1}$.

We will now consider the case

$$S_\rho = \begin{pmatrix} 1 & n \\ 0 & 1 \end{pmatrix}, \tag{4.181}$$

which is of infinite order. A local model for this monodromy can be defined by

$$\tau(t) = \frac{1}{2\pi i} n \log t. \tag{4.182}$$

Using the variable ρ defined in (4.163), we get, for the group $\mathcal{G}_\rho$ of automorphisms,

$$g(k, n_1, n_2) : (\rho, \lambda) \to (\rho + k, \lambda + n_1 n\rho + n_2), \tag{4.183}$$

and the local model for the elliptic fibration, out of the singular point,

$$(\mathcal{U}_\rho \times \mathbf{C})/\mathcal{G}_\rho, \tag{4.184}$$

i. e., fibers of the type of elliptic curves, with elliptic modulus $n\rho$. A simple way to think about these elliptic curves is in terms of cyclic unramified coverings [63]. Let us recall that a cyclic unramified covering, $\Pi : \hat{C} \to C$, of order n, of a curve C of genus g, is a curve $\hat{C}$ of genus

$$\hat{g} = ng + 1 - n. \tag{4.185}$$

Thus, for $g = 1$, we get $\hat{g} = 1$, for arbitrary n. Denoting by τ the elliptic modulus of C, in case $g = 1$, the elliptic modulus of $\hat{C}$ is given by

$$\hat{\tau} = n\tau. \tag{4.186}$$

Moreover, the generators $\hat{\alpha}$ and $\hat{\beta}$ of $H_1(\hat{C}; \mathbf{Z})$ are given in terms of the homology basis α, β of C as

$$\begin{aligned} \Pi\hat{\alpha} &= \alpha, \\ \Pi\hat{\beta} &= n\beta, \end{aligned} \tag{4.187}$$

with Π the projection $\Pi : \hat{C} \to C$. From (4.186) and (4.183), we can interpret the elliptic fibration (4.184) as one with elliptic fibers given by n-cyclic unramified coverings of a curve C with elliptic modulus ρ or, equivalently, $\frac{1}{2\pi i} \log t$. There exits a simple way to define a family of elliptic curves, with elliptic modulus given by $\frac{1}{2\pi i} \log t$, which is the plumbing fixture construction. Let D_0 be the unit disc around $t = 0$, and let C_0 be the Riemann sphere. Define two local coordinates, $z_a : \mathcal{U}_a \to D_0$, $z_b : \mathcal{U}_b \to D_0$, in disjoint neigbourhoods $\mathcal{U}_a$, $\mathcal{U}_b$, of two points P_a and P_b of C_0. Let us then define

$$\begin{aligned} W = \{(p,t) | t \in D_0, p \in C_0 - \mathcal{U}_a - \mathcal{U}_b,\ &\text{or } p \in \mathcal{U}_a, \text{ with } |z_a(p)| > |t|, \text{ or} \\ &p \in \mathcal{U}_b, \text{ with } |z_b(p)| > |t|\}, \end{aligned} \tag{4.188}$$

and let S be the surface

$$S = \{xy = t; (x, y, t) \in D_0 \times O_0 \times D_0\}. \tag{4.189}$$

We define the family of curves through the following identifications

$$\begin{aligned}(p_a, t) \in W \cap \mathcal{U}_a \times D_0 &\simeq (z_a(p_a), \frac{t}{z_a(p_a)}, t) \in S, \\ (p_b, t) \in W \cap \mathcal{U}_b \times D_0 &\simeq (\frac{t}{z_b(p_b)}, z_b(p_b), t) \in S.\end{aligned} \tag{4.190}$$

For each t we get a genus one curve, and at $t = 0$ we get a nodal curve by pinching the non zero homology cycles. The pinching region is characterized by

$$xy = t, \tag{4.191}$$

which defines a singularity of type A_0. The elliptic modulus of the curves is given by

$$\tau(t) = \frac{1}{2\pi i} \log t + C_1 t + C_2, \tag{4.192}$$

for some constants C_1 and C_2. We can use an appropiate choice of coordinate t, such that $C_1 = C_2 = 0$. The singularity at $t = 0$ is a singularity of type $\hat{A}_0$, in Kodaira's classification, corresponding to

$$S_\rho = \begin{pmatrix} 1 & 1 \\ 0 & 1 \end{pmatrix}. \tag{4.193}$$

Using now (4.186) and (4.192) we get, for the cyclic covering of order n, the result (4.182), and the group (4.183). The pinching region of the cyclic unramified covering is given by

$$xy = t^n, \tag{4.194}$$

instead of (4.191), i. e., for the surface defining the A_{n-1} singularity, $\mathbf{C}^2/\mathbf{Z}_n$. Now, we can proceed to the resolution of the singularity at $t = 0$. The resolution of the singularity (4.194) requires $n - 1$ exceptional divisors, $\Theta_1, \ldots, \Theta_{n-1}$. In addition, we have the rational curve Θ_0, defined by the complement of the node. Thus, we get, at $t = 0$,

$$C = \Theta_0 + \cdots + \Theta_{n-1}, \tag{4.195}$$

with $(\Theta_0, \Theta_1) = (\Theta_0, \Theta_{n-1}) = 1$, and $(\Theta_i, \Theta_{i+1}) = 1$, which is the $\hat{A}_{n-1}$ Dynkin diagram. The group of covering transformations of the nth order cyclic unramified covering is $\mathbf{Z}_n$, and the action over the components (4.195) is given by

$$\begin{aligned}\Theta_i &\to \Theta_{i+1}, \\ \Theta_{n-1} &\to \Theta_0.\end{aligned} \tag{4.196}$$

4.10 Singularities of Type $\hat{D}_{n+4}$.

This case is a combination of the two previous examples. Through the same reasoning as above, the group $\mathcal{G}_\rho$ is given, for

$$S_\rho = \begin{pmatrix} -1 & -n \\ 0 & -1 \end{pmatrix}. \tag{4.197}$$

by

$$g(k, n_1, n_2) : (\rho, \lambda) \to (k + \rho, (-1)^k(\lambda + n_1 n\rho + n_2)). \tag{4.198}$$

Using a new variable $\sigma^2 = t$, what we get is a set of irreducible components $\Theta_0, \dots \Theta_{2n}$, with the identifications $\Theta_i \to \Theta_{2n-i}$. In addition, we get the four fixed $\mathbf{Z}_2$ orbifold points described above. The singular fiber is then given by

$$\mathcal{C} = 2\Theta_0 + \cdots + 2\Theta_n + \Theta^1 + \Theta^2 + \Theta^3 + \Theta^4, \tag{4.199}$$

with the intersections of the $\hat{D}_{n+4}$ affine diagram. It is easy to see that in this case we also get

$$(\mathcal{C})^2 = 0. \tag{4.200}$$

Defing the genus of the singular fiber by $\mathcal{C}^2 = 2g - 2$, we conclude that $g = 1$, for all singularities of Kodaira type. Notice that for rational singularities, characterized by non affine Dynkin diagrams of ADE type [44], we get self intersection $\mathcal{C}^2 = -2$, which corresponds to genus equal zero.

4.11 Decompactification and Affinization.

The general framework in which we are working in order to get four dimensional $N = 1$ gauge theories is that of M-theory compactifications on elliptically fibered Calabi-Yau fourfolds, in the limit Vol $(E) = 0$, with E the elliptic fiber. As described above, we can interpolate between $N = 2$ supersymmetry in three dimensions, and $N = 1$ in four dimensions, by changing the radius R through

$$\text{Vol }(E) = \frac{1}{R}. \tag{4.201}$$

The three dimensional limit then corresponds to Vol $(E) \to \infty$, and the four dimensional to Vol $(E) \to 0$. Now, we will work locally around a singular fiber of Kodaira $\hat{A}\hat{D}\hat{E}$ type. As we know, for the Calabi-Yau fourfold X,

$$E \to X \stackrel{\Pi}{\to} B, \tag{4.202}$$

the locus C in B, where the fiber is singular, is of codimension one in B, i. e., of real dimension four. Let us now see what happens to the singular fiber in the three dimensional limit. In this case, we have Vol $(E) = \infty$. A possible way to represent this phenomenon is by simply extracting the point at infinity. In the case of $\hat{A}_{n-1}$ singularities, as described in previous subsection, taking out the point at infinity corresponds to decompactifying the irreducible component

Θ_0, that was associated with the curve itself. As was clear in this case, we then pass from the affine diagram, $\hat{A}_{n-1}$, to the non affine, A_{n-1}. More generally, as the elliptic fibration we are considering possesses a global section, we can select the irreducible component we are going to decompactify as the one intersecting with the basis of the elliptic fibration. When we decompactify, in the Vol $(E) = 0$ limit, what we are doing, at the level of the fiber, is precisely compactifying the extra irreducible component, which leads to the affine Dynkin diagram.

4.12 M-Theory Instantons and Holomorphic Euler Characteristic.

Using the results of reference [99] a vertical instanton in a Calabi-Yau fourfold, of the type (4.202), will be defined by a divisor D of X, such that $\Pi(D)$ is of codimension one in B, and with holomorphic Euler characteristic

$$\chi(D, \mathcal{O}_D) = 1. \tag{4.203}$$

It is in case (4.203) that we have two fermionic zero modes [99], and we can define a superpotential contribution associated to D. For N, the normal bundle to D in X, which is locally a complex line bundle on D, we define the $U(1)$ transformation

$$t \to e^{i\alpha} t, \tag{4.204}$$

with t a coordinate of the fiber of N. The two fermionic zero modes have $U(1)$ charge equal one half. Associated to the divisor D, we can define a scalar field ϕ_D that, together with Vol (D) defines the imaginary and real parts of a chiral superfield. Under $U(1)$ rotations (4.204), ϕ_D transforms as

$$\phi_D \to \phi_D + \chi(D)\alpha. \tag{4.205}$$

In three dimensions, this is precisely the transformation of the dual photon field as Goldstone boson [24]. However, transformation (4.205) has perfect sense, for vertical instantons, in the four dimensional decompactification limit.

Let us now consider an elliptically fibered Calabi-Yau fourfold, with singular fiber of $\hat{A}_{n-1}$ type, over a locus C of codimension one in B. We will assume that the singular fiber is constant over C. Moreover, in the geometrical engineering spirit, we will impose

$$h_{1,0}(C) = h_{2,0}(C) = 0 \tag{4.206}$$

and, thus, $\Pi_1(C) = 0$. This prevents us from having non trivial transformations on the fiber by going, on C, around closed loops, since all closed loops are contractible. In addition, we will assume, based on (4.206), that C is an Enriques surface. After impossing these assumptions, we will consider divisors D_i, with $i = 0, \ldots, n-1$, defined by the fibering over C, in a trivial way, of the irreducible components Θ_i of the $\hat{A}_{n-1}$ singularity [126]. Ussing the Todd representation of the holomorphic Euler class [137],

$$\chi(D) = \frac{1}{24}\int_{D_i} c_1(\Theta_i)c_2(C) \tag{4.207}$$

we get, for C an Enriques surface,

$$\chi(D_i) = 1. \tag{4.208}$$

Interpreting now the t variable (4.204) on the fiber of the normal bundle N of D in X as the t variable used in our previous description of Kodaira singularities of type $\hat{A}_{n-1}$, we can derive the transformation law, under the $\mathbf{Z}_n$ subgroup of $U(1)$, of the scalar fields ϕ_{D_i} associated to these divisors. Namely, from (4.196) we get

$$\mathbf{Z}_n : \phi_{D_i} \to \phi_{D_{i+1}}, \tag{4.209}$$

with the $\mathbf{Z}_n$ transformation being defined by

$$t \to e^{2\pi i/n}t. \tag{4.210}$$

Using now (4.205), we get

$$\mathbf{Z}_n : \phi_{D_i} \to \phi_{D_i} + \frac{2\pi}{n}. \tag{4.211}$$

Combining (4.209) and (4.211) we get, modulo 2π,

$$\phi_{D_j} = \frac{2\pi j}{n} + c, \tag{4.212}$$

with $j = 0, \ldots, n-1$, and c a constant independent of j.

Let us now consider the divisor $\mathcal{D}$ obtained by fibering over C the singular fiber $\mathcal{C} = \sum_{j=0}^{n-1}\Theta_j$, defined in (4.195). In this case we need to be careful in order to compute (4.207). If we naively consider the topological sum of components Θ_j in (4.207), we will get the wrong result $\chi(\mathcal{D}) = n$. This result would be correct topologically, but not for the holomorphic Euler characteristic we are interested in. In fact, what we should write in (4.207) for $\int c_1(\sum_{j=0}^{n-1}\Theta_j)$ is $2(1 - g(\sum_{j=0}^{n-1}\Theta_j))$, with g the genus of the cycle (4.195), as defined by $\mathcal{C}^2 = 2g - 2$, with $\mathcal{C}^2$ the self intrsection of the cycle (4.195) which, as for any other Kodaira singularity, is zero. Thus, we get $g = 1$, and [131]

$$\chi(\mathcal{D}) = 0. \tag{4.213}$$

We can try to intepret the result (4.213) in terms of the fermionic zero modes of each component Θ_i, and the topology of the cycle. In fact, associated to each divisor D_i we have, as a consequence of (4.208), two fermionic zero modes. In the case of the $\hat{A}_{n-1}$ singularity, we can soak up all zero modes inside the graph, as shown in Figure 1,

where from each node, representing one Θ_i, we have two fermionic zero mode lines. The soaking up of fermionic zero modes represented in the figure is an heuristic interpretation of the result (4.213).

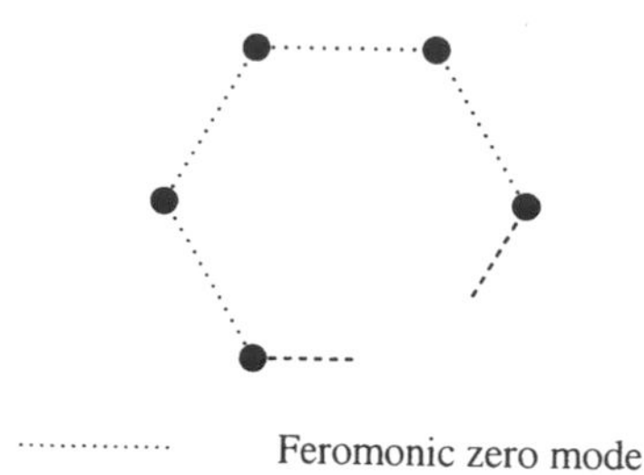

Fig. 4.3. Soaking up of zero modes for $\hat{A}_{n-1}$.

4.13 θ-Parameter and Gaugino Condensates.

We will, in this section, only consider singularities of $\hat{A}_{n-1}$ and $\hat{D}_{n+4}$ type. In both cases, and for each irreducible component Θ_i, we get a divisor D_i, with $\chi(D_i) = 1$. Associated to this divisor, we can get a superpotential term of the order [99]

$$\int d^2\theta e^{-(V(D_i))+i\phi_{D_i}}), \tag{4.214}$$

where $V(D_i)$ means the volume of the divisor D_i. As explained above, we are using vertical instanton divisors D_i, defined by a trivial fibering of Θ_i over the singular locus $C \subset B$, satisfying conditions (4.206). In order to get the four dimensional $N = 1$ limit, we will take the limit Vol $(E) = \frac{1}{R} \to 0$. Since the singular fibers are, topologically, the union of irreducible components (see (4.195) and (4.199)), we can write

$$\text{Vol }(\Theta_i) = \frac{1}{R\text{Cox}}, \tag{4.215}$$

with Cox the Coxeter number of the corresponding singularity, which equals the total number of irreducible components. Therefore, we will define Vol (D_i) as

$$\text{Vol }(D_i) = \lim_{R\to\infty} \text{Vol }(C)\frac{1}{R\text{Cox}}. \tag{4.216}$$

If we first consider the $N = 2$ supersymmetric three dimensional theory obtained by compactifying M-theory on the Calabi-Yau fourfold X, i. e., in the limit $R \to 0$, we know that only the divisor Θ_0, for the $\hat{A}_{n-1}$ case, is decompactified, passing from the affine diagram describing an elliptic singularity to the non affine diagram describing a rational, Artin like, singularity [44]. In that case, the volumes of the Θ_i components, for $i \neq 0$, are free parameters, corresponding to the Coulomb branch of the $N = 2$ three dimensional theory. In the three dimensional theory, the factor Vol (C) corresponds to the bare coupling constant in three dimensions,

$$\text{Vol }(C) = \frac{1}{g_3^2}, \tag{4.217}$$

and Vol $(D_i) = \frac{1}{g_3^2}\chi_i$, for $i \neq 0$, with χ_i the three dimensional Coulomb branch coordinates. In the four dimensional case, we must use (4.216), that becomes

$$\text{Vol}\ (D_i) = \lim_{R\to\infty} \frac{1}{g_3^2}\frac{1}{R\text{Cox}} = \frac{1}{g_4^2\text{Cox}}. \tag{4.218}$$

Let us now concentrate on the $\hat{A}_{n-1}$ case, where Cox $= n$. Using (4.214) we get the following superpotential for each divisor D_j,

$$\exp - \left(\frac{1}{g_4^2 n} + i\left(\frac{2\pi j}{n} + c\right)\right). \tag{4.219}$$

Let us now fix the constant c in (4.219). In order to do that, we will use the transformation rules (4.205). From the four dimensional point of view, these are the transformation rules with respect to the $U(1)_R$ symmetry. From (4.205) we get, that under $t \to e^{i\alpha}t$,

$$\sum_{i=0}^{\text{n-1}} \phi_{D_i} \to \sum_{i=0}^{\text{n-1}} \phi_{D_i} + n\alpha. \tag{4.220}$$

This is precisely the transformation rule under $U(1)_R$ of the $N = 1$ θ-parameter,

$$\theta \to \theta + n\alpha. \tag{4.221}$$

In fact, (4.220) is a direct consequence of the $U(1)$ axial anomaly equation: if we define θ as

$$\frac{\theta}{32\pi^2}F\tilde{F}, \tag{4.222}$$

the anomaly for $SU(n)$ is given by

$$\partial_\mu j_5^\mu = \frac{n}{16\pi^2}F\tilde{F}. \tag{4.223}$$

The factor 2, differing (4.222) from (4.223), reflects the fact that we are assigning $U(1)_R$ charge $\frac{1}{2}$ to the fermionic zero modes. Identifying the θ-parameter with the topological sum $\sum_{i=0}^{n-1}\phi_{D_i}$ we get that the constant c in (4.219) is simply

$$c = \frac{\theta}{n}, \tag{4.224}$$

so that we then finally obtain the superpotential

$$\exp - \left(\frac{1}{g_4^2 n} + i\left(\frac{2\pi j}{n} + \frac{\theta}{n}\right)\right) \simeq \Lambda^3 e^{2\pi i j/n} e^{i\theta/n}, \tag{4.225}$$

with $j = 0, \ldots, n-1$, which is the correct value for the gaugino condensate.

Let us now try to extend the previous argument to the $\hat{D}_{n+4}$ type of singularities. Defining again the four dimensional θ-parameter as the topological sum of ϕ_{D_i} for the whole set of irreducible components we get, for the cycle (4.199), the transformation rule

$$\theta \to \theta + \mathrm{Cox}\ .\alpha \tag{4.226}$$

where now the Coxeter for $\hat{D}_{n+4}$ is $2n+6$. Interpreting $\hat{D}_{n+4}$ as $O(N)$ gauge groups, with $N = 2n+8$, we get $\mathrm{Cox}(\hat{D}_{n+4}) = N-2$. Since θ is defined modulo 2π we get that for $\hat{D}_{n+4}$ singularities the value of ϕ_{D_i}, for any irreducible component, is

$$\frac{2\pi k}{N-2} + \frac{\theta}{N-2}, \tag{4.227}$$

with $k = 1, \ldots, N-2$. However, now we do not know how to associate a value of k to each irreducible component Θ_i of the $\hat{D}_{n+4}$ diagram. Using (4.227), we get a set of $N-2$ different values for the gaugino condensate for $O(N)$ groups:

$$\exp\left(-\frac{1}{g_4^2(N-2)} + i\left(\frac{2\pi k}{N-2} + \frac{\theta}{N-2}\right)\right), \tag{4.228}$$

with $k = 1, \ldots, N-2$. However, we still do not know how to associate to each Θ_i a particular value of k. A possibility will be associating consecutive values of k to components with non vanishing intersection; however, the topology of diagrams of type D prevents us from doing that globally. Notice that the problem we have is the same sort of puzzle we find for $O(N)$ gauge groups, concerning the number of values for $< \lambda\lambda >$, and the value of the Witten index, which in diagramatic terms is simply the number of nodes of the diagram. In order to unravel this puzzle, let us consider more closely the way fermionic zero modes are soaked up on a $\hat{D}_{n+4}$ diagram. We will use the cycle (4.199); for the components Θ^1 to Θ^4, associated to the $\mathbf{Z}_2$ orbifold points, we get divisors with $\chi = 1$. Now, for the components $2\Theta_0, \ldots, 2\Theta_n$ we get, from the Todd representation of the holomorphic Euler characteristic,

$$\chi = 4. \tag{4.229}$$

The reason for this is that the cycle 2Θ, with $\Theta^2 = -2$, has self intersection -8. Of course, (4.229) refers to the holomorphic Euler characteristic of the divisor obtained when fibering over C any of the cycles $2\Theta_i$, with $i = 0, \ldots, n$. Equation (4.229) implies 8 fermionic zero modes, with the topology of the soaking up of zero modes of the $\hat{D}_{n+4}$ diagram, as represented in Figure 2.

Notice that the contribution to χ of 2Θ is different form that of $(\Theta_1 + \Theta_2)$, with $(\Theta_1.\Theta_2) = 0$; namely, for the first case $\chi = 4$, and $\chi = 2$ for the second. For the $\hat{D}_{n+4}$ diagram, we can define: *i*) The Witten index tr $(-1)^F$, as the number of nodes, i. e., $5+n$; *ii*) The Coxeter number, which is the number of irreducible components, i. e., $2n+6$ and *iii*) The number of intersections as represented by the dashed lines in Figure 4, i. e., $8+4n$. From the point of view of the Cartan algebra, used to define the vacuum configurations in [22], we can only feel the number of nodes. The θ-parameter is able to feel the Coxeter

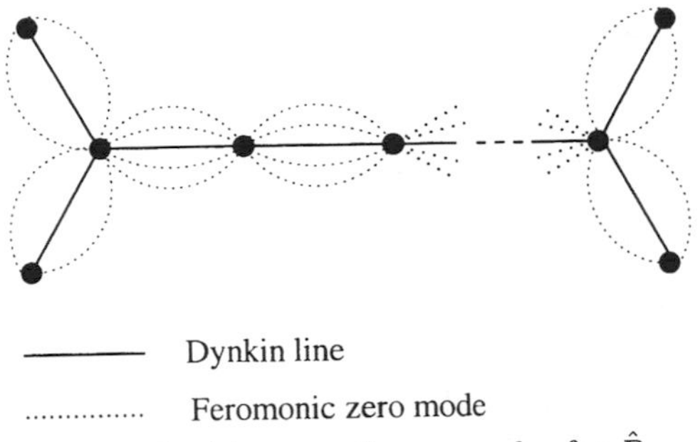

Fig. 4.4. Soaking up of zero modes for $\hat{D}_{n+4}$.

number; however, we now find a new structure related to the intersections of the graph. In the Witten index case, the nodes corresponding to cycles $2\Theta_i$, with $i = 0, \ldots, n$ contribute with one, in the number of $< \lambda\lambda >$ values with two, and in the number of intersections with four. This value four calls for an orientifold interpretation of these nodes. The topological definition of the θ-parameter implicitely implies the split of this orientifold into two cycles, a phenomena recalling the F-theory description [138] of the Seiberg-Witten splitting [35]. Assuming this splitting of the orientifold, the only possible topology for the soaking up of zero modes is the one represented in Figure 3, where the "splitted orientifold" inside the box is associated to four zero modes, corresponding to $\chi = 2$ for a cycle $\Theta_1 + \Theta_2$, with $\Theta_1.\Theta_2 = 0$.

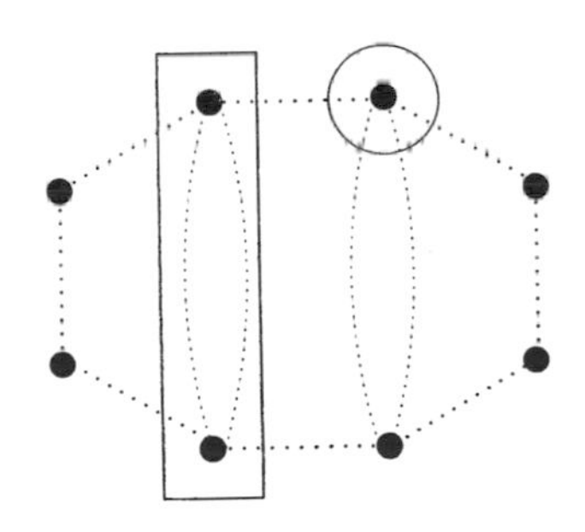

Fig. 4.5. Orientifold splitting.

On the other hand, each node surrounded by a circle in Figure 5 represents itself the disconnected sum of two non singular rational curves; thus, we represent each "orientifold" mode by four rational curves, with the intersections depicted inside the box of Figure 5. When we forget about internal lines in Figure 5, we get the cyclic $\mathbf{Z}_{2n+6} \equiv \mathbf{Z}_{N-2}$ structure of equation (4.228). It is clear that much more is necessary in order to reach a complete description of the $O(N)$ vacuum structure.

4.14 Domain Walls and Intersections.

The discussion in the previous section already raises the problem known as θ-puzzle. In fact, and discussing again only the $SU(n)$ case, the transformation law (4.221) together with the very definition fo the θ-angle as the topological sum $\sum_{i=0}^{n-1} \phi_{D_i}$ would imply that θ is the scalar field $\phi_{\mathcal{D}}$ of the 6-cycle associated to the $\hat{A}_{n-1}$ cycle, $\mathcal{C} = \sum_{i=0}^{n-1} \Theta_i$. On the basis of (4.205), this will be equivalent to saying that $\chi(\mathcal{D}) = n$, instead of zero. This is, in mathematical terms, the θ-puzzle. The mathematical solution comes from the fact that $\chi(\mathcal{D}) = 0$. In this section we will relate this result, on the value of the holomorphic Euler chareacteristic, to the appearance of domain walls [132, 133, 134]. To start with, let us consider a cycle $\mathcal{C} = \Theta_1 + \Theta_2$, with $(\Theta_1.\Theta_2) = 1$. The self intersection can be expressed as

$$(\mathcal{C}.\mathcal{C}) = -2 - 2 + 2, \tag{4.230}$$

where the -2 contributions come from Θ_1^2 and Θ_2^2, and the $+2$ comes from the intersection between Θ_1 and Θ_2. As usual, we can consider $\mathcal{C}$ trivially fibered on an Enriques surface. The holomorphic Euler chracteristic of the corresponding six cycle can be written as

$$\chi = \frac{1}{2}(-\mathcal{C}^2). \tag{4.231}$$

Using now the decomposition (4.230) we get two contributions of one, coming from the components Θ_1 and Θ_2, considered independently, and a contribution of -1 from the intersection term $+2$ in (4.230). In this sense, the intersection term can be associated to two fermionic zero modes, and net change of chiral charge oposite to that of the Θ_i components. When we do this for the cycle $\mathcal{C}$ of $\hat{A}_{n-1}$ singularities, we get that each intersection is soaking up two zero modes, leading to the result that $\chi(\mathcal{C}) = 0$. A graphical way to represent equation (4.230) is presented in Figure 4.

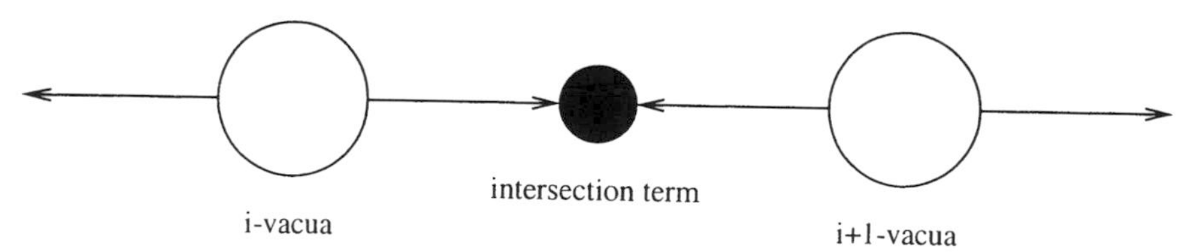

Fig. 4.6. Intersection term.

Now, we will wonder about the physical interpretation of the intersection terms leading to $\chi(\mathcal{C}) = 0$ for all Kodaira singularities. The simplest, and most natural answer, is certainly domain walls extending between different vacua, or values of $< \lambda\lambda >$.

From the point of view of zero mode counting, the "intersection term" behaves effectively as an anti-instanton with two fermionic zero modes. One of these fermionic zero modes, let us say $\psi_{j,j+1}$, is associated to the intersection of Θ_j with Θ_{j+1} and the other $\psi_{j+1,j}$ with the intersection of Θ_j and Θ_{j+1}. Thus, extending naively the computation done for irreducible components, the contribution of the black box in Figure 4 should be of the order

$$\Lambda^3 e^{2\pi ij/n}(1 - e^{2\pi i/n}). \tag{4.232}$$

In result (4.232), interpreted as the contribution of the intersection term, the most surprising fact is the appearance of Λ^3, since now we are geometrically considering simply a point; the factor Λ^3 in the computation of the gaugino condensate comes from the volume of the divisor. In the same way as we interpret M-theory instantons as fivebranes wrapped on the six-cycles used to define the instanton, we can think of the intersection terms as fivebranes wrapped on the cycle $C \times \{(\Theta_i.\Theta_{i+1})\}$, i. e., the product of the singular locus C and the intersection point. The fivebrane wrapped on this cycle defines, in four dimensions, a domain wall, let us say interwining between the vacua i, at $x_3 = +\infty$, and the vacua $i+1$, at $x_3 = -\infty$, where the coordinate x_3 is identified with the unwrapped direction. It is in this sense that we should use (4.232) to define the energy density, or tension, of the domain wall. In the four dimensional limit, Vol (E) goes zero as $\frac{1}{R}$; moreover, the local engineering approach works in the limit where the volume of the singular locus C is very large, so that we can very likely assume that intersection terms behave like (4.232), with Λ^3, but only in the four dimensional limit. Cyclicity of the $\hat{A}_{n-1}$ diagram allows us to pass from the j to the $j+1$ vacua in two different ways: $n-1$ steps, or a single one. The sum of both contributions should define the physical domain wall; thus the energy density will behave as

$$n\Lambda^3 |e^{2\pi ij/n}(1 - e^{2\pi i/n})|. \tag{4.233}$$

The extension of the previous argument to the case of $O(N)$ groups is certainly more involved, due to the topology of the $\hat{D}$ diagram, and the presence of orientifolds. It would certainly be interesting studying the interplay between orientifolds and domain walls in this case.

Finally, we will say some words on the QCD string. In reference [117], the geometry of QCD strings is intimately related to the topological fact that

$$H_1(Y/Z; \mathbf{Z}) = \mathbf{Z}_n, \tag{4.234}$$

where Σ is a rational curve associated to the configuration of fourbranes, and $Y = S^1 \times \mathbf{R}^5$ is the ambient space where Σ is embedded (see [117] for details). The QCD string is then associated to a partially wrapped membrane on a non trivial element of $H_1(Y/\Sigma; \mathbf{Z})$. Recall that $H_1(Y/\Sigma; \mathbf{Z})$ is defined by one-cycles in Y, with boundary on Σ. The previous discussion was done for $SU(N)$ gauge groups. Using our model of $\hat{A}_{n-1}$ singularities, described in section 2, the analog in our framework of (4.234) is equation (4.187). Then we can, in the same spirit as in reference [117], associate the QCD string to paths going from p_k to p_{k+1}, where p_k are the intersection points,

$$\Theta_k . \Theta_{k-1} = p_k . \tag{4.235}$$

Geometrically, it is clear that the tension of this QCD string is the square root of the domain wall tension. By construction, the QCD string we are suggesting here ends on domain walls, i. e., on intersection points.

To end up, let us include some comments on the existence of extra vacua, as suggested in [135]. It is known that the strong coupling computation of $< \lambda\lambda >$ does not coincide with the weak coupling computation; more precisely[136],

$$< \lambda\lambda >_{\mathrm{SC}} < < \lambda\lambda >_{\mathrm{WC}} . \tag{4.236}$$

In the framework of M-theory instanton computations, the numerical factors will depend in particular on the moduli of complex structures of the Calabi-Yau fourfold. In the strong coupling regime we must consider structures preserving the elliptic fibration structure and the Picard lattice. In the weak coupling regime, where the compuatation is performed in the Higgs phase, the amount of allowed complex structures contributing to the value of $< \lambda\lambda >$ is presumably larger. Obviously, the previous argument is only suggesting a possible way out of the puzzle (4.236).

Equally, at a very speculative level, the extra vacua, with no chiral symmetry breaking, could be associated to the cycle $\mathcal{D}$ defining the singular fiber, a cycle that we know leads to $\chi = 0$, and therefore does not produce any gaugino condensate. Notice that any other cycle with $\chi \neq 0$ will lead, if clustering is used, to some non vanishing gaugino condensates, so that $\mathcal{D}$ with $\chi = 0$ looks like a possible candidate to the extra vacua suggested in [135]. If this argument is correct this extra vacua will appears for any Kodaira singularity i. e. in any ADE $N = 1$ four dimensional gauge theory.

It is important to stress that the θ-puzzle is not exclusive of $N = 1$ gluodynamics. In the $N = 0$ case the Witten-Veneziano formula [139, 140] for the η' mass also indicates a dependence of the vacuum energy on θ in terms of $\frac{\theta}{N}$, which means a set of entangled "vacuum" states. In our approach to $N = 1$ the origin of this entanglement is due to the fact that $\chi = 0$ for the singular cycle. In fact, $\chi(\mathcal{D}) = 0$ means that the set of divisors D_i, plus the intersections, i. e., the domain walls, are invariant under $U(1)$, as implied by equation (4.205). If we naively think of something similar in $N = 0$ and we look for the origin of vacuum entanglement in intersections we maybe should think in translating the topology of intersections into topological properties of abelian proyection gauges [58].

A. M(atrix) Theory.

A.1 The Holographic Principle.

The holomorphic principle was originally suggested by 't Hooft in [141]. Let us work first in four dimensional spacetime. Let $S^{(2)}$ be a surface with the topology of two sphere in $\mathbf{R}^3$, and let us wonder about how many orthogonal quantum states can $S^{(2)}$ contain: we will find an upper bound for this number of states. In order to do this, we will use the Bekenstein-Hawking relation between the entropy and the horizon area of the black hole. Let us then call $\mathcal{N}$ the number of states inside $S^{(2)}$; the entropy can be defined as

$$\exp S = \mathcal{N}. \tag{A.1}$$

If S^2 is the horizon of a black hole, we have [142]

$$S \sim \frac{1}{4}\frac{A}{l_p^2}, \tag{A.2}$$

with A the area of the horizon, in Planck length units. Now, let us translate all the physical information contained inside the $S^{(2)}$ surface, in terms of states of q-bits, defined as quantum systems of two states. The number n of q-bits we need is given by

$$N = 2^n. \tag{A.3}$$

Using (A.1) to (A.3) we then get

$$n = \frac{1}{4\ln 2}\frac{A}{l_p^2}, \tag{A.4}$$

which is essentially the number of cells, of area l_p^2, covering the surface $S^{(2)}$. What we learn from this is that all three dimensional physics inside $S^{(2)}$ can be described using states of q-bits, living on the two dimensional surface $S^{(2)}$. We will call these q-bits the holographic degrees of freedom. What we need now is the two dimensional dynamics governing these two dimensional degrees of freedom, able to reproduce, in holographic projection, the three dimensional physics taking place $S^{(2)}$. We can even consider, instead of $S^{(2)}$, an hyperplane of dimension two, dividing space into two regions. The extension, to this extreme situation, of the holomorphic principle, will tell us that the $3+1$ dynamics can be described in terms of some $2+1$ dynamics for the holomorphic degrees of freedom living on the hypersurface.

This picture of the holographic principle allows to introduce M(atrix) theory [143] as the holographic projection of M-theory. In this case, we will pass from eleven dimensional to ten dimensional physics. What we will need, in order to formulate M(atrix) theory, will be

i) An explicit definition of the holographic projection.
ii) Identifying the holomorphic degrees of freedom.

iii) Providing a ten dimensional dynamics for these degrees of freedom.

The conjectured answer in [143] to *i*), *ii*) and *iii*) are

i) The infinite momentum frame.
ii) D-0branes as degrees of freedom.
iii) The dynamics is implemented through the worldvolume of the lagrangian of D-0branes.

These set of conjectures define M(atrix) theory at present. The idea of the infinite momentum frame is boosting in the eleventh direction in eleven dimensional spacetime, in such a way that p_{11}, the eleventh component of the momentum, becomes larger than any scale in the problem. In this frame, we associate, with an eleven dimensional massles system of momentum $\mathbf{p} = (p_{11}, p_{\perp})$, a ten dimensional galilean system with mass p_{11}, and energy

$$E = \frac{p_{\perp}^2}{2p_{11}}. \tag{A.5}$$

If we introduce an infrared cut off, by compactifying the eleventh dimension on a circle S^1, of radius R, the p_{11} is measured in units of $\frac{1}{R}$. Then,

$$p_{11} = \frac{n}{R}. \tag{A.6}$$

We will interpret n as the number of partons which are necessary to describe a system with value of p_{11} given by (A.6). These partons are the ten dimensional degrees of freedom we are going to consider as holographic variables.

Using (A.4), we can define the ten dimensional size of an eleven dimensional massless particle, in eleven dimensions, with some given p_{11}. In fact, $n = p_{11}R$ is the number of holographic degrees of freedom which, by (A.4), means that

$$\frac{r^9}{l_p^9} \sim p_{11}R, \tag{A.7}$$

and the radius r, characterizing the size, will be $(p_{11}R)^{1/9}l_p$. Now, we can look for objects in ten dimensions with mass equal to the mass of a parton, i. e., $\frac{1}{R}$. Natural candidates are D-0branes. From this, it seems natural to conjecture that the worldvolume dynamics of D-0branes will be a good candidate for the holographic description of M-theory.

M(atrix) theory would hence simply be defined as the worldvolume theory of D-0branes. As for any other type of D-branes, this worldvolume theory is defined as the dimensional reduction down to $0+1$ dimensions of ten dimensional Yang-Mills with $N=1$ supersymmetry. If we consider a set of N D-0branes, we have to introduce matrices X^i, with $i = 1, \ldots, 9$. As usual, the diagonal part of this matrices can be interpreted in terms of the classical positions of the N D-0branes, and the off diagonal terms as representing the exchange of open strings. Thus, the worldvolume lagrangian we get for

N D-0branes is $U(N)$ Yang-Mills quantum mechanics. Using units in which $\alpha' = 1$, the bosonic part of this lagrangian is simply

$$\mathcal{L} = \frac{1}{2g}[\ \mathrm{tr}\dot{X}^i\dot{X}^i - \frac{1}{2}\ \mathrm{tr}[X^i, X^j]^2], \tag{A.8}$$

in units where $l_p = 1$, and with g the string coupling constant. In this definition we have simply consider D-0branes in a ten dimensional type IIA string theory. However, we know that D-0branes are in fact Kaluza-Klein modes of an eleven dimensional theory named M-theory. In the eleven dimensional spacetime, the D-0branes have a momentum p_{11} given by

$$p_{11} = \frac{1}{R}, \tag{A.9}$$

with R the radius of the eleventh dimension. The way to relate (A.8) to the physics of the partons defined in the infinite momentum frame is observing that the kinetic term in (A.8) coincides with the equation (A.5) for p_{11} given in (A.9). In fact, using the relation

$$R = gl_s, \tag{A.10}$$

and choosing units where $l_s = 1$, we notice that (A.8) is precisely the galilean lagrangian for particles of mass $\frac{1}{g}$. Thus, we will interpret the worldvolume dynamics of D-0branes, (A.8), as the infinite momentum frame of the M-theory D-0branes.

Our main task now will consist in deriving the brane spectrum directly from the M(atrix) lagrangian (A.8), interpreting the different branes as collective excitations of D-0branes. In order to achive this, it will be necessary to work in the $N \to \infty$ of (A.8). Using the relation

$$l_s = g^{-1/3} l_p, \tag{A.11}$$

between the string length, and the Plank scale, we can pass to Plank units by defining $Y = \frac{X}{g^{1/3}}$. In Y variables, and with $l_p = 1$, we get

$$\mathcal{L} = \mathrm{tr}\left[\frac{1}{2R}D_tY^iD_tY^i - \frac{1}{4}R[Y^i, Y^j]^2\right], \tag{A.12}$$

where $D_t = \partial_t + iA$, with A equal the A_0 piece of the ten dimensional Yang-Mills theory. In order to get (A.8), going to the temporal gauge $A_0 = 0$ is all what is needed.

Now, some of the ingredients introduced in chapter I will be needed; namely, the matrices P and Q defined in (1.87). In terms of this basis of matrices, any matrix Z can be written as

$$Z = \sum_{n,m=1}^{N} z_{n,m} P^n Q^m. \tag{A.13}$$

Taking into account that

$$PQ = QPe^{2\pi i/N}, \tag{A.14}$$

we can define

$$P = e^{i\hat{p}}, \quad Q = e^{i\hat{q}}, \tag{A.15}$$

with

$$[\hat{p}, \hat{q}] = \frac{2\pi i}{N}. \tag{A.16}$$

Replacing (A.15) in (A.13) we get

$$Z = \sum_{n,m} z_{n,m} e^{in\hat{p}} e^{im\hat{q}}, \tag{A.17}$$

which looks like the Fourier transform of a function $Z(p,q)$. The only difference is that this function is defined on a quantum phase space defined by $\hat{p}$ and $\hat{q}$ variables satisfying (A.16). In the $N \to \infty$ limit, we can interpret this quantum space as classical, since in this limit $[\hat{p}, \hat{q}] = 0$. Thus, in the $N \to \infty$ limit, the matrices can be replaced by functions $Z(p,q)$, as defined by (A.17). The matrix operations become, in this limit,

$$\begin{aligned} \mathrm{tr} Z \quad &\to \quad N \int Z(p,q) dp dq, \\ [X,Y] \quad &\to \quad \frac{1}{N} [\partial_q X \partial_p Y - \partial_q Y \partial_p X], \end{aligned} \tag{A.18}$$

i. e., the conmutator becomes the Poisson bracket. Now, we can use (A.18) in (A.12); what we then get is

$$\mathcal{L} = \frac{p_{11}}{2} \left[\int dp dq \dot{Y}^i(p,q) \dot{Y}^i(p,q) - \frac{1}{P_{11}} \int dp dq [\partial_q Y^i \partial_p Y^j - \partial_q Y^j \partial_p Y^i] \right], \tag{A.19}$$

where $p_{11} = \frac{N}{R}$. The interest of (A.19) is that this result coincides with the eleven diemensional lagrangian for the eleven dimensional supermembrane in the light cone frame. Notice that i in (A.19) goes from 1 to 9, which can be interpreted as the transversal directions to the supermembrane worldvolume. The previous result is alraedy a good indication of the consistency of M(atrix) theory as a microscopic description of M-theory. Next, we will try to define toroidal compactifications of (A.8).

A.2 Toroidal Compactifications.

The definition of toroidal compactifications [145] of M(atrix) theory is quite simple. We will consider the worldvolume of lagrangian of D-0branes, starting with ten dimensional $N = 1$ supersymmetric Yang-Mills in $\mathbf{R}^9 \times S^1$. In order to clear up the procedure, we will keep all indices for a while, so that we will write

$$X^i_{k,l} \tag{A.20}$$

for the matrix X^i. The indices k and l will hence label different D-0branes. Now, if we force D-0branes to live in $\mathbf{R}^9 \times S^1$, and interpret S^1 as

$$\mathbf{R}/\Gamma, \tag{A.21}$$

with Γ a one dimensional lattice defined by a vector $\mathbf{e} = 2\pi R$, we can think of copies of each D-0brane, parametrized by integers n, depending on the cell of $\mathbf{R}/\Gamma$ where they are. Then, (A.20) should be changed to

$$X^i_{k,m;l,m}. \tag{A.22}$$

We can now forget about the indices k and l, to write X^i_{nm}, where n and m are integers. The lagrangian (A.8) then becomes

$$\mathcal{L} = \frac{1}{2g}[\mathrm{tr}\dot{X}^i_{mn}\dot{X}^i_{mn} + \frac{1}{2}\mathrm{tr}(X^i_{mq}X^j_{qn} - X^j_{mq}X^i_{qn})(X^i_{nr}X^j_{rm} - X^j_{nr}X^i_{rm})]. \tag{A.23}$$

Now, we should imposse symmetry with with respect to the action of Γ, which implies

$$\begin{aligned} X^i_{mn} &= X^i_{m-1\,n-1} \quad i > 1, \\ X^1_{mn} &= X^1_{m-1\,n-1} \quad m \neq n, \\ X^1_{mn} &= 2\pi R\mathbf{I} + X^1_{m-1\,n-1}. \end{aligned} \tag{A.24}$$

The meaning of (A.24) is that the coordinate X^1 is periodic, so that the difference in X^1 for n D-0branes, and $n+1$ D-0branes, is simply the length of the compactified direction. Using (A.24), matrices can be simply labelled by one index, $X^i_{0,n} \equiv X^i_n$, and the lagrangian (A.23) becomes

$$\mathcal{L} = \frac{1}{2g}[\sum_{i=1}^{9} \mathrm{tr}\dot{X}^i_n \dot{X}^i_{-n} - \sum_{j=2}^{9} \mathrm{tr}S^j_n(S^j_n)^+ - \frac{1}{2}\sum_{j,k=2}^{9} \mathrm{tr}T^{jk}_n(T^{jk}_n)^+], \tag{A.25}$$

where

$$\begin{aligned} S^j_n &= \sum_q([X^1_q, X^j_{n-q}]) - 2\pi R n X^j_n, \\ T^{jk}_n &= \sum_q [X^j_q, X^k_{n-q}]. \end{aligned} \tag{A.26}$$

Once we get lagrangian (A.21), we can compare it with the worldvolume lagrangian for D-1branes. In fact, for D-1branes the worldvolume lagrangian is $1+1$ dimensional super Yang-Mills theory, with gauge fields A^1 and A^0, and matter fields Y^j (with $j = 2, \dots, 9$) in the adjoint representation. We can then work in the temporal gauge, fixing $A^0 = 0$. On the other hand, performing T-duality on S^1 takes form D-0branes to D-1branes. Hence, on the dual S^1 the worldvolume lagrangian for D-1branes should coincide with that of D-0branes in $\mathbf{R}^9 \times S^1$, i. e, with lagrangian (A.25). The D-1brane worldvolume lagrangian in the dual S^1, with radius $R' = \frac{1}{2\pi R}$,

$$\mathcal{L} = \int dx \frac{dt}{2\pi R'} \frac{1}{2g} [\text{ tr } \dot{Y}^i \dot{Y}^i + \text{ tr } \dot{A}^1 \dot{A}^1 + \frac{1}{2} \text{ tr } [Y^i, Y^j]^2 - \text{ tr } [\partial_1 Y^i - i[A^1, Y^i]]^2, \tag{A.27}$$

can be compared with (A.26) if we just interpret X_n^i as the Fourier modes of $Y^i(x)$, and X_n^1 as the Fourier modes of $A^1(x)$:

$$\begin{aligned} A^1(x) &= \sum_n e^{inx/R'} X_n^1, \\ Y^i(x) &= \sum_n e^{inx/R'} X_n^i. \end{aligned} \tag{A.28}$$

Hence, we can readily induce the following result: M(atrix) theory compactified on T^d is equivalent to $d+1$ supersymmetric Yang-Mills on the dual $\hat{T}^d \times \mathbf{R}$, with $\mathbf{R}$ standing for the time direction, and the supersymmetric Yang-Mills theory defined through dimensional reduction from $N=1$ ten dimensional Yang-Mills theory. This is a surprising result, connecting M(atrix) compactifications with Yang-Mills theories, a relation with far reaching consequences, some of which we will consider in what follows.

A.3 M(atrix) Theory and Quantum Directions.

We can then represent M(atrix) theory, compactified on T^d, as supersymmetric Yang-Mills theory defined on $\hat{T}^d \times \mathbf{R}$, i. e., as the worldvolume lagrangian of d D-branes wrapped on the dual torus, $\hat{T}^d$. Let us then work out some simple cases. We will first compactify M(atrix) on T^4, which will be an interesting case concerning the U-duality symmetry.

Let L_i be the lengths of T^4. The dual torus $\hat{T}$ will then be defined with sides of length

$$\Sigma_i = \frac{l_s^2}{L_i}, \tag{A.29}$$

with l_s the string length. In terms of the eleven dimensional Planck scale, l_P, we have

$$\frac{l_P^3}{R} = l_s^2, \tag{A.30}$$

and therefore

$$\Sigma_i = \frac{l_p^3}{L_i R}. \tag{A.31}$$

Let us now consider the infinite momentum frame energy of a state with one unit of p_{11}, and one unit of momentum in some internal direction, L_i,

$$E = \frac{p_\perp}{2P_{11}} = \frac{R}{2L_i^2}. \tag{A.32}$$

This state corresponds, in supersymmetric Yang-Mills, to a gauge configuration with a non trivial Wilson line $A(C_i)$ (recall that in the toroidal compactification the compactified components X^i behave as Yang-Mills fields. This non trivial Wilson line means a flux through C_i. This energy is given by

$$\frac{g_{SYM}^2 \Sigma_i^2}{\Sigma_1 \Sigma_2 \Sigma_3 \Sigma_4}. \tag{A.33}$$

Identiying (A.32) and (A.33) we get

$$g_{SYM}^2 = \frac{R}{2L_i^2} \frac{\Sigma_1 \Sigma_2 \Sigma_3 \Sigma_4}{\Sigma_i^2}. \tag{A.34}$$

Using (A.31) we get

$$g_{SYM}^2 = \frac{R^3 \Sigma_1 \Sigma_2 \Sigma_3 \Sigma_4}{2l_p^6} = \frac{l_p^6}{2L_1 L_2 L_3 L_4 R}, \tag{A.35}$$

which means that g^2, as expected in $4+1$ dimensions, has units of length.

From the definition of M(atrix) compactifications, we expect that M(atrix) on T^4 will reproduce type IIA string theory on T^4 that, has been derived in chapter III, is invariant under the U-duality group, $Sl(5, \mathbf{Z})$. Thus, our task is to unravel this U-duality invariance, considering supersymmetric Yang-Mills on $\hat{T}^4 \times \mathbf{R}$. From (A.35), we observe a clear $Sl(4, \mathbf{Z})$ invariance of the gauge theory. These transformations exchange all Σ_i, leaving their product invariant. In order to extend this symmetry to $Sl(5, \mathbf{Z})$, an extra dimension Σ_5 needs to be defined. A way to do this is using as such direction the coupling constant itself in $4+1$ directions that, as can be clearly seen from (A.35), has dimensions of length. In this way, we can think that M(atrix) on T^4 is described by a $5+1$ dimensional theory, with space dimensions a torus T^5, of dimensions Σ_i, with $i = 1, 2, 3, 4$, and $\Sigma_5 = \frac{l_p^6}{L_1 L_2 L_3 L_4 R}$. This is exactly the same picture we have in M-theory, understood as the strong coupled limit of type IIA string theory. There, we associated the RR D-0branes with Kaluza-Klein modes of the extra dimension. In the gauge theory context we should look for objects in $4+1$ dimensions, that can be interpreted as Kaluza-Klein modes of the extra dimension required by U-duality. As candidates to these states, we can use instantons. Instantons are associated with the Π_3 homotopy group of the gauge group so that, in $4+1$ dimensions, they look like particles. Moreover, their mass is given by $\frac{1}{g^2}$, with the gauge coupling constant (recall that $\frac{1}{g^2}$ is the action for the instanton in $3+1$ dimensions). Therefore, using (A.35), we get the desired result, namely that instantons ar the Kaluza-Klein modes of the extra dimension.

We can, in fact, try to understand what kind of dynamics is playing the role here, using string theory language. The supersymmetric Yang-Mills theory on T^4, with gauge group $U(N)$, can be interpreted as the worldvolume lagrangian for N fourbranes of type IIA, wrapped around T^4. In M-theory, we can interpret this fourbranes as fivebranes partially wrapped in around the internal eleventh dimension. When we move to strong coupling, we open the extra direction, and we effectively get a $5+1$ dimensional gauge theory. If this is the correct picture, we can check it by comparison of the mass of the instanton and the expected mass of the wrapped around T^4 and the internal eleventh dimension. The energy of the fivebrane would then be

$$E = \frac{L_1 L_2 L_3 L_4 R}{l_p^6}, \tag{A.36}$$

which is exactly the mass of the instanton,

$$\frac{1}{g^2} \equiv \frac{1}{\hat{g}} \frac{L_1 L_2 L_3 L_4 R}{l_p^6}. \tag{A.37}$$

In order to understand the effect described above, it would be convenient to discuss briefly the scales entering the theory. Using relations (A.32) to (A.35), and $gl_s = R$ we get, for generic dimension d,

$$g_{SYM}^2 = \frac{R^{3-d} l_s^{3d-6}}{L_i^d g^{d-3}}. \tag{A.38}$$

It is clear form (A.38) that for $d \leq 3$ the limit of string coupling constant equal zero gives a weak coupled supersymmetric Yang-Mills theory. However, a barrier appears in $d = 4$. In fact, for $d \geq 4$ the limit $g \to 0$ leads to strong coupling in the field theory. One of these strong copling effects is the generation of the quantum dimension needed for U-duality.

Acknowledgments

This work is partially supported by European Community grant ERBFM-RXCT960012, and by grant AEN-97-1711.

References

1. P. A. M. Dirac, "Quantized Singularities in the Electromagnetic Field", Proc. Roy. Soc. Lond. **A133** (1931), 60.
2. T. T. Wu and C. N. Yang, "Concept of Nonintegrable Phase Factors and Global Formulation of Gauge Fields", Phys. Rev. **D12** (1975), 3845.
3. P. Goddard and D. Olive, Prog. Rep. Phys. **41** (1978), 1357.
4. P. Goddard, J. Nuyts and D. Olive, "Gauge Theories and Magnetic Charge", Nucl. Phys. **B125** (1977), 1.
5. C. Montonen and D. Olive, "Magnetic Monopoles as Gauge Particles ?", Phys. Lett. **B72** (1977), 177.

6. H. Georgi and S. Glashow, "Unified Weak and Electromagnetic Interactions Without Neutral Currents", Phys. Rev. Lett. **28** (1972), 1494.
7. G. 't Hooft, "Magnetic Monopoles in Unified Gauge Theories", Nucl. Phys. **B79** (1974), 276.
   A. M. Polyakov, "Particle Spectrum in the Quantum Field Theory", JETP Lett. **20** (1974), 194.
8. M. K. Prasad and C. M. Sommerfeld, "An Exact Classical Solution for the 't Hooft Monopole and the Julia-Zee Dyon", Phys. Rev. Lett. **35** (1975), 760.
9. E. B. Bogomolny, "Stability of Classical Solutions", Sov. J. Nucl. Phys. **24** (1976), 449.
10. A. A. Belavin, A. M. Polyakov, A. S. Swartz and Y. S. Tyupkin, "Pseudoparticle Solutions of the Yang-Mills Equations", Phys. Lett. **B59** (1975), 85.
11. G. 't Hooft, "Computation of the Quantum Effects due to a Four Dimensional Pseudoparticle", Phys. Rev. **D14** (1977), 3432.
12. R. Jackiw and C. Rebbi, "Degree of Freedom in Pseudoparticle Physics", Phys. Lett. **B67** (1977), 189.
13. M. F. Atiyah N. Hitchin and I. M. Singer, "Self-Duality in Four Dimensional Riemannian Geometry", Proc. Roy. Soc. Lond. **A362** (1978), 475.
14. M. F. Atiyah and I. M. Singer, "Index of Elliptic Operators I", Ann. Math. **87** (1968), 485.
    M. F. Atiyah and G. B. Segal, "Index of Elliptic Operators II", Ann. Math. **87** (1968), 531.
    M. F. Atiyah and I. M. Singer, "Index of Elliptic Operators III", Ann. Math. **87** (1968), 546.
    M. F. Atiyah and I. M. Singer, "Index of Elliptic Operators IV", Ann. Math. **93** (1971), 119.
    M. F. Atiyah and I. M. Singer, "Index of Elliptic Operators V", Ann. Math. **87** (1971), 139.
15. G. 't Hooft, "Symmetry Breaking Through Bell-Jackiw Anomalies", Phys. Rev. Lett. **37** (1976), 8.
16. R. Jackiw and C. Rebbi, "Vacuum Periodicity in a Yang-Mills Quantum Theory", Phys. Rev. Lett. **37** (1976), 172.
17. C. G. Callan, R. Dashen and D. J. Gross, "Towards a Theory of the Strong Interactions", Phys. Rev. **D17** (1978), 2717.
18. E. Witten, "Dyons af Charge $e\theta/2\pi$", Phys. Lett. **B86** (1979), 283.
19. V. A. Rubakov, "Adler-Bell-Jackiw Anomaly and Fermionic-Number Breaking in the Presence of a Magnetic Monopole", Nucl. Phys. **B203** (1982), 311.
20. G. 't Hooft, "A Property of Electric and Magnetic Charges in Non Abelian Gauge Theories", Nucl. Phys. **B153** (1979), 141.
21. G. 't Hooft, "Some Twisted Self-Dual Solutions for the Yang-Mills Equations on a Hypertorus", Commun. Math. Phys. **81** (1981), 267.
22. E. Witten, "Constraints on Supersymmetry Breaking", Nucl. Phys. **B202** (1982), 253.
23. A. M. Polyakov, "Quark Confinement and Topology of Gauge Groups", Nucl. Phys. **B120** (1977), 429.
24. I. Affleck, J. A. Harvey and E. Witten, "Instantons and Supersymmetry Breaking in 2 + 1 Dimensions", Nucl. Phys. **B206** (1982), 413.
25. C. Callias, "Axial Anomalies and Index Theorems on Open Spaces", Comm. Math. Phys. **62** (1978), 213.
26. J. Wess and J. Bagger, "*Supersymmetry and Supergravity*", Princeton University Press, Princeton, 1984.

27. V. A. Novikov, M. A. Shifman, A. I. Vainshtein and V. I. Zakharov, "Exact Gell-Mann-Low Function of Supersymmetric Yang-Mills Theories From Instanton Calculus", Nucl. Phys. **B229** (1983), 381.
28. V. A. Novikov, M. A. Shifman, A. I. Vainshtein and V. I. Zakharov, "Instanton Effects in Supersymmetric Theories", Nucl. Phys. **B229** (1983), 407.
29. M. A. Shifman, A. I. Vainshtein and V. I. Zakharov, "On Gluino Condensation in Supersymmetric Gauge Theories. $SU(N)$ and $O(N)$ Gauge Groups", Nucl. Phys. **B296** (1988), 445.
30. D. Amati, K. Konishi, Y. Meurice, G. C. Rossi and G. Veneziano, "Non Perturbative Aspects in Supersymmetric Gauge Theories", Phys. Rep. **162** (1988), 169.
31. I. Affleck. M. Dine and N. Seiberg, "Dynamical Supersymmetry Breaking in Supersymmetric QCD", Nucl. Phys. **B241** (1984), 493.
32. I. Affleck. M. Dine and N. Seiberg, "Dynamical Supersymmetry Breaking in Four-Dimensions and its Phenomenological Implications", Nucl. Phys. **B256** (1985), 557.
33. M. A. Shifman and A. I. Vainshtein, "On Gluino Condensation in Supersymmetric Gauge Theories, $SU(N)$ and $O(N)$ Gauge Groups", Nucl. Phys. **B296** (1988), 445.
34. E. Cohen and C. Gómez, "Chiral Symmetry Breaking in Supersymmetric Yang-Milss", Phys. Rev. Lett. **52** (1984), 237.
35. N. Seiberg and E. Witten, "Electric-Magnetic Duality, Monopole Condensation and Confinement in $N = 2$ Supersymmetric Yang-Mills Theory", Nucl. Phys. **B426** (1994), 19.
36. N. Seiberg and E. Witten, "Monopoles, Duality and Chiral Symmetry Breaking in $N = 2$ Supersymmetric QCD", Nucl. Phys. **B431** (1994), 484.
37. N. Seiberg and E. Witten, "Gauge Dynamics and Compactification to Three Dimensions", **hep-th/9607163**.
38. M. Shifman and A. I. Vainshtein, "On Holomorphic Dependence and Infrared Effects in Supersymmetric Gauge Theories", Nucl. Phys. **B359** (1991), 571.
39. N. Seiberg, "Supersymmetry and Nonperturbative Beta Functions", Phys. Lett. **B206** (1988), 75.
40. K. G. Wilson and J. Kogut, "The Renormalization Group and the ϵ-Expansion", Phys. Rep. **12** (1974), 75.
41. L. Adler and W. A. Bardeen, "Absence of Higher Order Corrections in the Anomalous Axial Vector divergence Equation", Phys. Rev. **182** (1969), 1517.
42. M. Atiyah and N. Hitchin, "*The Geometry and Dynamics of Magnetic Monopoles*", Princeton University Press, Princeton, 1988.
43. V. I. Arnold, "*Singularity Theory*", London Math. Soc. Lecture Note Series **53**, Cambridge University Press, Cambridge, 1981.
44. M. Artin, "On Isolated Rational Singularities of Surfaces", Amer. J. Math. **88** (1966), 129.
45. K. Kodaira, Ann. Math. **77**, 3 (1963), 563.
46. A. Klemm, W. Lerche, S. Theisen and S. Yankielowicz, "Simple Singularities and $N = 2$ Supersymmetric Yang-Mills", Phys. Lett. **B344** (1995), 169.
47. P. Argyres and A. Faraggi, "The Vacuum Structure and Spectrum of $N = 2$ Supersymmetric $SU(N)$ Gauge Theory", Phys. Rev. Lett. **73** (1995), 3931.
48. A. Hanany and Y. Oz, "On the Quantum Moduli Space of Vacua of $N = 2$ Supersymmetric $SU(N_c)$ Gauge Theories", Nucl. Phys. **B452** (1995), 283.
49. P. Argyres, M. Plesser and A. Shapere, "The Coulomb Phase of $N = 2$ Supersymmetric QCD", Phys. Rev. Lett. **75** (1995), 1699.
50. U. Danielsson and B. Sundborg, "The Moduli Space and Monodromies of $N = 2$ Supersymmetric $SO(2r+1)$ Yang-Mills Theory", Phys. Lett. **B358** (1995), 273.

51. A. Brandhuber and K. Landsteiner, "On the Monodromies of $N = 2$ Supersymmetric Yang-Mills Theory with Gauge Group $SO(2n)$, Phys. Lett. **B358** (1995), 73.
52. K. Intriligator and N. Seiberg, "Lectures on Supersymmetric Gauge Theories and Electric-Magnetic Duality", Nucl. Phys. Proc. Supl. **45** (1996).
53. C. Gómez and R. Hern'andez, "Electric-Magnetic Duality and Effective Field Theories", **hep-th/9510023**; published in "**Advanced School on Effective Theories**", F. Cornet and M. J. Herrero eds. World Scientific (1997).
A. Bilal, "Duality in $N = 2$ SUSY $SU(2)$ Yang-Mills Theory: A Pedagogical Introduction to the Work of Seiberg and Witten", **9601007**.
W. Lerche, "Introduction to Seiberg-Witten Theory and its Stringy Origin", Nucl. Phys. Proc. Supl. **B55** (1997), 83.
L. Álvarez-Gaumé and F. Zamora, "Duality in Quantum Field Theory (and String Theory)", **hep-th/9709180**.
54. O. J. Ganor, D. R. Morrison and N. Seiberg, "Branes, Calabi-Yau Spaces, and Toroidal Compactification of the $N = 1$ Six-Dimensional E_8 Theory", Nucl. Phys. **B487** (1997), 93.
55. Mandelstam, "Vortices and Quark Confinement in Non Abelian Gauge Theories", Phys. Rep. **23** (1976), 245.
56. G. 't Hooft, "On the Phase Transition Towards Permanent Quark Confinement", Nucl. Phys. **B138** (1978), 1.
57. H. B. Nielsen and P. Olesen, "Vortex Line Models for Dual Strings", Nucl. Phys. **B61** (1973), 45.
58. G. 't Hooft, "Topology of the Gauge Condition and New Confinement Phases in Non Abelian Gauge Theories", Nucl. Phys. **B190** (1981), 455.
59. M. B. Green, J. H. Schwarz and E. Witten, "*Superstring Theory*", Cambridge University Press, Cambridge, 1987.
60. J. Polchinski, "What is String Theory?", **hep-th/9411028**.
61. C. Vafa, "Lectures on Strings and Dualities", **hep-th/9702201**.
62. E. Kiritsis, "Introduction to Superstring Theory", **hep-th/9709062**.
63. J. D. Fay, "Theta Functions on Riemann Surfaces", Lecture Notes in Mathematics, **352**, Springer-Verlag, 1973.
64. K. S. Narain, "New Heterotic String Theories in Uncompactified Dimensions < 10", Phys. Lett. **B169** (1986), 41; K. S. Narain, M. H. Samadi and E. Witten, "A Note on the Toroidal Compactification of Heterotic String Theory", Nucl. Phys. **B279** (1987), 369.
65. For reviews see A. Giveon, M. Porrati and E. Rabinovici, "Target Space Duality in String Theory", Phys. Rep. **244** (1994), 77.
E. Álvarez, L. Álvarez-Gaumé and Y. Lozano, "An Introduction to T-Duality in String Theory", **hep-th/9410237**.
66. P. Griffiths and J. Harris, "*Principles of Algebraic Geometry*", Wiley-Interscience, 1978.
67. P. A. Aspinwall, "$K3$ Surfaces and String Duality", **hep-th/9611137**.
68. U. Persson, Lecture Notes in Mathematics, 1124, Springer-Verlag.
69. N. Seiberg, "Observations on the Moduli Space of Superconformal Field Theories", Nucl. Phys. **B303** (1988), 286.
70. B. Greene and M. R Plesser, "Duality in Calabi-Yau Moduli Space", Nucl Phys. **B338** (1990), 15.
P. Candelas, M. Lynker and R. S. Schimmrigk, "Calabi-Yau Manifolds in Weighted P(4)", Nucl. Phys. **B341** (1990), 383.
P. Candelas, X. de la Ossa, P. Green and L. Parkes, "A Pair of Calabi-Yau Manifolds as an Exactly Soluble Superconformal Theory", Nucl. Phys. **B359** (1991), 21.

V. Batyrev, "Dual Polyhedra and Mirror Symmetry for Calabi-Yau Hypersurfaces in Toric Varieties", **alg-geom/9410003**.
For a series of references, see S. -T. Yau ed., "Essays on Mirror Manifolds", International Press, Hong-Kong, 1992.
71. I. V. Dolgachev, "Mirror Symmetry for Lattice Polarized $K3$-Surfaces", **alg-geom/9502005**.
72. J. Polchinski and Y. Cai, "Consistency of Open Superstring Theories", Nucl. Phys. **B296** (1988), 91.
M. Green, "Space-Time Duality and Dirichlet String Theory", Phys. Lett. **B266** (1991), 325.
J. Polchinski, "Dirichlet Branes and Ramond-Ramond Charges", Phys. Rev. Lett. **75** (1995), 4724.
J. Polchinski, S. Chaudhuri and C. V. Johnson, "Notes on D-Branes", **hep-th/9602052**, and "TASI Lectures on D-Branes", **hep-th/9611050**.
73. J. C. Paton and H. M. Chan, "Generalized Veneziano Model with Isospin", Nucl. Phys. **B10** (1969), 516.
74. N. J. Hitchin, A. Karlhede, U. Lindström, M. Rocek, "Hyperkähler Metrics and Supersymmetry", Comm. Math. Phys. **108** (1987), 535.
75. J. Cardy and E. Rabonovici, "Phase Structure of $Z(P)$ Models in the Presence of a Theta Parameter", Nucl. Phys. **B205** (1982), 1.
76. A. Font, L. Ibañez, D. Lüst and F. Quevedo, "Strong-Weak Coupling Duality and Non-Perturbative Effects in String theory", Phys. Lett. **B249** (1990), 35.
77. A. Sen, "Strong-Weak Coupling Duality in Four Dimensional Field theory", Int. J. Mod. Phys. **A9** (1994), 3707.
78. C. Hull and P. Townsend, "Unity of Superstring Dualities", Nucl. Phys. **B438** (1995), 109.
79. M. Duff, "Strong/Weak Coupling Duality from the Dual String", Nucl. Phys. **B442** (1995), 47.
80. P. Townsend, "The Eleven-Dimensional Supermembrane Revisited", Phys. Lett. **B350** (1995), 184.
81. E. Witten, "String Theory Dynamics in Various Dimensions", Nucl. Phys. **B443** (1995), 85.
82. A. Sen, "String-String Duality Conjecture in Six Dimensions and Charged Solitonic Strings", Nucl. Phys. **B450** (1995), 103.
83. J. Harvey and A. Strominger, "The Heterotic String is a Soliton", Nucl. Phys. **B449** (1995), 535.
84. B. Greene, D. Morrison and A. Strominger, "Black Hole Condensation and the Unification of String Vacua", Nucl. Phys. **B451** (1995), 109.
85. C. Vafa and E. Witten, "A One-Loop Test Of String Duality", Nucl. Phys. **B447** (1995), 261.
86. C. Hull and P. Townsend, "Enhanced Gauge Symmetries in Superstring Theories", Nucl. Phys. **B451** (1995), 525.
87. J. Schwarz, "An $Sl(2, \mathbf{Z})$ Multiplet of Type IIB Superstrings", Phys. Lett. **B360** (1995), 13.
88. D. J. Gross, J. A. Harvey E. Martinec and R. Rohm, "Heterotc String", Phys. Rev. Lett. **54** (1985), 502; "Heterotic String Theory (I). The Free Heterotic String.", Nucl. Phys. **B256** (1985), 253; "Heterotic String Theory (II). The Interacting Heterotic String.", Nucl. Phys. **B267** (1986), 75.
89. S. Katz, D. R. Morrison and M. R. Plesser, "Enhanced Gauge Symmetry in Type II String Theory", Nucl. Phys. **477** (1996), 105.
90. A. Klemm and P. Mayr, "Strong Coupling Singularities and Non-abelian Gauge Symmetries in $N = 2$ String Theory", Nucl. Phys. **B469**, 37. (1996),

91. P. S. Aspinwall, "Enhanced Gauge Symmetries and $K3$ Surfaces", Phys. Lett. **B357** (1995), 329..
92. P. S. Aspinwall and J. Louis, "On the Ubiquity of $K3$ Fibrations in String Duality", Phys. Lett. **B369** (196), 233..
93. A. Klemm, W. Lerche and P. Mayr,"$K3$-Fibrations and Heterotic-Type II String Duality", Phys. Lett. **B357** (1995), 313.
94. C. Vafa, "Evidence for F-Theory", Nucl. Phys. **B469** (1996), 403.
95. D. R. Morrison and C. Vafa, "Compactifications of F-Theory on Calabi-Yau Threefolds. I.", Nucl. Phys. **B473** (1996), 74.
96. D. R. Morrison and C. Vafa, "Compactifications of F-Theory on Calabi-Yau Threefolds. II.", Nucl. Phys. **B476** (1996), 437.
97. S. Kachru and C. Vafa, "Exact Results for $N = 2$ Compactifications of Heterotic Strings", Nucl. Phys. **450**, (1995), 69.
98. M. J. Duff, P. Howe, T. Inami and K. S. Stelle, "Superstrings in $D = 10$ from Supermembranes in $D = 11$", Phys. Lett. **B191** (1987), 70.
M. J. Duff, R. Minasian and J. T. Liu, "Duality Rotations in Membrane Theory", Nucl. Phys. **B347** (1990), 394.
M. J. Duff, R. Minasian and J. T. Liu, "Eleven Dimensional Origin of String/String Duality: A One Loop Test", Nucl. Phys. **B452** (1995), 261.
99. E. Witten, "Non-Perturbative Superpotentials in String Theory", Nucl. Phys. **B474** (1996), 343.
100. A. Hanany and E. Witten, "Type IIB Superstrings, BPS Monopoles and Three Dimensional Gauge Dynamics", Nucl. Phys. **B492** (1997), 152.
101. S. Elitzur, A. Giveon and D. Kutasov, "Branes and $N = 1$ Duality in String Theory", Phys. Lett. **B400** (1997), 269.
102. J. de Boer, K. Hori, Y. Oz and Z. Yin, "Branes and Mirror Symmetry in $N = 2$ Supersymmetric Gauge Theories in Three Dimensions", **hep-th/9702154**.
103. E. Witten, "Solutions of Four-Dimensional Fields Theories Via M-Theory", Nucl. Phys. **B500** (1997), 3.
104. N. Evans, C. V. Johnson and A. D. Shapere, "Orientifolds, Branes and Duality of $4D$ Field Theories", **hep-th/9703210**.
105. S. Elitzur, A. Giveon, D. Kutasov, E. Ravinovici and A. Schwimmer, "Brane Dynamics and $N = 1$ Supersymmetric Gauge Theory", **hep-th/9704104**.
106. J. L. F. Barbon, "Rotated Branes and $N = 1$ Duality", Phys. Lett. **B402** (1997), 59.
107. J. Brodie and A. Hanany, "Type IIA Superstrings, Chiral Symmetry, and $N = 1$ $4D$ Gauge Theory Dualities", **hep-th/9704043**.
108. A. Brandhuber, J. Sommenschein, S. Theisen and S. Yanckielowicz, "Brane Configurations and $4D$ Field Theories", **hep-th/9704044**.
109. O. Aharony and A. Hanany, "Branes, Superpotentials and Superconformal Fixed Points", **hep-th/9704170**.
110. R. Tartar, "Dualities in $4D$ Theories with Product Gauge Groups from Brane Configurations", **hep-th/9704198**.
111. I. Brunner and A. Karch, "Branes and Six Dimensional Fixed Points", **hep-th/9705022**.
112. B. Kol, "5d Field Theories and M Theory", **hep-th/9705031**.
113. A. Marshakov, M. Martellini and A. Morozov, "Insights and Puzzles from Branes: $4d$ SUSY Yang-Mills from $6d$ Models", **hep-th/9706050**.
114. K. Landsteiner, E. López and D. A. Lowe, "$N = 2$ Supersymmetric Gauge Theories, Branes and Orientifolds", **hep-th/9705199**.
115. A. Brandhuber, J. Sommenschein, S. Theisen and S. Yanckielowicz, **hep-th/9705232**.

116. K. Hori, H. Ooguri and Y. Oz, "Strong Coupling Dynamics of Four Dimensional $N = 1$ Gauge Theory fron M theory Fivebrane" **hep-th/9706082**.
117. E. Witten, "Branes and the Dynamics of QCD", **hep-th/9706109**.
118. A. Strominger, S. -T. Yau and E. Zaslow, "Mirror Symmetry is T-Duality", Nucl. Phys. **B479** (1996), 243.
119. D. R. Morrison, "The Geometry Underlying Mirror Symmetry", **alg-geom/9608006**.
120. R. Donagi and E. Witten, "Supersymmetric Yang-Mills and Integrable Systems", Nucl. Phys. **B460** (1996), 299.
121. N. Hitchin, "The Self-Duality Equations on a Riemann Surface", Proc. London Math. Soc. **55** (1987), 59; "Stable Bundles and Integrable Systems", Duke Math. J. **54** (1987), 91.
122. S. Kachru, A. Klemm, W. Lerche P. Mayr and C: Vafa, "Non Perturbative Results on the Point Particle Limit of $N = 2$ Heterotic String Compactifications", Nucl. Phys. **B459** (1996), 537.
123. C. Gómez, R. Hernández and E. López, "S-Duality and the Calabi-Yau Interpretation of the $N = 4$ to $N = 2$ Flow", Phys. Lett. **B386** (1996), 115.
124. A. Klemm, W. Lerche, P. Mayr, C. Vafa and N. Warner, "Self-Dual Strings and $N = 2$ Supersymmetric Field Theory", Nucl. Phys. **B477**, (1996), 746.
125. C. Gómez, R. Hernández and E. López, "$K3$-Fibrations and Softly Broken $N = 4$ Supersymmetric Gauge Theories", **hep-th/9608104**.
126. S. Katz and C. Vafa, "Geometric Engineering of Quantum Field Theories", Nucl. Phys. **B497** (1997), 173, and "Geometric Engineering of $N = 1$ Quantum Field Theories", Nucl. Phys. **B497** (1997), 196.
127. S. Katz, A. Klemm, C. Vafa, "Geometric Engineering of Quantum Field Theories", Nucl. Phys. **B497** (1997), 173
128. M. Bershadsky, A. Johansen, T. Pantev, V. Sadov and C. Vafa, "F-Theory, Geometric Engineering and $N=1$ Dualities", **hep-th/9612052**.
129. A. Gorskii, I. Krichever, A. Marshakov, A. Mironov and A. Morozov, "Integrability and Exect Seiberg-Witten Solution", Phys. Lett. **B355** (1995), 466.
130. E. Martinec and N. P. Warner, "Integrable Systems and Supersymmetric Gauge Theories", Nucl. Phys. **B459** (1996), 97.
131. C. Gómez, "Elliptic Singularities, θ-Puzzle and Domain Wall", **hep-th/9711074**.
132. G. Dvali and M. Shifman, "Domain Walls in Strongly Coupled Theories", Phys. Lett. **B396** (1997), 64.
133. A. Kovner, M. Shifman and A. Smilga,"Domain Walls in Supersymmetric Yang-Mills Theories", **hep-th/9706089**.
134. A. Smilga and A. Vaselov, "Complex BPS Domain Walls and Phase Transition in Mass in Supersymmetric QCD", **hep-th/9706217** and "Domain Walls Zoo in Supersymmetric QCD", **hep-th/9710123**.
135. A. Kovner and M. Shifman, "Chirally Symmetric Phase of Supersymmetric Gluodynamics", **hep-th/9702174**.
136. V. A. Novikov, M. A. Shifman, A. I. Vainshtein and V. I. Zakharov, "Supersymmetric Instanton Calculus (Gauge Theories with Matter)", Nucl. Phys. **B260** (1985), 157.
137. W. Fulton, "*Intersection Theory*", Springer-Verlag, 1980.
138. A. Sen, "F-Theory and Orientifolds", Nucl. Phys. B475 (1996), 562.
139. E. Witten, "Current Algebra Theorems for the $U(1)$ 'Goldstone Boson' ", Nucl. Phys. B156 (1979), 269.
140. G. Veneziano, "$U(1)$ without Instantons", Nucl. Phys. B159 (1979), 213.
141. G. 't Hooft, "Dimensional Reduction in Quantum Gravity", **gr-qc/9310026**.
142. J. D. Bekenstein, "Black Holes and Entropy", Phys. Rev. **D7** (1973), 2333.

143. T. Banks, W. Fischler, S. H. Shenker and L. Susskind, "M-Theory as a Matrix Model: A Conjecture", Phys. Rev. **D55** (1997), 5112.
144. T. Banks, "Matrix Theory", **hep-th/9710231**.
145. W. Taylor, "D-Brane Field Theory on Compact Spaces", Phys.Lett. **B394** (1997), 283.
O. J. Ganor, S. Ramgoolam, W. Taylor IV, "Branes, Fluxes and Duality in M(atrix)-Theory", Nucl. Phys. **B492** (1997), 191.
146. M. Rozali, "Matrix Theory and U-Duality in Seven Dimensions", Phys. Lett. **B400** (1997), 260.
147. S. Sethi and L. Susskind, "Rotational Invariance in the M(atrix) Formulation of Type IIB Theory", Phys. Lett. **B400** (1997), 265.
148. A. Sen, "D0 Branes on T^n and Matrix Theory", **hep-th/9709220**.
149. N. Seiberg, "Why is the Matrix Model Correct?", **hep-th/9710009**.

q-Hypergeometric Functions and Representation Theory

Vitaly Tarasov

St.Petersburg Branch of Steklov Mathematical Institute

Introduction

Multidimensional hypergeometric functions are well known to be closely related to the representation theory of Kac-Moody Lie algebras and quantum groups, see [27], [35]. The basic way to connect these subjects goes via integral representation for solutions of the Knizhnik-Zamolodchikov (KZ) equations.

There are three the most essential points on this way. First, one consideres the twisted de Rham complex associated with a one-dimensional local system and its cohomology and homology groups, the pairing between the twisted differential forms and cycles producing multidimensional hypergeometric functions. The hypergeometric functions obey a system of differential equations which describes periodic sections of the Gauss-Manin connection associated with the local system.

The cohomology groups with coefficients in the local system admit a natural description in terms of the representation theory of Kac-Moody algebras, and it turns out that the system of difference equations for the hypergeometric functions induced by the Gauss-Manin connection can be identified with the KZ equation appearing in the representation theory.

The homology groups with coefficients in the local system have a natural description in terms of the representation theory of quantum groups. This reduces the problem of calculating monodromies of the KZ equation to a geometric problem of computing certain relations between twisted cycles. As a result, one can get a geometric proof of the Kohno-Drinfeld theorem that the monodromies of the KZ equation associated with a semisimple Lie algebra $\mathfrak{g}$ are given by the R-matrices associated with the quantum group $U_q(\mathfrak{g})$, see [35].

The quantized Knizhnik-Zamolodchikov (qKZ) equation is a difference analogue of the KZ equation and it turns into the KZ equation in a suitable limit. The qKZ equations had been introduced in [9] as equations for matrix elements of vertex operators of quantum affine algebras. An important special case of the qKZ equation had been considered earlier in [26] as equations for form factors in massive integrable models of quantum field theory. Later the qKZ equations were derived as equations for correlation functions in lattice integrable models, cf. [15] and references therein. Integral representation for solutions of the qKZ equation are closely related to diagonalization of the transfer-matrix of the corresponding lattice integrable model by the algebraic Bethe ansatz method [32].

Integral representation for solutions of the qKZ equation are studied in detail only in the simplest $\mathfrak{sl}_2$ case. But the form of the results suggests that it is plausible they can be naturally extended to the general case.

It is shown in [30], [31] that the geometric picture for the KZ equation can be naturally quantized and its quantization connects multidimensional q-hypergeometric functions with the representation theory of affine quantum groups, i.e. Yangians, quantum affine algebras and elliptic quantum groups. One can define the twisted difference de Rham complex associated with a one-dimensional discrete (difference) local system, and the corresponding cohomology and homology groups, the pairing between them producing q-hypergeometric functions. There is a difference analogue of the Gauss-Manin connection, the q-hypergeometric functions obeying a system of difference equations which describes periodic sections of the discrete Gauss-Manin connection associated with the discrete local system.

Both the cohomology and homology groups with coefficients in a discrete local system have a natural description in terms of the representation theory of affine quantum groups. This allows to identify the system of difference equations for the q-hypergeometric functions induced by the discrete Gauss-Manin connection with the qKZ equation, and to express the transition functions between asymptotic solutions of the qKZ equation via suitable R-matrices, which is analogous to the Kohno-Drinfeld theorem on the monodromies of the KZ equation. It is remarkable that in the difference case both the cohomology and homology groups can be represented as certain spaces of functions very similar to each other. So the quantization brings up more symmetry between "differential forms" and "cycles".

There are three versions of the above mentioned geometric construction, the rational, trigonometric and elliptic one. For the rational case considered in [31], q-hypergeometric functions are given by multidimensional integrals of Mellin-Barnes type, the qKZ equation is written in terms of the rational R-matrices, intertwiners of modules over the Yangian $Y(\mathfrak{sl}_2)$, and the transition functions are computed via the trigonometric R-matrices, intertwiners of modules over the quantum loop algebra $U_q(\widetilde{\mathfrak{sl}_2})$.

The trigonometric case is studied in [30]. In this case the qKZ equations is written in terms of the trigonometric R-matrices and the transition functions are expressed via the dynamical elliptic R-matrices, intertwiners of modules over the elliptic quantum group $E_{\rho,\gamma}(\mathfrak{sl}_2)$.

In the elliptic case, see [10], [11], the corresponding system of difference equations is some modification of the qKZ equation, the quantized Knizhnik-Zamolodchikov-Bernard ($qKZB$) equation. Its solutions are given by elliptic generalizations of q-hypergeometric functions. For the elliptic $qKZB$ equation one consideres a monodromy problem for the hypergeometric solutions instead of calculating the transition functions between asymptotic solutions;

for the $qKZB$ equation with the elliptic modulus ρ and the step α the monodromy problem produces the $qKZB$ equation with the elliptic modulus α and the step ρ. In a sense, this means that in the elliptic case we reach complete symmetry between cohomology and homology spaces.

In these lectures we review the relation of the q-hypergeometric functions and the representation theory of quantum groups via the qKZ equation in the trigonometric case. To avoid introducing a lot of notions and notations in advance we start the exposition from relatively elementary constructions formulating in Section3. a remarkable identity satisfied by multidimensional q-hypergeometric functions, the hypergeometric Riemann identity [28], see Theorem 3.1. In Section 4. we outline the proof of the hypergeometric Riemann identity. In particular, we write down a system of difference equations for q-hypergeometric functions and give asymptotics of the q-hypergeometric functions in a suitable asymptotic zone. In the sequel we explain the geometric origin of the system of difference equations as a periodic section equation for a discrete Gauss-Manin connection, as well as expose the relation with the representation theory of quantum loop algebra $U_q(\widetilde{\mathfrak{sl}_2})$ and the elliptic quantum group $E_{\rho,\gamma}(\mathfrak{sl}_2)$.

Technically, the relation with the representation theory is described via the so-called tensor coordinates. The trigonometric version of the tensor coordinates transforms the difference equations associated with the discrete Gauss-Manin connection to the qKZ equation, while the elliptic version of the tensor coordinates is responsible for expressing the transition functions between asymptotic solutions of the qKZ equation via the dynamical elliptic R-matrices.

In addition to the qKZ equation we consider the dual qKZ equation, solutions of the qKZ equation taking values in a tensor product of $U_q(\mathfrak{sl}_2)$-modules, and solutions of the dual qKZ equation taking values in the dual space. Using the geometric picture of the qKZ equation we identify the spaces of solutions of the qKZ and the dual qKZ equations with the tensor product of the corresponding modules over the elliptic quantum group $E_{\rho,\gamma}(\mathfrak{sl}_2)$ and its dual space, respectively. In this context the hypergeometric Riemann identity means that the hypergeometric solutions of the qKZ and dual qKZ equations transform the natural pairing of the spaces of solutions to the natural pairing of the target spaces; namely, given respective solutions Ψ and Ψ^* of the qKZ and dual qKZ equations we have that

$$\langle \mathit{Value}\Psi^*, \mathit{Value}\Psi \rangle_{\mathit{target\ spaces}} = \langle \Psi^*, \Psi \rangle_{\mathit{spaces\ of\ solutions}} .$$

In particular, one can say that the hypergeometric Riemann identity is a deformation of Gaudin-Korepin's formula for norms of the Bethe vectors [17], cf. [32].

We preface the discussion of the multidimensional case with the one-dimensional examples, both the differential and difference one. They are instructive for understanding the main ideas in a simpler context. In these examples we also pay attention to another appearance of the hypergeometric Riemann identity which is ignored in the multidimensional case. Namely, we show that the hypergeometric Riemann identity is an analogue of the Riemann bilinear relation for the (co)homology groups with coefficients in a one-dimensional local system and the dual local system. In other words, we obtain difference analogues of intersection forms of cohomology classes and of homology classes in the differential case. The deformation of the Riemann bilinear relation for the hyperelliptic Riemann surfaces was obtained in [25].

Though at first glance, the hypergeometric Riemann identity is the main topic of the lectures, in fact it plays a role of a stem which can be depicted using only things of a common mathematical knowledge, and which allows us to introduce more sophisticated themes in the process of its proof. At the moment the proof of the hypergeometric Riemann identity includes virtually all essential results concerning the geometric picture of the qKZ equation, so it serves well as an entering point to the subject.

The results presented in the paper were obtained as a part of the joint project developed by the author together with A. Varchenko /, University of North Carolina at Chapel Hill, and G. Felder, ETH Zürich. I am very grateful to them for fruitful collaboration and valuable discussions. I thank organizers of the CIME summer school at Cetraro for the kind opportunity to present the lectures, and all participants of the school for their interest.

1. One-dimensional differential example

In this section we consider an example of the Riemann bilinear relation for period matrices of the twisted one-dimensional de Rham (co)homologies. This example was considered in [4].

Let $z_1, \dots, z_n$ be pairwise distinct complex numbers. Fix noninteger complex numbers $\lambda_1, \dots, \lambda_n$ such that $\sum\limits_{m=1}^{n} \lambda_m \notin \mathbb{Z}$. Consider a one-dimensional local system

$$\Phi(t; z_1, \dots, z_n) = \prod_{m=1}^{n} (t - z_m)^{\lambda_m}.$$

The local system determines the twisted differential d_Φ and the twisted boundary operator ∂_Φ. Namely, for a function $f(t)$ we have

$$d_\Phi f = df + f\,\Phi^{-1} d\Phi = df + f \sum_{m=1}^{n} \lambda_m\, \omega_m$$

where

$$\omega_m = \frac{dt}{t - z_m}, \qquad m = 1, \dots, n,$$

and for a contour γ the twisted boundary $\partial_\Phi \gamma$ is such that for any rational function $g(t)$

$$\int_\gamma \Phi\, d_\Phi\, g \;=\; g|_{\partial_\Phi \gamma}\,.$$

Let $\widehat{\mathcal{F}}_{cl}$ be the space of rational functions in one variable which are regular in $\mathbb{C}\backslash\{z_1,\dots z_n\}$ and let $\mathcal{F}_{cl}\subset\widehat{\mathcal{F}}_{cl}$ be the subspace of functions with at most simple poles and vanishing at infinity.

Consider the holomorphic de Rham complex $(\Omega^\bullet(\widehat{\mathcal{F}}_{cl}), d_\Phi\,)$ on $\mathbb{C}\setminus\{z_1,\dots z_n\}$. For a form $\omega\in\Omega^1(\widehat{\mathcal{F}}_{cl})$ denote by $\lfloor\omega\rfloor_\Phi\in H^1(\Omega^\bullet, d_\Phi\,)$ its cohomology class.

The following statement is quite straightforward.

Proposition 1.1. $\dim H^1(\Omega^\bullet, d_\Phi\,) \;=\; n-1$. *Moreover,* $\lfloor\Omega^1(\mathcal{F}_{cl})\rfloor_\Phi \;=\; H^1(\Omega^\bullet, d_\Phi\,)$.

It is clear that the differential forms $\omega_1,\dots,\omega_n$ form a basis in $\Omega^1(\mathcal{F}_{cl})$ and

$$d_\Phi\, 1 \;=\; \sum_{m=1}^{n}\lambda_m\,\omega_m\,. \tag{1.1}$$

Set $\rho_m=\omega_m-\omega_{m+1}$, $m=1,\dots,n-1$.

Corollary 1.1. *The cohomology classes* $\lfloor\rho_1\rfloor_\Phi\,,\dots,\lfloor\rho_{n-1}\rfloor_\Phi$ *form a basis in* $H^1(\Omega^\bullet, d_\Phi\,)$.

A contour γ defines a linear functional $\langle\gamma,\cdot\rangle_\Phi$ on the space of differential forms $\Omega^1(\widehat{\mathcal{F}}_{cl})$ by the rule:

$$\langle\gamma,\omega\rangle_\Phi \;=\; \int_\gamma \Phi\,\omega\,.$$

If γ is a twisted cycle, i.e. $\partial_\Phi\,\gamma \;=\; 0$, then the functional $\langle\gamma,\cdot\rangle_\Phi$ can be considered as an element of the space $H_1(\Omega^\bullet, d_\Phi\,)$ of linear functionals on the cohomology space $H^1(\Omega^\bullet, d_\Phi\,)$. Below we give examples of twisted cycles; namely, we will describe certain important twisted cycles $\gamma_1,\dots,\gamma_{n-1}$.

To simplify notations, from now on we assume that $z_1,\dots,z_n$ are real and $z_1<,\dots,<z_n$. We denote

$$\xi_m \;=\; \exp(2\pi i\lambda_m)\,,\qquad\qquad m=1,\dots,n\,.$$

Fix a small positive number ε. Let $\Delta_1,\dots,\Delta_{n-1}$ be the following intervals:

$$\Delta_m=[z_m+\varepsilon, z_{m+1}-\varepsilon]$$

oriented from $z_m+\varepsilon$ to $z_{m+1}-\varepsilon$, and let $\Sigma_1^\pm,\dots,\Sigma_n^\pm$ be the counterclockwise oriented circles:

$$\Sigma_m^\pm=\{t\mid |t-z_m|=\varepsilon\}$$

starting at $z_m\pm\varepsilon$, respectively. Set

$$\gamma_m \;=\; \Delta_m+\frac{1}{1-\xi_{m+1}}\,\Sigma_{m+1}^-\,\frac{1}{1-\xi_m}\,\Sigma_m^+\,, \tag{1.2}$$

that is

$$\langle \gamma_m, \omega \rangle_\Phi = \int_{\Delta_m} \Phi > \omega + \frac{1}{1-\xi_{m+1}} \int_{\Sigma^-_{m+1}} \Phi > \omega - \frac{1}{1-\xi_m} \int_{\Sigma^+_m} \Phi > \omega \,,$$

and we fix branches of the functionΦ by the following prescriptions:

$$\begin{aligned} -\pi \le |\arg(t-z_k)| \le 0\,, \qquad & t \in \Delta_l\,, \qquad & k,l = 1,\dots,n\,, \\ |\arg(t-z_k)| < \pi/2\,, \qquad & t \in \Sigma^\pm_l\,, \qquad & 1 \le k < l \le n\,, \\ |\pi + \arg(t-z_k)| < \pi/2\,, \qquad & t \in \Sigma^\pm_l\,, \qquad & 1 \le l < k \le n\,, \\ 0 \le \arg(t-z_k) < 2\pi\,, \qquad & t \in \Sigma^+_k\,, \qquad & k = 1,\dots,n\,, \\ -\pi \le \arg(t-z_k) < \pi\,, \qquad & t \in \Sigma^-_k\,, \qquad & k = 1,\dots,n\,. \end{aligned}$$

It is easy to see that $\gamma_1,\dots,\gamma_{n-1}$ are twisted cycles: $\partial_\Phi \gamma_m = 0$, $m = 1,\dots,n-1$, and the functionals $\langle \gamma_1, \cdot \rangle \dots, \langle \gamma_{n-1}, \cdot \rangle_\Phi$ do not depend on ε. Moreover, the functionals $\langle \gamma_1, \cdot \rangle \dots, \langle \gamma_{n-1}, \cdot \rangle_\Phi$ do not change under coherent smooth deformations of $\Delta_1,\dots,\Delta_{n-1}$, $\Sigma^\pm_1,\dots,\Sigma^\pm_n$ in $\mathbb{C} \setminus \{z_1,\dots z_n$; namely, we assume that at any moment of the deformation there are pairwise distinct points $z^\pm_1,\dots,z^\pm_n \in \mathbb{C} \setminus \{z_1,\dots z_n$ such that each Δ_m is an arc going from z^+_m to z^-_{m+1} and each $\Sigma^\pm_m$ is a loop going around z_m counterclockwise starting and ending at $z^\pm_m$, respectively. Originally, we have $z^\pm_m = z_m \pm \varepsilon$, $m = 1,\dots,n$.

Remark 1.1. In what follows abusing notations we usually do not distinguish between a differential form ω and its cohomology class $\lfloor \omega \rfloor_\Phi$ as well as a cycle γ and the corresponding functional $\langle \gamma, \cdot \rangle_\Phi$, if it causes no confusion.

Proposition 1.2. *The twisted cycles $\gamma_1,\dots,\gamma_{n-1}$ form a basis in the homology space $H_1(\Omega^\bullet, d_\Phi)$.*

Consider the dual local system

$$\Phi'(t; z_1,\dots,z_n) = \prod_{m=1}^{n} (t-z_m)^{-\lambda_m} = \frac{1}{\Phi(t; z_1,\dots,z_n)}\,. \tag{1.3}$$

obtained via replacing $\lambda_1,\dots,\lambda_n$ by $-\lambda_1,\dots,-\lambda_n$. We define accordingly the twisted cycles $\gamma'_1,\dots,\gamma'_{n-1}$:

$$\gamma'_m = \Delta_m - \frac{\xi_{m+1}}{1-\xi_{m+1}} \Sigma^-_{m+1} + \frac{\xi_m}{1-\xi_m} \Sigma^+_m\,, \tag{1.4}$$

$$\langle \gamma'_m, \omega \rangle_{\Phi'} = \int_{\Delta_m} \Phi'\omega - \frac{\xi_{m+1}}{1-\xi_{m+1}} \int_{\Sigma^-_{m+1}} \Phi'\omega + \frac{\xi_m}{1-\xi_m} \int_{\Sigma^+_m} \Phi'\omega\,, \tag{1.5}$$

using the same intervals $\Delta_1,\dots,\Delta_{n-1}$ and circles $\Sigma^\pm_1,\dots,\Sigma^\pm_n$ as before.

The intersection numbers $\gamma_l \circ \gamma'_m$ can be calculated in a simple geometric way [20]; they are given below:

$$\gamma_m \circ \gamma'_m = \frac{(1-\xi_m)\xi_{m+1}}{(1-\xi_m)\,(1-\xi_{m+1})}\,, \qquad m=1,\dots,n\,, \tag{1.6}$$

$$\gamma_{m-1} \circ \gamma'_m = -\frac{1}{1-\xi_m}\,, \tag{1.7}$$

$$\gamma_m \circ \gamma'_{m-1} = -\frac{\xi_m}{1-\xi_m}\,, \qquad m=1,\dots,n-1\,, \tag{1.8}$$

$$\gamma_l \circ \gamma'_m = 0\,, \qquad l,m=1,\dots,n\,, \qquad |l-m|>1\,. \tag{1.9}$$

Notice that the matrix $g=(\gamma_l \circ \gamma'_m)^n_{l,m=1}$ is nondegenerate. Formulae 1.6 completely determine the intersection form

$$\circ\,:\,H_1(\Omega^\bullet, d_\Phi\,) \otimes H_1(\Omega^\bullet, d_{\Phi'}\,) \to \mathbb{C}\,. \tag{1.10}$$

It is known that there is a natural nondegenerate pairing

$$\bullet\,:\,H^1(\Omega^\bullet, d_\Phi\,) \otimes H^1(\Omega^\bullet, d_{\Phi'}\,) \to \mathbb{C}, \tag{1.11}$$

called the intersection form. Let us extend it to a pairing

$$\bullet\,:\,\Omega^1(\widehat{\mathcal{F}}_{cl}) \otimes \Omega^1(\widehat{\mathcal{F}}_{cl}) \to \mathbb{C}$$

by requirement

$$d_\Phi\, f \bullet g = f \bullet d_{\Phi'}\, g = 0\,. \tag{1.12}$$

The restriction $\bullet\,:\,\Omega^1(\mathcal{F}_{cl}) \otimes \Omega^1(\mathcal{F}_{cl}) \to \mathbb{C}$ was explicitly described in [4] and the intersection numbers of the differential forms $\rho_1,\dots,\rho_{n-1}$ were calculated therein:

$$\rho_m \bullet \rho_m = \frac{\lambda_m + \lambda_{m+1}}{\lambda_m\lambda_{m+1}}\,, \qquad m=1,\dots,n\,, \tag{1.13}$$

$$\rho_{m-1} \bullet \rho_m = \rho_m \bullet \rho_{m-1} = -\frac{1}{\lambda_m}\,, \qquad m=1,\dots,n-1\,, \tag{1.14}$$

$$\rho_l \bullet \rho_m = 0\,, \qquad l,m=1,\dots,n\,, \qquad |l-m|>1\,. \tag{1.15}$$

It is remarkable that the intersection numbers of the forms $\rho_1,\dots,\rho_{n-1}$ can be obtained by rational degeneration from the intersection numbers for the cycles $\gamma_1,\dots,\gamma_{n-1},\gamma'_1,\dots,\gamma'_{n-1}$. We will see a similar feature in the difference case, where it will be even more transparent.

Using relations 1.1, 1.12 and 1.13 one can find the intersection numbers of the forms $\omega_1,\dots,\omega_n$:

$$\omega_m \bullet \omega_m = \frac{1}{\lambda_m} - \frac{1}{\lambda_1 <,\dots,+\lambda_n}\,, \qquad m=1,\dots,n\,, \tag{1.16}$$

$$\omega_l \bullet \omega_m = -\frac{1}{\lambda_1 <,\dots,+\lambda_n}\,, \qquad l,m=1,\dots,n\,, \qquad l \neq m\,. \tag{1.17}$$

Now we can formulate the Riemann bilinear relation for the twisted de Rham (co)homologies.

Theorem 1.1. *[4] For any differential forms $\omega,\omega' \in \Omega^1(\widehat{\mathcal{F}}_{cl})$ the following relation hold:*

$$2\pi i(\omega \bullet \omega') = \sum_{l,m=1}^{n-1} (g^{-1})_{lm}\,\langle \gamma_l, \omega\rangle_\Phi\,\langle \gamma'_m, \omega'\rangle_{\Phi'}\,. \tag{1.18}$$

2. One-dimensional difference example

Let $\mathbb{C}^\times = \mathbb{C} \setminus \{0\}$. Fix $p \in \mathbb{C}^\times$ such that $|p| < 1$. Set $p^{\mathbb{Z}} = \{p^s \mid s \in \mathbb{Z}\}$. Let

$$(u)_\infty = (u;p)_\infty = \prod_{s=0}^{\infty} (1 - p^s u)$$

and let $\theta(u) = (u)_\infty (pu)_\infty (p)_\infty$ be the Jacobi theta-function.

For any vector space V we denote by V^* the space of linear functionals on V.

Fix nonzero complex numbers $x_1, \ldots, x_n$, $y_1, \ldots, y_n$. We will use the following compact notations : $x = (x_1, \ldots, x_n)$, $y = (y_1, \ldots, y_n)$. We assume that

$$x_l / y_m \notin p^{\mathbb{Z}}, \qquad x_l / x_m \notin p^{\mathbb{Z}}, \qquad y_l / y_m \notin p^{\mathbb{Z}},$$

for any $l, m = 1, \ldots, n$, $l \neq m$ in the second and third cases, and say that $x_1, \ldots, x_n$, $y_1, \ldots, y_n$ are generic if the above conditions hold.

Let $\widehat{\mathcal{F}}[x;y]$ be the space of rational functions in one variable with at most simple poles at points

$$p^s x_m, \qquad p^{-s-1} y_m, \qquad m = 1, \ldots, n, \qquad s \in \mathbb{Z}_{\geq \nu},$$

and any pole at zero. Let $\mathcal{F}[x], \mathcal{F}'[x] \subset \widehat{\mathcal{F}}[x;y]$ be the subspaces of functions regular in $\mathbb{C}P^1 \setminus \{x_1, \ldots, x_n\}$ and vanishing at zero or at infinity, respectively. It is clear that $\dim \mathcal{F}[x] = \dim \mathcal{F}'[x] = n$.

Fix $\alpha \in \mathbb{C}^\times$. For any function $f(t)$ set

$$D[x;y;\alpha] f(t) = \varphi(t;x;y;\alpha) f(pt) - f(t) \tag{2.1}$$

where

$$\varphi(t;x;y;\alpha) = \alpha \prod_{m=1}^{n} \frac{1 - t/y_m}{1 - t/x_m} . \tag{2.2}$$

We call functions of the form $D[x;y;\alpha]f$ the *twisted total differences.*

It is easy to see that the space $\widehat{\mathcal{F}}[x;y]$ is invariant with respect to the operator $D[x;y;\alpha]$. So we have a *discrete local system* on $\mathbb{C}^\times$ with the functional space $\widehat{\mathcal{F}}[x;y]$ and the connection coefficient $\varphi(\cdot;x;y;\alpha)$, see [30]. The top cohomology space $H^1[x;y;\alpha]$ of the discrete local system is canonically isomorphic to the quotient space $\widehat{\mathcal{F}}[x;y] / D[x;y;\alpha] \widehat{\mathcal{F}}[x;y]$, see [30]. In this paper we take the equality

$$H^1[x;y;\alpha] = \widehat{\mathcal{F}}[x;y] \,/\, D[x;y;\alpha] \widehat{\mathcal{F}}[x;y]$$

for the definition of the space $H^1[x;y;\alpha]$, and denote by

$$\lfloor \cdot \rfloor : \widehat{\mathcal{F}}[x;y] \to H^1[x;y;\alpha]$$

the canonical projection.

Say that α is generic if $\alpha \notin p^{\mathbb{Z}}$ and $\alpha \prod_{m=1}^{n} x_m/y_m \notin p^{\mathbb{Z}}$.

Proposition 2.1. *Let α, x, y be generic. Then $\dim H^1[x;y;\alpha] = n$. Moreover,*

$$\lfloor \mathcal{F}[x] \rfloor = \lfloor \mathcal{F}'[x] \rfloor = H^1[x;y;\alpha] .$$

The proof is straightforward.

The homology space $H_1[x;y;\alpha]$ of linear functionals on $H^1[x;y;\alpha]$ can be viewed as a subspace of $(\widehat{\mathcal{F}}[x;y])^*$ spanned by functionals annihilating the twisted total differences. We construct such functionals using the hypergeometric integral described below.

For any $z \in \mathbb{C}^\times$ denote by $\mathcal{C}[z]$ the counterclockwise oriented small circle around z. We assume that the circles $\mathcal{C}[z]$ and $\mathcal{C}[z']$ are outside each other if $z \neq z'$. Take a positive number r such that

$$r \neq |p^s x_m| , \qquad r \neq |p^s y_m| , \qquad m = 1, \ldots, n , \qquad s \in \mathbb{Z} .$$

Let $\mathbb{T}_r$ be the circle $\{ t \in \mathbb{C} \mid |t| = r \}$ oriented counterclockwise. Set

$$\widetilde{\mathbb{T}}_r[x;y] = \mathbb{T}_r + \sum_{m=1,\, s\in\mathbb{Z},\, |p^s x_m|>r}^{n} \mathcal{C}[p^s x_m] - \sum_{m=1,\, s\in\mathbb{Z},\, |p^s y_m|<r}^{n} \mathcal{C}[p^s y_m] . \qquad (2.3)$$

Let $\mathbb{U}[x;y] = \mathbb{C}^\times \setminus \{ p^s x_m , p^s y_m \mid m = 1, \ldots, n, s \in \mathbb{Z} \}$.

Lemma 2.1. *For any r, r' the contours $\widetilde{\mathbb{T}}_r[x;y]$ and $\widetilde{\mathbb{T}}_{r'}[x;y]$ are homologous in $\mathbb{U}[x;y]$.*

The proof is straightforward.

For any function $f(t)$ holomorphic in $\mathbb{U}[x;y]$ set

$$Int[x;y](f) = \frac{1}{2\pi i} \int_{\widetilde{\mathbb{T}}_r[x;y]} f(t) \frac{dt}{t} , \qquad (2.4)$$

if the integral converges. It is easy to see that the integral in the right hand side does not depend on r due to Lemma 2.1. Moreover, if a function f_p is given by $f_p(t) = f(pt)$, then for the same reason

$$Int[x;y](f_p) = Int[x;y](f) . \qquad (2.5)$$

Let $\mathcal{F}_{ell}[x;\alpha]$ be the space of functions $f(t)$ such that the product $f(t)\prod_{m=1}^{n} \theta(t/x_m)$ is a holomorphic function on $\mathbb{C}^\times$ and

$$f(pt) = \alpha f(t) .$$

One can see that $\dim \mathcal{F}_{ell}[x;\alpha] = n$, say by Laurent series.

Let $\Phi(t;x;y)$ be the following function:

$$\Phi(t;x;y) = \prod_{m=1}^{n} \frac{(t_a/x_m)_\infty}{(t_a/y_m)_\infty}. \tag{2.6}$$

It solves a difference equation

$$\Phi(pt;x;y) = \Phi(t;x;y) \prod_{m=1}^{n} \frac{1-t/y_m}{1-t/x_m}, \tag{2.7}$$

cf. (2.2). We call $\Phi(t;x;y)$ the it phase function. For any $w \in \widehat{\mathcal{F}}[x;y]$ and $W \in \mathcal{F}_{ell}[x;\alpha]$ set

$$I\,[x;y;\alpha]\,(w,W) = Int[x;y]\,\big(w\,W\,\Phi(\cdot;x;y)\big)\,. \tag{2.8}$$

We call the integral of this form the *hypergeometric integral.* Notice that for given functions w and W there is only a finite number of small circles in the definition (2.3) of the integration contour $\widetilde{\mathbb{T}}_r$ containing a pole of the integrand $\Phi(\cdot;x;y)\,w\,W$ inside and, therefore, contributing to the hypergeometric integral $Int[x;y]\,(\Phi(\cdot;x;y)\,w\,W)$.

The hypergeometric integral defines the *hypergeometric pairing*

$$\hat{I}\,[x;y;\alpha] \,:\, \widehat{\mathcal{F}}[x;y] \otimes \mathcal{F}_{ell}[x;\alpha] \to \mathbb{C}\,, f \otimes g \mapsto I\,[x;y;\alpha](f,g)\,. \tag{2.9}$$

We will also consider the corresponding linear map

$$\hat{I}\,[x;y;\alpha] \,:\, \mathcal{F}_{ell}[x;\alpha] \to \big(\widehat{\mathcal{F}}[x;y]\big)^*,$$

denoting it by the same letter.

Proposition 2.2. *For any $w \in \widehat{\mathcal{F}}[x;y]$ and $W \in \mathcal{F}_{ell}[x;\alpha]$ we have*

$$I\,[x;y;\alpha]\,(D[x;y;\alpha]\,w,W) = 0\,.$$

Proof . Formulae (2.1), (2.2) and (2.7) imply that

$$\Phi(t)\;D[\alpha]w(t)\;W(t) = \Phi(pt)\;w(pt)\;W(pt) - \Phi(t)\;w(t)\;W(t)\,.$$

Hence the statement follows from (2.8) and (2.5).

The last proposition means that $\hat{I}\,[x;y;\alpha]$ maps the space $\mathcal{F}_{ell}[x;\alpha]$ into the homology space $H_1[x;y;\alpha]$.

Theorem 2.1. *[30] For generic α, x, y the map*

$$\hat{I}\,[x;y;\alpha] \,:\, \mathcal{F}_{ell}[x;\alpha] \to H_1[x;y;\alpha]$$

is an isomorphism of vector spaces.

The theorem follows from Proposition [2.2] and Corollary [A.1] for $\ell = 1$.

To obtain a picture of the dual discrete local system we interchange the parameters $x_1, \dots, x_n$ with $y_1, \dots, y_n$ and replace α by α^{-1} in all definitions.

Define the *Shapovalov pairing* $S[x;y] : \mathcal{F}[x] \otimes \mathcal{F}'[y] \to \mathbb{C}$ by the rule:

$$S[x;y]\,(f,g) \;=\; Int[x;y]\,(fg)\,, \qquad f \in \mathcal{F}[x]\,, \quad g \in \mathcal{F}'[y]\,.$$

It is easy to check that

$$S[x;y]\,(f,g) \;=\; \sum_{m=1}^{n} \operatorname{Res}\,\big(f(t)\,g(t)\,t^{-1}\,dt\big)|_{t=x_m} \;= \tag{2.10}$$

$$=\; -\sum_{m=1}^{n} \operatorname{Res}\,\big(f(t)\,g(t)\,t^{-1}\,dt\big)|_{t=y_m}\,. \tag{2.11}$$

Similarly, we define the *elliptic Shapovalov pairing*

$$\mathcal{S}_{ell}[x;y;\alpha] \;:\; \mathcal{F}_{ell}[x;\alpha] \otimes \mathcal{F}_{ell}[y;\alpha^{-1}] \;\to\; \mathbb{C}\,, \tag{2.12}$$

$$\mathcal{S}_{ell}[x;y;\alpha]\,(f,g) \;=\; \sum_{m=1}^{n} \operatorname{Res}\,\big(f(t)\,g(t)\,t^{-1}\,dt\big)|_{t=x_m}\,, \tag{2.13}$$

$$f \in \mathcal{F}_{ell}[x;\alpha]\,, \quad g \in \mathcal{F}_{ell}[y;\alpha^{-1}]\,. \tag{2.14}$$

Since $f(pt)\,g(pt) = f(t)\,g(t)$, one can check that

$$\mathcal{S}_{ell}[x;y;\alpha]\,(fg) \;=\; -\sum_{m=1}^{n} \operatorname{Res}\,\big(f(t)\,g(t)\,t^{-1}\,dt\big)|_{t=y_m}\,. \tag{2.15}$$

We will also consider the Shapovalov pairings $S[x;y]$ and $\mathcal{S}_{ell}[x;y;\alpha]$ as linear maps

$$S[x;y] \;:\; \mathcal{F}'[y] \;\to\; \big(\mathcal{F}[x]\big)^{*}\,, \tag{2.16}$$

$$\mathcal{S}_{ell}[x;y;\alpha] \;:\; \mathcal{F}_{ell}[y;\alpha^{-1}] \;\to\; \big(\mathcal{F}_{ell}[x;\alpha]\big)^{*}\,. \tag{2.17}$$

Proposition 2.3. *[30] Let x, y be generic. Then the Shapovalov pairing $S[x;y]$ is nondegenerate. Moreover, for generic α the Shapovalov pairing $\mathcal{S}_{ell}[x;y;\alpha]$ is nondegenerate.*

Proof. We prove the nondegeneracy of the pairing $\mathcal{S}_{ell}$. The proof for the pairing S is similar. Consider functions $g_1, \dots, g_n \in \mathcal{F}_{ell}[x;\alpha]$:

$$g_m(t) \;=\; \frac{\theta(\alpha^{-1}t/x_m)}{\theta(t/x_m)}\,,$$

and functions $g'_1, \dots, g'_n \in \mathcal{F}_{ell}[y;\alpha^{-1}]$:

$$g'_m(t) \;=\; \frac{\theta(\tilde{\alpha}\,t/x_m)}{\theta(t/y_m)} \prod_{l=1,\,l\neq m}^{n} \frac{\theta(t/x_l)}{\theta(t/y_l)}$$

where $\widetilde{\alpha} = \alpha \prod_{m=1}^{n} x_m/y_m$. It is easy to see that $\mathcal{S}_{ell}(g_l, g'_m) = 0$ unless $l = m$ and

$$\mathcal{S}_{ell}(g_m, g'_m) = \frac{\theta(\alpha^{-1})\,\theta(\widetilde{\alpha})}{\theta'(1)} \prod_{l=1,\, l\neq m}^{n} \theta(x_m/x_l) \prod_{k=1}^{n} \frac{1}{\theta(x_m/y_k)}$$

where $\theta'(1) = \frac{d}{du}\theta(u)|_{u=1} = -\,(p)^3$. The last formulae clearly imply the claim.

Consider the following restrictions of the hypergeometric pairing $\hat{I}$:

$$I\,[x;y;\alpha] : \mathcal{F}[x] \otimes \mathcal{F}_{ell}[x;\alpha] \to \mathbb{C}, \qquad I'[x;y;\alpha] : \mathcal{F}'[y] \otimes \mathcal{F}_{ell}[y;\alpha^{-1}] \to \mathbb{C},$$

and the corresponding linear maps

$$I\,[x;y;\alpha] : \mathcal{F}_{ell}[x;\alpha] \to \left(\mathcal{F}[x]\right)^*, \qquad I'[x;y;\alpha] : \mathcal{F}_{ell}[y;\alpha^{-1}] \to \left(\mathcal{F}'[y]\right)^*,$$

denoting them by the same letters.

Theorem [2.2] describes an abstract form of the hypergeometric Riemann identity. It is an analogue of Theorem 1.1 in the differential case.

Theorem 2.2. *The following diagramm is commutative:*

$$\begin{array}{ccc}
\mathcal{F}_{ell}[x;\alpha] & \xrightarrow{\;I\,[x;y;\alpha]\;} & \left(\mathcal{F}[x]\right)^* \\
\Big\downarrow {\scriptstyle \left(\mathcal{S}_{ell}[x;y;\alpha]\right)^*} & & \Big\downarrow {\scriptstyle -\left(S\,[x;y]\right)} \\
\left(\mathcal{F}_{ell}[y;\alpha^{-1}]\right)^* & \xleftarrow[\;\left(I'[x;y;\alpha]\right)^*\;]{} & \mathcal{F}'[y]
\end{array}$$

As we will see in the multidimensional case, the proof of the hypergeometric Riemann identity does not involve Theorem 2.1. So, one gets an independent proof of Theorem 2.1 using Propositions 2.3 and Theorem 2.2.

Let α be generic. Then by Proposition Theorem 2.1 we have isomorphisms

$$\mathcal{F}[x] \approx H^1[x;y;\alpha]\,, \qquad \mathcal{F}'[y] \approx H^1[y;x;\alpha]\,,$$

$$\mathcal{F}_{ell}[x;\alpha] \approx H_1[x;y;\alpha]\,, \qquad \mathcal{F}_{ell}[y;\alpha^{-1}] \approx H_1[y;x;\alpha]\,,$$

which allow us to consider the pairing $S[x;y]$ as a nondegenerate pairing of the cohomology spaces and the pairing $\mathcal{S}_{ell}[x;y]$ as a nondegenerate pairing of the homology spaces, see Proposition 2.3. Moreover, Theorem 2.2 means that these pairings are difference analogues of the intersection forms (1.11) and (1.10), respectively.

To obtain an analytic form of the hypergeometric Riemann identity similar to the formula (1.18) we have to pick up bases of the vector spaces mentioned in the commutative diagram.

For any $m = 1, \ldots, n$ introduce the following functions:

$$w_m(t;x;y) = \frac{t}{t-x_m} \prod_{1\le l<m} \frac{t-y_l}{t-x_l} \in \mathcal{F}[x]\,, \tag{2.18}$$

$$w'_m(t;x;y) = \frac{y_m}{t-y_m} \prod_{1\le l<m} \frac{t-x_l}{t-y_l} \in \mathcal{F}'[y]\,,$$

$$W_m(t;x;y;\alpha) = \frac{\theta(\alpha_m^{-1} t/x_m)}{\theta(t/x_m)} \prod_{1\le l<m} \frac{\theta(t/y_l)}{\theta(t/x_l)} \in \mathcal{F}_{ell}[x;\alpha]\,,$$

$$W'_m(t;x;y;\alpha) = W(t;y;x;\alpha^{-1}) \in \mathcal{F}_{ell}[y;\alpha^{-1}]\,,$$

where $\alpha_m = \alpha \prod_{1\le l<m} x_l/y_l$.

Lemma 2.2. *Let x, y be generic. Then*

a) the functions $w_1(\cdot;x;y), \dots, w_n(\cdot;x;y)$ form a basis in the space $\mathcal{F}[x]$;
b) the functions $w'_1(\cdot;x;y), \dots, w'_n(\cdot;x;y)$ form a basis in the space $\mathcal{F}'[y]$;
c) The functions $W_1(\cdot;x;y;\alpha), \dots, W_n(\cdot;x;y;\alpha)$ form a basis in the space $\mathcal{F}_{ell}[x;\alpha]$ provided $\alpha \prod_{1\le l\le m} x_l/y_l \notin p^{\mathbb{Z}}$, $m = 1, \dots, n-1$.

Proof. We will give a proof of claim c). The proofs of claims a) and b) are similar.

Since $\dim \mathcal{F}_{ell} = n$ it suffices to show that the functions $W_1, \dots, W_n$ are linear independent. Observe that $W_m(y_l) = 0$ for $l < m$. Hence, the functions $W_1, \dots, W_n$ are linear independent if $\prod_{m=1}^{n} W_m(y_m) \ne 0$, which holds provided

$$\alpha \prod_{1\le l\le m} x_l/y_l \notin p^{\mathbb{Z}}, \qquad m = 1, \dots, n\,. \tag{2.19}$$

Similarly, res $W_m(x_l) = 0$ for $l > m$. Hence, the functions $W_1, \dots, W_n$ are linear independent if $\prod_{m=1}^{n} \operatorname{res} W_m(x_m) \ne 0$, which holds provided

$$\alpha \prod_{1\le l\le m} x_l/y_l \notin p^{\mathbb{Z}}, \qquad m = 0, \dots, n-1\,. \tag{2.20}$$

Since either (2.19) or (2.20) holds under the assumptions of the lemma, claim c) is proved.

Lemma 2.3. *$S(w_l, w'_m) = 0$ and $\mathcal{S}_{ell}(W_l, W'_m) = 0$ unless $l = m$. Moreover,*

$$S(w_m, w'_m) = \frac{y_m}{x_m - y_m}\,,$$

$$\mathcal{S}_{ell}(W_l, W'_m) = \frac{\theta(\alpha_m^{-1})\, \theta(\alpha_m\, x_m/y_m)}{\theta'(1)\, \theta(x_m/y_m)}$$

where $\alpha_m = \alpha \prod_{1\le l<m} x_l/y_l$ and $\theta'(1) = \frac{d}{du}\theta(u)|_{u=1} = -\,(p)^3_\infty$.

Proof. We prove the statement for the pairing $\mathcal{S}_{ell}$. The proof for the pairing S is similar. Observe that

$$\operatorname{Res}\,\bigl(W_l(t)\,W'_m(t)\bigr)|_{t=x_k} = 0 \qquad \text{unless} \qquad m < k < l\,,$$

hence, $\mathcal{S}_{ell}(W_l, W'_m) = 0$ unless $m < l$, see (2.12). Similarly,

$$\operatorname{Res}\,\bigl(W_l(t)\,W'_m(t)\bigr)|_{t=y_k} = 0 \qquad \text{unless} \qquad l < k < m\,,$$

hence, $\mathcal{S}_{ell}(W_l, W'_m) = 0$ unless $l < m$, see (2.15). This proves the first part of the statement. Moreover, from the above consideration we also have that

$$\mathcal{S}_{ell}(W_m, W'_m) \;=\; \operatorname{Res}\,\bigl(W_m(t)\,W'_m(t)\bigr)|_{t=x_m}$$

and the second part of the statement follows from the straightforward calculation.

Set $N_m \;=\; \dfrac{(p)^3_\infty\,\theta(x_m/y_m)}{\theta(\alpha_m^{-1})\,\theta(\alpha_m\,x_m/y_m)}$ where $\alpha_m = \alpha\prod_{1\le l<m} x_l/y_l$.
Theorem 2.2 is equivalent to each of the following formulae:

$$S(f,g) \;=\; \sum_{m=1}^{n} N_m\, I(f, W_m)\, I'(g, W'_m)$$

for any $f \in \mathcal{F}[x]$ and $g \in \mathcal{F}'[y]$, and

$$\mathcal{S}_{ell}(f,g) \;=\; \sum_{m=1}^{n}(1 - x_m/y_m)\, I(w_m, f)\, I'(w_m, g)$$

for any $f \in \mathcal{F}_{ell}[x;\alpha]$ and $g \in \mathcal{F}_{ell}[y;\alpha^{-1}]$. Set

$$X(t,u) \;=\; \sum_{m=1}^{n}(1 - x_m/y_m) w_m(t) w'_m(u)\,, \mathcal{X}_{ell}(t,u) \;=\; \sum_{m=1}^{n} N_m W_m(t) W'_m(u)\,. \tag{2.21}$$

Then we have

$$S(f,g) \;=\; I_t \otimes I_u\,\bigl(f(t)\,g(u), \mathcal{X}_{ell}(t,u)\bigr)$$

for any $f \in \mathcal{F}[x]$, $g \in \mathcal{F}'[y]$, and

$$\mathcal{S}_{ell}(f,g) \;=\; I_t \otimes I_u\,\bigl(X(t,u), f(t)\,g(u)\bigr)$$

for any $f \in \mathcal{F}_{ell}[x;\alpha]$, $g \in \mathcal{F}_{ell}[y;\alpha^{-1}]$. Here subscripts in $I_t \otimes I_u$ indicate the integration variables for the hypergeometric integrals.

There are explicit formulae for the functions $X(t,u)$ and $\mathcal{X}_{ell}(t,u)$.

Lemma 2.4.

$$X(t,u) = \frac{t}{t-u}\left(1-\prod_{m=1}^{n}\frac{(t-y_m)\,(u-x_m)}{(t-x_m)\,(u-y_m)}\right),$$

$$\mathcal{X}_{ell}(t,u) = \frac{(p)_\infty^3}{\theta(t/u)}\left(\frac{\theta(\alpha^{-1}\,t/u)}{\theta(\alpha^{-1})} - \frac{\theta(\widetilde{\alpha}^{-1}\,t/u)}{\theta(\widetilde{\alpha}^{-1})}\prod_{m=1}^{n}\frac{\theta(t/y_m)\,\theta(u/x_m)}{\theta(t/x_m)\,\theta(u/y_m)}\right)$$

where $\widetilde{\alpha} = \alpha \prod_{m=1}^{n} x_m/y_m$.

Proof. We prove the second formula of the lemma. The proof of the first formula is similar.

It suffices to prove the formula for $n = 1$. The general case easily follows from this particular case by induction with respect to n.

Let $n = 1$ and denote by $\widetilde{X}(t,u)$ the right hand side of the second formula. For any u the function $\widetilde{X}(\cdot,u)$ belongs to $\mathcal{F}_{ell}[x;\alpha]$. So $\widetilde{X}(\cdot,u)$ is proportional to W_1, cf. Lemma 2.2, the proportionality coefficient depending on u:

$$\widetilde{X}(t,u) = W_1(t)\,f(u)\,.$$

Substituting $t = y_1$ into the above equality one gets $f(u) = N_1\,W_1'(u)$. The lemma is proved.

Define a pairing $\widehat{S}[x;y;\alpha] : \widehat{\mathcal{F}}[x;y] \otimes \widehat{\mathcal{F}}[y;x] \to \mathbb{C}$ by the rule:

$$\widehat{S}(f,g) = I_t \otimes I_u\left(f(t)\,g(u), \mathcal{X}_{ell}(t,u)\right)$$

for $f \in \widehat{\mathcal{F}}[x;y]$, $g \in \widehat{\mathcal{F}}[y;x]$. Then we have that

$$\widehat{S}[x;y;\alpha]\left(D[x;y;\alpha]\,f,g\right) =$$

$$\widehat{S}[x;y;\alpha](f, D[x;y;\alpha^{-1}]\,g) = 0$$

for any $f \in \widehat{\mathcal{F}}[x;y]$, $g \in \widehat{\mathcal{F}}[y;x]$, and the restriciton of $\widehat{S}[x;y;\alpha]$ on $\mathcal{F}[x] \otimes \mathcal{F}'[y]$ coincides with the Shapovalov pairing $S[x;y]$. Therefore, the pairing $\widehat{S}[x;y]$ can be considered as a difference analogue of the intersection form

$$\bullet : \Omega^1(\widehat{\mathcal{F}}_{cl}) \otimes \Omega^1(\widehat{\mathcal{F}}_{cl}) \to \mathbb{C}$$

of the differential forms.

Notice that the spaces $\mathcal{F}[x]$ and $\mathcal{F}'[x]$ can be considered as degenerations of the space $\mathcal{F}_{ell}[x;\alpha]$ as $p \to 0$ and then $\alpha \to 0$ and $\alpha \to \infty$, respectively. The same degeneration connects the the function W_m with the functions w_m and w_m' as well as their Shapovalov pairings. This fact generalizes a similar observation for the intersections numbers $\gamma_l \circ \gamma'_m$ and $\rho_l \bullet \rho_m$ in the differential case.

In this section we are not striving for absolute similarity of formulae in the differential and difference examples. In fact, the similarity can be improved by a suitable modifications of words. One can also consider more sophisticated situation taking a special value of the parameter α, for instance, $\alpha = 1$ or $\alpha = \prod_{m=1}^{n} x_m/y_m$. But we will not do it in these lectures.

3. The hypergeometric Riemann identity

In the rest of the paper we consider the multidimensional difference case. Unlike the one-dimensional case we will start with introducing certain finite-dimensional spaces of functions called the hypergeometric spaces with pairings between them being given by the q-hypergeometric integrals, and formulate the hypergeometric Riemann identity. Then we will describe relations of the hypergeometric spaces and q-hypergeometric integrals to the (co)homology groups with coefficients in a discrete local system and periodic sections of the discrete Gauss-Manin connection. Further we will concentrate on the relation of the q-hypergeometric integrals with the representation theory of the quantum loop algebra $U'_q(\widetilde{\mathfrak{gl}_2})$ and the elliptic quantum group $E_{\rho,\gamma}(\mathfrak{sl}_2)$, the relation skipped from discussion in the one-dimensional difference example.

Though the exposition will be basically independent of the one-dimensional examples considered in the previous two sections we are going to use sometime the one-dimensional difference example as a helpful illustration of the general story. We omit almost all the proofs for the multidimensional case referring a reader to [30]. But to show the main technical ideas we will give some proofs for the one-dimensional case. It also seems to be instructive as an introduction to [30], since the one-dimensional example is not specially considered there.

The results of Sections 3. and 4. concerning the hypergeometric Riemann identity, see Theorems 3.1, 4.1, are due to [28].

3.1 Basic notations

Here we describe basic notations which are used all over the paper. Some of them were introduced already in Section 2., but we will mention them again, in order to make the rest of the paper selfcontained.

Let $\mathbb{C}^\times = \mathbb{C} \setminus \{0\}$. Fix $p \in \mathbb{C}^\times$ such that $|p| < 1$. For any $\mathbb{X} \subset \mathbb{Z}$ set $p^{\mathbb{X}} = \{ p^s \mid s \in \mathbb{X} \}$.

Let $(u)_\infty = (u;p)_\infty = \prod_{s=0}^{\infty}(1 - p^s u)$ and let $\theta(u) = (u)_\infty (pu)_\infty (p)_\infty$ be the Jacobi theta-function.

Fix a nonnegative integer ℓ. Take nonzero complex numbers η, $x_1, \dots, x_n$, $y_1, \dots, y_n$ called *parameters*. Say that the parameters are generic if for any $r = 0, \dots, \ell - 1$ and any $k, m = 1, \dots, n$, we have

$$\eta^{r+1} \notin p^{\mathbb{Z}}, \qquad \eta^{\pm r} x_k / x_m \notin p^{\mathbb{Z}}, \qquad \eta^{\pm r} y_k / y_m \notin p^{\mathbb{Z}} \qquad \text{for } k \neq m, \tag{3.1}$$
$$\eta^{\pm r} x_k / y_m \notin p^{\mathbb{Z}}.$$

All over the paper we assume that the parameters are generic, unless otherwise stated.

For any function$f(t_1, \dots, t_\ell)$ and any permutation $\sigma \in \mathbf{S}_\ell$ set

$$[\, f\,]_\sigma (t_1, \dots, t_\ell) = f(t_{\sigma_1}, \dots, t_{\sigma_\ell}) \prod_{1 \le a < b \le \ell} \sigma_a > \sigma_b \frac{t_{\sigma_b} - \eta\, t_{\sigma_a}}{\eta\, t_{\sigma_b} - t_{\sigma_a}}\,, \tag{3.2}$$

$$[\![\, f\,]\!]_{\sigma(t_1, \dots, t_\ell)} = f(t_{\sigma_1}, \dots, t_{\sigma_\ell}) \prod_{1 \le a < b \le \ell} \sigma_a > \sigma_b \frac{\eta\, \theta(\eta^{-1} t_{\sigma_b} / t_{\sigma_a})}{\theta(\eta\, t_{\sigma_b} / t_{\sigma_a})}\,. \tag{3.3}$$

Each of the formulae defines an action of the symmetric group $\mathbf{S}_\ell$.

For any $\mathfrak{l} = (\mathfrak{l}_1, \dots, \mathfrak{l}_\mathfrak{n}) \in \mathbb{Z}^\mathfrak{n}_{\ge 0}$ we set

$$\mathfrak{l}^\mathfrak{m} = \mathfrak{l}_1 <, \dots, + \mathfrak{l}_\mathfrak{m}\,, \qquad \mathfrak{m} = 1, \dots, \mathfrak{n}\,. \tag{3.4}$$

Let
$$\mathcal{Z}^n_\ell = \{\, \mathfrak{l} \in \mathbb{Z}^\mathfrak{n}_{\ge 0} \mid \mathfrak{l}_1 <, \dots, + \mathfrak{l}_\mathfrak{n} = \ell \,\}.$$

For any $\mathfrak{l}, \mathfrak{m} \in \mathcal{Z}^\mathfrak{n}_\ell$, $\mathfrak{l} \neq \mathfrak{m}$, we say that

$$\mathfrak{l} \ll \mathfrak{m} \qquad \text{if} \qquad \mathfrak{l}^\mathfrak{k} \le \mathfrak{m}^\mathfrak{k} \qquad \text{for any} \quad \mathfrak{k} = 1, \dots, \mathfrak{n} - 1\,, \tag{3.5}$$

which defines a partial ordering on the set $\mathcal{Z}^n_\ell$.

Remark 3.1. If we identify $\mathfrak{l} \in \mathbb{Z}^\mathfrak{n}_{\ge 0}$ with a partition $\mathfrak{l}^\mathfrak{n} <, \dots, \ge \mathfrak{l}^1$, then the introduced ordering on $\mathcal{Z}^n_\ell$ coincides with the inverse dominance ordering for the corresponding partitions of ℓ.

For any $x \in \mathbb{C}^n$ and $\mathfrak{l} \in \mathcal{Z}^\mathfrak{n}_\ell$ we define the point $x \triangleright \mathfrak{l}\,[\eta] \in \mathbb{C}^{\times \ell}$ as follows:

$$x \triangleright \mathfrak{l}\,[\eta] = (\eta^{1-\mathfrak{l}_1} \mathfrak{x}_1, \eta^{2-\mathfrak{l}_1} \mathfrak{x}_1, \dots, \mathfrak{x}_1, \qquad \eta^{1-\mathfrak{l}_2} x_2, \dots, x_2, \tag{3.6}$$
$$\dots, \qquad \eta^{1-\mathfrak{l}_\mathfrak{n}} x_n, \dots, x_n)\,.$$

For a function$f(t_1, \dots, t_\ell)$ and a point $t^\star = (t_1^\star, \dots, t_\ell^\star)$ we define a multiple residue $\operatorname{Res}\big(f(t)\,(dt/t)^\ell\,\big)|_{t=t^\star}$ by B

$$\operatorname{Res}\big(f(t)\,(dt/t)^\ell\,\big)|_{t=t^\star} = \tag{3.7}$$
$$= \operatorname{Res}\big(\dots \operatorname{Res}\big(f(t_1, \dots, t_\ell)\,(dt_\ell/t_\ell)\big)|_{t_\ell = t_\ell^\star} \cdots (dt_1/t_1)\big)|_{t_1 = t_1^\star}\,.$$

We often use in the paper the following compact notations:

$$t = (t_1, \dots, t_\ell)\,, \qquad x = (x_1, \dots, x_n)\,, \qquad y = (y_1, \dots, y_n)\,.$$

For any vector space V we denote by V^* the dual vector space, and for a linear operator A we denote by A^* the dual operator.

In this paper we extensively use results from [30]. The parameters $x_1,\dots,x_n$, $y_1,\dots,y_n$ in this paper correspond to the parameters $\xi_1,\dots,\xi_n$, $z_1,\dots,z_n$ in [30] as follows:

$$x_m = \xi_m z_m\,, \qquad y_m = \xi_m^{-1} z_m\,, \qquad m = 1,\dots,n\,. \tag{3.8}$$

3.2 The hypergeometric integral

Let $\widetilde{\Phi}(t;x;y;\eta)$ be the following function:

$$\widetilde{\Phi}(t;x;y;\eta) = \prod_{m=1}^{n}\prod_{a=1}^{\ell} \frac{1}{(x_m/t_a)_\infty\,(t_a/y_m)_\infty} \prod_{a,b=1\,a\neq b}^{\ell} \frac{1}{(\eta^{-1}t_a/t_b)_\infty}\,.$$

For any function$f(t_1,\dots,t_\ell)$ holomorphic in $\mathbb{C}^{\times\ell}$ we define below the *hypergeometric integral* $Int(f\,\widetilde{\Phi})$.

Assume that $|\eta|>1$ and $|x_m|<1$, $|y_m|>1$, $m=1,\dots,n$. Then we set

$$Int[x;y;\eta](f\,\widetilde{\Phi}) = \frac{1}{(2\pi i)^\ell\,\ell\,!}\int_{\mathbb{T}^\ell} f(t)\,\widetilde{\Phi}(t)\,(dt/t)^\ell \tag{3.9}$$

where $(dt/t)^\ell = \prod_{a=1}^{\ell} dt_a/t_a$ and $\mathbb{T}^\ell = \{\,t\in\mathbb{C}^\ell \mid |t_1|=1,\,\dots,|t_\ell|=1\,\}$. B We define $Int[x;y;\eta](f\,\widetilde{\Phi})$ for arbitrary values of the parameters η, $x_1,\dots,x_n$, $y_1,\dots,y_n$ by the analytic continuation with respect to the parameters.

Proposition 3.1. *For generic values of the parameters x,y,η, see (3.1), the hypergeometric integral $Int[x;y;\eta](f\,\widetilde{\Phi})$ is well defined and is a holomorphic function of the parameters.*

Proof. For generic values of the parameters x,y,η, the singularities of the integrand $f(t)\,\widetilde{\Phi}(t)$ are at most at the following hyperplanes:

$$t_a = 0\,, \qquad t_a = p^s x_m\,, \qquad t_a = p^s y_m\,, \qquad t_a = p^{-s}\eta\,t_b\,, \tag{3.10}$$

$a,b=1,\dots,\ell$, $a\neq b$, $m=1,\dots,n$, $s\in\mathbb{Z}_{\geq 0}$. The number of edges (nonempty intersections of the hyperplanes) of configuration (3.10) and dimensions of the edges are always the same for nonzero generic values of the parameters. Therefore, the topology of the complement in $\mathbb{C}^\ell$ of the union of the hyperplanes (3.10) does not change if the parameters are nonzero generic.

The rest of the proof is similar to the proof of Theorem 5.7 in [31].

It is clear from the proof of Proposition [3.1] that for generic values of the parameters the hypergeometric integral $Int[x;y;\eta](f\,\widetilde{\Phi})$ can be represented as an integral

$$Int[x;y;\eta](f\,\widetilde{\Phi}) = \frac{1}{(2\pi i)^\ell\,\ell!}\int_{\widetilde{\mathbb{T}}^\ell[x;y;\eta]} f(t)\,\widetilde{\Phi}(t;x;y;\eta)\,(dt/t)^\ell \tag{3.11}$$

where $\widetilde{\mathbb{T}}^\ell[x;y;\eta]$ is a suitable deformation of the torus $\mathbb{T}^\ell$ which does not depend on f.

It is easy to see that for $\ell = 1$ the present definition of $Int(f\,\widetilde{\Phi})$ is equivalent to the definition (2.4).

Remark 3.2. In what follows we are using the hypergeometric integrals $Int(f\,\widetilde{\Phi})$ only for symmetric functions f which have a certain particular form. In this case the hypergeometric integrals coincide with the symmetric A-type Jackson integrals, cf. Appendix B..

3.3 The hypergeometric spaces and the hypergeometric pairing

Let $\mathcal{F}[x;\eta;\ell]$ be the space of rational functions $f(t_1,\dots,t_\ell)$ such that the product

$$f(t_1,\dots,t_\ell)\prod_{a=1}^{\ell} t_a^{-1}\prod_{m=1}^{n}\prod_{a=1}^{\ell}(t_a - x_m)\prod_{1\le a<b\le\ell}\frac{\eta\,t_a - t_b}{t_a - t_b} \tag{3.12}$$

is a symmetric polynomial of degree less than n in each of the variables $t_1,\dots,t_\ell$. Elements of the space $\mathcal{F}[x;\eta;\ell]$ are invariant with respect to the action (3.2) of the symmetric group $\mathbf{S}_\ell$. Set

$$\mathcal{F}'[x;\eta;\ell] = \{\,f(t_1,\dots,t_\ell)\mid t_1\dots t_\ell\,f(t_1,\dots,t_\ell)\in\mathcal{F}[x;\eta^{-1};\ell]\,\}\,. \tag{3.13}$$

The spaces $\mathcal{F}$ and $\mathcal{F}'$ are called the *trigonometric hypergeometric spaces.*

Remark 3.3. There are quite a few motivations to call the spaces $\mathcal{F}$ and $\mathcal{F}'$ trigonometric, though no trigonometric functions will appear actually in the paper. To describe the origin of the name *trigonometric* let us change the variables:

$$\eta = \exp(\epsilon\,h)\,,\qquad t_a = \exp(\epsilon\,u_a)\,,\qquad x_m = \exp(\epsilon\,\xi_m)\,. \tag{3.14}$$

Then elements of the spaces $\mathcal{F}$ and $\mathcal{F}'$ are rational functions of exponentials, i.e. trigonometric functions. In the limit $\epsilon\to 0$ both the trigonometric hypergeometric spaces degenerate into the space of rational functions $\varphi(u_1,\dots,u_\ell)$ such that the product

$$\varphi(u_1,\dots,u_\ell)\prod_{m=1}^{n}\prod_{a=1}^{\ell}(u_a - \xi_m)\prod_{1\le a<b\le\ell}\frac{u_a - u_b + h}{u_a - u_b}$$

is a symmetric polynomial of degree less than n in each of the variables $u_1, \dots, u_\ell$. The last space is called the *rational hypergeometric space*, see [31].

Fix $\alpha \in \mathbb{C}^\times$. Let $\mathcal{F}_{ell}[x;\eta\,;\alpha;\ell\,]$ be the space of functions $g(t_1,\dots,t_\ell)$ such that

$$g(t_1,\dots,t_\ell)\prod_{m=1}^{n}\prod_{a=1}^{\ell}\theta(t_a/x_m)\prod_{1\le a<b\le\ell}\frac{\theta(\eta\,t_a/t_b)}{\theta(t_a/t_b)} \tag{3.15}$$

is a symmetric holomorphic function of $t_1,\dots,t_\ell$ in $\mathbb{C}^{\times\ell}$ and

$$g(t_1,\dots,pt_a,\dots,t_\ell) = \alpha\,\eta^{2-2a}\,g(t_1,\dots,t_\ell)\,. \tag{3.16}$$

Elements of the space $\mathcal{F}_{ell}[x;\eta\,;\alpha;\ell\,]$ are invariant with respect to the action (3.3) of the symmetric group $\mathbf{S}_\ell$. Set

$$\mathcal{F}'_{ell}[x;\eta\,;\alpha;\ell\,] = \mathcal{F}_{ell}[x;\eta^{-1};\alpha^{-1};\ell\,]\,.$$

The spaces $\mathcal{F}_{ell}$ and $\mathcal{F}'_{ell}$ are called the *elliptic hypergeometric spaces.*

Remark 3.4. The parameter α here is related to the parameter κ in [30]: $\alpha = \kappa\,\eta^{\ell-1}\prod_{m=1}^{n}\xi_m^{-1}$, cf. (3.8).

Proposition 3.2. *[30] For any α, η, $x_1,\dots,x_n$ we have that*

$$\dim\mathcal{F}[x;\eta\,;\ell\,] = \dim\mathcal{F}'[x;\eta\,;\ell\,] = \dim\mathcal{F}_{ell}[x;\eta\,;\alpha;\ell\,] = (\,n\,)+\ell-1n-1\,.$$

The statement follows from Lemma [A.2].

Proof. In what follows we do not indicate explicitly all arguments for the hypergeometric spaces and related maps if it causes no confusion. The suppressed arguments are supposed to be the same for all the spaces and maps involved.

Remark 3.5. The trigonometric hypergeometric spaces can be considered as degenerations of the elliptic hypergeometric spaces as $p\to 0$ and then $\alpha\to 0$. In this limit the function$\theta(v)$ turns into $(1-v)$ and the spaces $\mathcal{F}_{ell}[\alpha]$ and $\mathcal{F}'_{ell}[\alpha]$ degenerate into the spaces $\mathcal{F}$ and $\mathcal{F}'$, respectively. Because of this correspondence we use two slightly different versions of the trigonometric hypergeometric spaces.

Another degeneration of the elliptic hypergeometric space to the trigonometric one occurs after the change of variables (3.14) supplemented by

$$p = \exp(\epsilon\,\rho)\,, \qquad \alpha = \exp(\mu)\,, \qquad 0\le \operatorname{Im}\,\mu < 2\pi\,.$$

Then in the limit $\epsilon\to 0$ the function$\theta(e^{\epsilon\,u})$ is to be replaced by $\sin(\pi u/\rho)$ and the space $\mathcal{F}_{ell}[\alpha]$ turns into the space of functions $\psi(u_1,\dots,u_\ell)$ such that

$$\psi(u_1,\dots,u_\ell)\,\exp\bigl((\pi i\,n-\mu)\sum_{a=1}^{\ell} u_a/\rho\bigr)\;\times$$

$$\times\;\prod_{m=1}^{n}\prod_{a=1}^{\ell}\sin\bigl(\pi(u_a-\xi_m)/\rho\bigr)\prod_{1\le a<b\le\ell}\frac{\sin\bigl(\pi(u_a-u_b+h)/\rho\bigr)}{\sin\bigl(\pi(u_a-u_b)/\rho\bigr)}$$

is a symmetric polynomial of degree less than n in each of the variables $e^{2\pi i u_1},\dots,e^{2\pi i u_\ell}$. This is another version of the trigonometric hypergeometric space, see [31].

Let $\Phi(t;x;y;\eta)$ be the following function:

$$\Phi(t;x;y;\eta)\;=\;\prod_{m=1}^{n}\prod_{a=1}^{\ell}\frac{(t_a/x_m)}{(t_a/y_m)}\prod_{1\le a<b\le\ell}\frac{(\eta\,t_a/t_b)}{(\eta^{-1}t_a/t_b)}\,. \tag{3.17}$$

We call $\Phi(t;x;y;\eta)$ the *phase function.* Notice that

$$\Phi(t;y;x;\eta^{-1})\;=\;\bigl(\Phi(t;x;y;\eta)\bigr)^{-1}.$$

The hypergeometric integral (3.11) induces the *hypergeometric pairings* of the trigonometric and elliptic hypergeometric spaces:

$$\begin{aligned}
I\,[x;y;\eta\,;\alpha]\;&:\;\mathcal{F}[x;\eta\,]\otimes\mathcal{F}_{ell}[x;\eta\,;\alpha]\;\to\;\mathbb{C}\,,\\
f\otimes g\;&\mapsto\;Int[x;y;\eta\,]\bigl(fg\,\Phi(\cdot\,;x;y;\eta)\bigr)\,,\\
I'[x;y;\eta\,;\alpha]\;&:\;\mathcal{F}'[y;\eta\,]\otimes\mathcal{F}'_{ell}[y;\eta\,;\alpha]\;\to\;\mathbb{C}\,,\\
f\otimes g\;&\mapsto\;Int[y;x;\eta^{-1}]\bigl(fg\,\Phi(\cdot\,;y;x;\eta^{-1})\bigr)\,.
\end{aligned} \tag{3.18}$$

We also consider these pairings as linear maps denoting them by the same letters:

$$\begin{aligned}
I\,[x;y;\eta\,;\alpha]\;&:\;\mathcal{F}_{ell}[x;\eta\,;\alpha]\;\to\;\bigl(\mathcal{F}[x;\eta\,]\bigr)^*,\\
I'[x;y;\eta\,;\alpha]\;&:\;\mathcal{F}'_{ell}[y;\eta\,;\alpha]\;\to\;\bigl(\mathcal{F}'[y;\eta\,]\bigr)^*.
\end{aligned} \tag{3.19}$$

Remark 3.6. In this paper we multiply the hypergeometric pairings by an additional factor $\dfrac{1}{(2\pi i)^\ell\,\ell\,!}$ compared with the hypergeometric pairing in [[30]].

Proposition 3.3. *[[30]] Let the parameters x,y,η be generic. Assume that*

$$\alpha\ne p^s\eta^r\,,\,\alpha\,\eta^{2-2\ell}\prod_{m=1}^{n}x_m/y_m\ne p^{-s-1}\eta^{-r}\,,\,r=0,\dots,\ell-1\,,\,s\in\mathbb{Z}_{\ge0}\,.$$

Then the hypergeometric pairing $I\,[x;y;\eta\,;\alpha]$ is nondegenerate/.

The statement follows from Corollary [A.1].

Corollary 3.1. *Let the parameters x, y, η be generic. Assume that*

$$\alpha \neq p^{-s-1}\eta^r \,, \alpha\,\eta^{2-2\ell} \prod_{m=1}^{n} x_m/y_m \neq p^s\eta^{-r} \,, r = 0, \ldots, \ell-1 \,, s \in \mathbb{Z}_{\geq 0} .$$

Then the hypergeometric pairing $I'[x; y; \eta\,; \alpha]$ is nondegenerate/.

Proof. Let π be the following map: $\pi : f(t_1, \ldots, t_\ell) \mapsto t_1 \ldots t_\ell\, f(t_1, \ldots, t_\ell)$. Then the next diagram is commutative

$$\begin{array}{ccc} \mathcal{F}_{ell}[y; \eta^{-1}; p^{-1}\alpha^{-1}] & \xrightarrow{I\,[y;x;\eta^{-1};p^{-1}\alpha^{-1}]} & \big(\mathcal{F}[y; \eta^{-1}]\big)^* \\ \Big\downarrow{\scriptstyle \pi} & & \Big\downarrow{\scriptstyle \pi^*} \\ \mathcal{F}'_{ell}[y; \eta\,; \alpha] & \xrightarrow{I'[x;y;\eta\,;\alpha]} & \big(\mathcal{F}'[y; \eta\,]\big)^* \end{array} \tag{3.20}$$

and the vertical arrows are invertible, which proves the statement.

3.4 The Shapovalov pairings

For any function$f(t_1, \ldots, t_\ell)$ set

$$\mathit{Res}\,[x; \eta\,](f) \;=\; \sum_{\mathfrak{m} \in \mathcal{Z}_\ell^n} \mathrm{Res}\,\big(f(t_1, \ldots, t_\ell)\,(dt/t)^\ell\,\big)\big|_{t = x \triangleright \mathfrak{m}[\eta]} \tag{3.21}$$

where the points $x \triangleright \mathfrak{l}\,[\eta\,] \in \mathbb{C}^{\times \ell}$, $\mathfrak{l} \in \mathcal{Z}_\ell^n$, are defined by (3.6) and the multiple residue is defined by (3.7).

Lemma 3.1. *For any $f \in \mathcal{F}[x; \eta\,; \ell\,]$ and $g \in \mathcal{F}'[y; \eta\,; \ell\,]$ we have*

$$\mathit{Int}\,[x; y; \eta\,](fg) \;=\; \mathit{Res}\,[x; \eta\,](fg) \;=\; (-1)^\ell \mathit{Res}\,[y; \eta^{-1}](fg)\,.$$

The statement is equivalent to Lemma C.8 in [[30]].

Lemma 3.2. *For any $f \in \mathcal{F}_{ell}[x; \eta\,; \alpha; \ell\,]$ and $g \in \mathcal{F}'_{ell}[y; \eta\,; \alpha; \ell\,]$ we have*

$$\mathit{Res}\,[x; \eta\,](fg) \;=\; (-1)^\ell \mathit{Res}\,[y; \eta^{-1}](fg)\,.$$

The statement is equivalent to Lemma C.3 in [[30]].

We define the *Shapovalov pairings* of the trigonometric and elliptic hypergeometric spaces as follows:

$$\begin{aligned} &S\,[x; y; \eta\,] \;:\; \mathcal{F}[x; \eta\,; \ell\,] \otimes \mathcal{F}'[y; \eta\,; \ell\,] \;\to\; \mathbb{C}, \\ &f \otimes g \;\mapsto\; \mathit{Res}\,[x; \eta\,](fg)\,, \\ &S_{ell}[x; y; \eta\,; \alpha] \;:\; \mathcal{F}_{ell}[x; \eta\,; \alpha; \ell\,] \otimes \mathcal{F}'_{ell}[y; \eta\,; \alpha; \ell\,] \;\to\; \mathbb{C}, \\ &f \otimes g \;\mapsto\; \mathit{Res}\,[x; \eta\,](fg)\,. \end{aligned} \tag{3.22}$$

We also consider these pairings as linear maps, denoting them by the same letters:

$$S[x;y;\eta] : \mathcal{F}'[y;\eta] \to (\mathcal{F}[x;\eta])^*, \tag{3.23}$$

$$\mathcal{S}_{ell}[x;y;\eta;\alpha] : \mathcal{F}'_{ell}[y;\eta;\alpha] \to (\mathcal{F}_{ell}[x;\eta;\alpha])^*.$$

Proposition 3.4. *For generic x, y, η the Shapovalov pairing $S[x;y;\eta]$ is nondegenerate/.*

The statement follows from either Lemma [4.1] or Proposition [A.3].

Proposition 3.5. *Let the parameters x, y, η be generic. Assume that*

$$\alpha\,\eta^{-r} \notin p^{\mathbb{Z}}, \qquad \alpha\,\eta^{r+2-2\ell} \prod_{m=1}^{n} x_m/y_m \notin p^{\mathbb{Z}}, \qquad r = 0, \dots, \ell - 1.$$

Then the Shapovalov pairing $\mathcal{S}_{ell}[x;y;\eta;\alpha]$ is nondegenerate/.

The statement follows from Proposition [A.3].

3.5 The hypergeometric Riemann identity

Now we formulate the main result of this section, the hypergeometric Riemann identity which involves both the hypergeometric and Shapovalov pairings. We prove this result in the next section.

Theorem 3.1. *Let the parameters x, y, η be generic. Then the following diagram is commutative:*

$$\begin{array}{ccc}
\mathcal{F}_{ell}[x;\eta;\alpha] & \xrightarrow{\quad I[y;x;\eta;\alpha] \quad} & (\mathcal{F}[x;\eta])^* \\
{\scriptstyle (\mathcal{S}_{ell}[x;y;\eta;\alpha])^*}\big\downarrow & & \big\downarrow{\scriptstyle (-1)^{\ell}\,(S[x;y;\eta])^{-1}} \\
(\mathcal{F}'_{ell}[y;\eta;\alpha])^* & \xrightarrow{\quad (I'[x;y;\eta;\alpha])^* \quad} & \mathcal{F}'[y;\eta]
\end{array}$$

This theorem generalizes Theorem [2.2] in the one-dimensional case.

Given bases of the hypergeometric spaces, Theorem [3.1] translates into bilinear relations for the corresponding hypergeometric integrals. In the next section we describe an important example of the bases — the bases given by the weight functions (4.1) – (4.3).

4. Tensor coordinates on the hypergeometric spaces and the hypergeometric maps

In this section we give a proof of the hypergeometric Riemann identity. We introduce the bases of weight functions in the trigonometric and elliptic hypergeometric spaces, and using these bases define the tensor coordinates on the hypergeometric spaces, see (4.7). Furthermore, we introduce the hypergeometric maps, see (4.11), and formulate an equivalent form of the hypergeometric Riemann identity, see Theorem [4.1]. To prove the identity we use the fact that the hypergeometric maps satisfy certain systems of difference equations, cf. (4.19), (4.20), and then study asymptotics of the hypergeometric maps in a suitable asymptotic zone.

4.1 Bases of the hypergeometric spaces

For any $\mathfrak{l} \in \mathcal{Z}_\ell^{\mathfrak{n}}$ define the functions $w_\mathfrak{l}$ and $W_\mathfrak{l}$ by the formulae:

$$w_\mathfrak{l}(t;x;y;\eta) = \tag{4.1}$$

$$= \prod_{m=1}^{n}\prod_{s=1}^{\mathfrak{l}_m} \frac{1-\eta}{1-\eta^s} \sum_{\sigma\in\mathbf{S}_\ell} \Bigl[\prod_{m=1}^{n}\prod_{a\in\Gamma_m}\Bigl(\frac{t_a}{t_a-x_m}\prod_{1\le l<m}\frac{t_a-y_l}{t_a-x_l}\Bigr)\Bigr]_\sigma,$$

$$W_\mathfrak{l}(t;x;y;\eta\,;\alpha) = \prod_{m=1}^{n}\prod_{s=1}^{\mathfrak{l}_m}\frac{\theta(\eta)}{\theta(\eta^s)} \times \tag{4.2}$$

$$\times \sum_{\sigma\in\mathbf{S}_\ell}\Bigl[\Bigl[\prod_{m=1}^{n}\prod_{a\in\Gamma_m}\Bigl(\frac{\theta(\eta^{2a-2}\alpha_m^{-1}t_a/x_m)}{\theta(t_a/x_m)}\prod_{1\le l<m}\frac{\theta(t_a/y_l)}{\theta(t_a/x_l)}\Bigr)\Bigr]\Bigr]_\sigma$$

where $\Gamma_m = \{1+\mathfrak{l}^{m-1},\dots,\mathfrak{l}^m\}$ and $\alpha_m = \alpha\prod_{1\le l<m}x_l/y_l$, $m=1,\dots,n$. Set

$$w'_\mathfrak{l}(t;x;y;\eta) = \prod_{m=1}^{n} y_m^{\mathfrak{l}_m}\prod_{a=1}^{\ell} t_a^{-1}\,w_\mathfrak{l}(t;y;x;\eta^{-1}), \tag{4.3}$$
$$W'_\mathfrak{l}(t;x;y;\eta\,;\alpha) = W_\mathfrak{l}(t;y;x;\eta^{-1};\alpha^{-1}).$$

The functions $w_\mathfrak{l}$, $w'_\mathfrak{l}$ and $W_\mathfrak{l}$, $W'_\mathfrak{l}$ are called the *trigonometric* and *elliptic weight functions*, respectively. The weight functions in the one-dimensional case are given by (2.18).

Proposition 4.1. *[30] For generic x, y, η the functions $\{\,w_\mathfrak{l}(t;x;y;\eta)\,\}_{\mathfrak{l}\in\mathcal{Z}_\ell^\mathfrak{n}}$ form a basis in the trigonometric hypergeometric space $\mathcal{F}[x;\eta]$.*

The statement follows from Proposition [A.1].

Corollary 4.1. *For generic x, y, η the functions $\{\,w'_\mathfrak{l}(t;x;y;\eta)\,\}_{\mathfrak{l}\in\mathcal{Z}_\ell^\mathfrak{n}}$ form a basis in the trigonometric hypergeometric space $\mathcal{F}'[y;\eta]$.*

Proposition 4.2. *[30] Let the parameters x, y, η be generic. Assume that*

$$\alpha\,\eta^{-r}\textstyle\prod_{1\le l\le m} x_l/y_l \notin p^{\mathbb{Z}}, \qquad m = 1, \dots, n-1, \quad r = 0, \dots, 2\ell - 2\,. \tag{4.4}$$

Then the functions $\{\, W_{\mathfrak{l}}(t; x; y; \eta\,; \alpha) \,\}_{\mathfrak{l}\in\mathcal{Z}_\ell^n}$ form a basis in the elliptic hypergeometric space $\mathcal{F}_{ell}[x; \eta\,; \alpha]$.

The statement follows from Proposition [A.2].

Corollary 4.2. *Under the assumptions of Proposition [4.2] the functions $\{\, W'_{\mathfrak{l}}(t; x; y; \eta\,; \alpha) \,\}_{\mathfrak{l}\in\mathcal{Z}_\ell^n}$ form a basis in the elliptic hypergeometric space $\mathcal{F}'_{ell}[x; \eta\,; \alpha]$.*

The bases $\{\, w_{\mathfrak{l}} \,\}$ and $\{\, w'_{\mathfrak{l}} \,\}$ of the trigonometric hypergeometric space are biorthogonal with respect to the Shapovalov pairing (3.22), and the same holds for the bases $\{\, W_{\mathfrak{l}} \,\}$, $\{\, W'_{\mathfrak{l}} \,\}$ of the elliptic hypergeometric spaces, see the next lemma.

Lemma 4.1. *[30]*

$$S\,(w_{\mathfrak{l}}, w'_{\mathfrak{m}}) \;=\; \delta_{\mathfrak{l}\,\mathfrak{m}} \prod_{m=1}^{n} \prod_{s=0}^{\mathfrak{l}_m - 1} \frac{(1-\eta)\,\eta^s\, y_m}{(1-\eta^{s+1})\,(x_m - \eta^s y_m)}\,, \tag{4.5}$$

$$S_{ell}(W_{\mathfrak{l}}, W'_{\mathfrak{m}}) \;=\; \delta_{\mathfrak{l}\,\mathfrak{m}} \prod_{m=1}^{n} \prod_{s=0}^{\mathfrak{l}_m - 1} \frac{\eta^s\,\theta(\eta)\,\theta(\eta^s \alpha_{\mathfrak{l},m}^{-1})\,\theta(\eta^{1-s-\mathfrak{l}_m} \alpha_{\mathfrak{l},m}\, x_m/y_m)}{\theta'(1)\,\theta(\eta^{s+1})\,\theta(\eta^{-s} x_m/y_m)}$$

where $\alpha_{\mathfrak{l},m} = \alpha \prod_{1\le j<m} \eta^{-2\mathfrak{l}_j} x_j/y_j$ *and* $\theta'(1) = \dfrac{d}{du}\theta(u)|_{u=1} \;=\; -\,(p)^3_\infty$.

Proof. One can see from the definition of the weight functions that

$$\begin{aligned} \operatorname{Res}\,\bigl(w_{\mathfrak{l}}(t; x; y; \eta)\,(dt/t)^\ell\,\bigr)|_{t=x\triangleright\mathfrak{n}[\eta]} &= 0 \quad \text{unless } \mathfrak{l} \ll \mathfrak{n} \text{ or } \mathfrak{l} = \mathfrak{n}, \\ w_{\mathfrak{l}}(t; x; y; \eta)|_{t=y\triangleright\mathfrak{n}[\eta^{-1}]} &= 0 \quad \text{unless } \mathfrak{l} \gg \mathfrak{n} \text{ or } \mathfrak{l} = \mathfrak{n}, \end{aligned} \tag{4.6}$$

and similarly,

$$\begin{aligned} \operatorname{Res}\,\bigl(w'_{\mathfrak{m}}(t; x; y; \eta)\,(dt/t)^\ell\,\bigr)|_{t=y\triangleright\mathfrak{n}[\eta^{-1}]} &= 0 \quad \text{unless } \mathfrak{m} \ll \mathfrak{n} \text{ or } \mathfrak{m} = \mathfrak{n}, \\ w'_{\mathfrak{m}}(t; x; y; \eta)|_{t=x\triangleright\mathfrak{n}[\eta]} &= 0 \quad \text{unless } \mathfrak{m} \gg \mathfrak{n} \text{ or } \mathfrak{m} = \mathfrak{n}. \end{aligned}$$

Therefore, by (3.22) and (3.21) we have $S(w_{\mathfrak{l}}, w'_{\mathfrak{m}}) = 0$ unless $\mathfrak{l} \ll \mathfrak{m}$ or $\mathfrak{l} = \mathfrak{m}$, while using in addition Lemma [3.1] we obtain that $S(w_{\mathfrak{l}}, w'_{\mathfrak{m}}) = 0$ unless $\mathfrak{l} \gg \mathfrak{m}$ or $\mathfrak{l} = \mathfrak{m}$. Moreover, we get

$$S(w_{\mathfrak{l}}, w'_{\mathfrak{l}}) \;=\; \operatorname{Res}\,\bigl(w_{\mathfrak{l}}(t)\; w'_{\mathfrak{l}}(t)\;(dt/t)^\ell\,\bigr)|_{t=x\triangleright\mathfrak{l}[\eta]}\,,$$

and the straightforward calculation of the residue yields the right hand side of the first formula (4.5). The proof of the second formula is similar.

4.2 Tensor coordinates and the hypergeometric maps

Let $V = \bigoplus_{\mathfrak{m}\in\mathcal{Z}_\ell^n} \mathbb{C}\, v_{\mathfrak{m}}$ and let $V^* = \bigoplus_{\mathfrak{m}\in\mathcal{Z}_\ell^n} \mathbb{C}\, v^*_{\mathfrak{m}}$ be the dual space. Denote by $\langle\,,\rangle$ the canonical pairing: $\langle v^*_{\mathfrak{l}}, v_{\mathfrak{m}}\rangle = \delta_{\mathfrak{l}\,\mathfrak{m}}$.

Introduce the *tensor coordinates* on the hypergeometric spaces, cf. [30]. They are the following linear maps:

$$
\begin{aligned}
B[x;y;\eta] : V^* &\to \mathcal{F}[x;\eta]\,, & B_{ell}[x;y;\eta\,;\alpha] : V^* &\to \mathcal{F}_{ell}[x;\eta\,;\alpha]\,,\\
v^*_{\mathfrak{m}} &\mapsto w_{\mathfrak{m}}(t;x;y;\eta)\,, & v^*_{\mathfrak{m}} &\mapsto W_{\mathfrak{m}}(t;x;y;\eta\,;\alpha)\,,\\
B'[x;y;\eta] : V^* &\to \mathcal{F}'[y;\eta]\,, & B'_{ell}[x;y;\eta\,;\alpha] : V^* &\to \mathcal{F}'_{ell}[y;\eta\,;\alpha]\,,\\
v^*_{\mathfrak{m}} &\mapsto w'_{\mathfrak{m}}(t;x;y;\eta)\,, & v^*_{\mathfrak{m}} &\mapsto W'_{\mathfrak{m}}(t;x;y;\eta\,;\alpha)\,.
\end{aligned}
\tag{4.7}
$$

Under the assumptions of Propositions [4.1] and [4.2] the tensor coordinates are isomorphisms of the respective vector spaces.

Remark 4.1. The tensor coordinates used in this paper differ from the tensor coordinates in [30] by normalization factors.

The tensor coordinates and the Shapovalov pairings (3.23) induce bilinear forms on the spaces V and V^*:

$$
\begin{aligned}
(\,,)[x;y;\eta]\, : V\otimes V &\to \mathbb{C}, & (\!(\,,)\!)[x;y;\eta\,;\alpha]\, : V^*\otimes V^* &\to \mathbb{C},\\
(u,v) &= \langle B'^{\,-1} S^{\,-1} B^{*-1} u, v\rangle\,, & (\!(u,v)\!) &= (-1)^\ell \langle u, (B_{ell})^* \mathcal{S}_{ell}\, B'_{ell}\, v\rangle\,.
\end{aligned}
\tag{4.8}
$$

We omit the common arguments in the second line. The explicit formulae for these pairings are:

$$
(v_{\mathfrak{l}}, v_{\mathfrak{m}}) \;=\; \delta_{\mathfrak{l}\,\mathfrak{m}} \prod_{m=1}^{n} \prod_{s=0}^{\mathfrak{l}_m-1} \frac{(1-\eta^{s+1})\,(x_m-\eta^s y_m)}{(1-\eta)\,\eta^s\, y_m}\;, \tag{4.9}
$$

$$
(\!(v^*_{\mathfrak{l}}, v^*_{\mathfrak{m}})\!) \;=\; \delta_{\mathfrak{l}\,\mathfrak{m}} \prod_{m=1}^{n} \prod_{s=0}^{\mathfrak{l}_m-1} \frac{\eta^s\,\theta(\eta)\,\theta(\eta^s\alpha^{-1}_{\mathfrak{l},\mathfrak{m}})\,\theta(\eta^{1-s-\mathfrak{l}_m}\alpha_{\mathfrak{l},\mathfrak{m}}\, x_m/y_m)}{\theta(\eta^{s+1})\,\theta(\eta^{-s}x_m/y_m)\,(p)^3_\infty} \tag{4.10}
$$

where $\alpha_{\mathfrak{l},\mathfrak{m}} = \alpha\prod_{1\le j<m}\eta^{-2\mathfrak{l}_j} x_j/y_j$, cf. Lemma [4.1]. These formulae imply the next proposition.

Proposition 4.3. *Let the parameters x,y,η be generic. Then the form $(\,,)$ is nondegenerate/. The form $(\!(\,,)\!)$ is nondegenerateprovided that*

$$
\alpha\,\eta^{-r} \notin p^{\mathbb{Z}}, \qquad \alpha\,\eta^{r+2-2\ell} \prod_{m=1}^{n} x_m/y_m \notin p^{\mathbb{Z}}, \qquad r = 0,\dots,\ell-1\,,
$$

and

$$
\alpha\,\eta^{-r}\textstyle\prod_{1\le l\le m} x_l/y_l \notin p^{\mathbb{Z}}, \qquad m = 1,\dots,n-1\,, \quad r = 0,\dots,2\ell-2\,.
$$

Remark 4.2. Let the parameters q, γ, ρ be defined by

$$\eta = e^{-4\pi i\gamma}, \qquad q = e^{-2\pi i\gamma}, \qquad p = e^{2\pi i\rho}.$$

In Section [7.] we identify the space V with either a weight subspace in a tensor product of $U_q(\mathfrak{sl}_2)$ Verma modules, or the dual space to the weight subspace. Under these identifications the form $(\,,\,)$ transforms to the canonical pairing of the weight subspace and its dual space.

In Section [8.] we do similar identifications of the space V with either a weight subspace in a tensor product of evaluation Verma modules over the elliptic quantum group $E_{\rho,\gamma}(\mathfrak{sl}_2)$, or the dual space to the weight subspace. Then the form $(\!(\,,\,)\!)$ turns into the canonical pairing.

Notice that notations in Sections [7.], [8.] differ slightly from notations used here. In particular, we explicitly distinguish there the vector spaces associated with the tensor coordinates of different types.

Consider the following linear maps:

$$\begin{aligned} \bar I\,[x;y;\eta\,;\alpha] &: V^* \to V\,, & \bar I'[x;y;\eta\,;\alpha] &: V^* \to V\,, \\ \bar I &= B^*\, I\, B_{\mathrm{ell}}\,, & \bar I' &= (B')^*\, I'\, B'_{\mathrm{ell}}\,, \end{aligned} \tag{4.11}$$

where I and I' are given by (3.19) and we omit the common arguments in the second line. We call $\bar I$ and $\bar I'$ the *hypergeometric maps*.

Theorem [3.1] is equivalent to the following statement.

Theorem 4.1. *Let the parameters x, y, η be generic. Then the hypergeometric maps $\bar I\,[x;y;\eta\,;\alpha]$, $\bar I'[x;y;\eta\,;\alpha]$ respect the forms $(\,,\,)[x;y;\eta]$, $(\!(\,,\,)\!)[x;y;\eta\,;\alpha]$. That is for any $u, v \in V^*$ we have*

$$(\!(u,v)\!)[\alpha] = \big(\bar I\,[\alpha]\,u, \bar I'[\alpha]\,v\big)\,.$$

4.3 Difference equations for the hypergeometric maps

Let $L_m[x;y;\eta]$ and $L'_m[x;y;\eta]$, $k = 1, \ldots, n$, be linear operators acting in the trigonometric hypergeometric spaces $\mathcal F[x;y;\eta]$ and $\mathcal F'[x;y;\eta]$, respectively. The operators are defined by their actions on the bases of the trigonometric weight functions:

$$L_k[x;y;\eta\,;\alpha]\, w_{\mathfrak l}(\cdot\,;x;y;\eta) = \Big(\alpha\,\eta^{1-\ell}\prod_{m=1}^{n} x_m/y_m\Big)^{\mathfrak l_{\mathfrak k}} w_{{}^k\mathfrak l}(\cdot\,;{}^k x;{}^k y;\eta)\,, \tag{4.12}$$

$$L'_k[x;y;\eta\,;\alpha]\, w'_{\mathfrak l}(\cdot\,;x;y;\eta) = \Big(\alpha\,\eta^{1-\ell}\prod_{m=1}^{n} x_m/y_m\Big)^{-\mathfrak l_{\mathfrak k}} w'_{{}^k\mathfrak l}(\cdot\,;{}^k x;{}^k y;\eta)\,,$$

where

$$^k\mathfrak l = (\mathfrak l_{\mathfrak k+1}, \ldots, \mathfrak l_n, \mathfrak l_1, \ldots, \mathfrak l_{\mathfrak k})\,, \tag{4.13}$$

$$^k x = (x_{k+1}, \ldots, x_n, x_1, \ldots, x_k)\,, \qquad ^k y = (y_{k+1}, \ldots, y_n, y_1, \ldots, y_k)\,.$$

Recall that $\mathfrak{l}^{m} = \mathfrak{l}_1 <, \dots, +\mathfrak{l}_m$, $m = 1, \dots, n$, and in particular $\mathfrak{l}^{n} = \ell$.

Using the tensor coordinates we introduce operators $K_m, K'_m \in \mathrm{End}\,(V)$, $m = 1, \dots, n$, by the formulae:

$$K_m[x;y;\eta\,;\alpha] = \left(\left(B[x;y;\eta]\right)^{-1} L_m[x;y;\eta\,;\alpha]\, B[x;y;\eta]\,\right)^{*}, \quad (4.14)$$

$$K'_m[x;y;\eta\,;\alpha] = \left(\left(B'[x;y;\eta]\right)^{-1} L'_m[x;y;\eta\,;\alpha]\, B'[x;y;\eta]\,\right)^{*}. \quad (4.15)$$

We also define operators $M_m[x;y;\eta\,;\alpha] \in \mathrm{End}\,(V^*)$, $m = 1, \dots, n$:

$$M_m[x;y;\eta\,;\alpha]\, v^*_{\mathfrak{l}} = \mu_{\mathfrak{l},\mathrm{m}}[x;y;\eta\,;\alpha]\, v^*_{\mathfrak{l}}\,, \mu_{\mathfrak{l},\mathrm{m}} = \Big(\alpha\,\eta^{1-\mathfrak{l}^{m}} \prod_{1\le j\le m} x_j/y_j\Big)^{-\mathfrak{l}^{m}}. \quad (4.16)$$

Let $Q^h_1, \dots, Q^h_n$ be the multiplicative shift operators acting on functions of $x_1, \dots, x_n$, $y_1, \dots, y_n$:

$$Q^h_m f(x_1, \dots, x_n; y_1, \dots, y_n) = \quad (4.17)$$

$$= f(hx_1, \dots, hx_m, x_{m+1}, \dots, x_n; hy_1, \dots, hy_m, y_{m+1}, \dots, y_n)\,. \quad (4.18)$$

Set $Q_m = Q^p_m$, $m = 1, \dots, n$.

Theorem 4.2. *[30] The hypergeometric map $\bar{I}\,[x;y;\eta\,;\alpha]$ satisfies the following system of difference equations:*

$$Q_m \bar{I}\,[x;y;\eta\,;\alpha] = K_m[x;y;\eta\,;\alpha]\, \bar{I}\,[x;y;\eta\,;\alpha]\, M_m[x;y;\eta\,;\alpha]\,, \quad (4.19)$$

$m = 1, \dots, n$.

Corollary 4.3. *The hypergeometric map $\bar{I}'[x;y;\eta\,;\alpha]$ satisfies the following system of difference equations:*

$$Q_m \bar{I}'[x;y;\eta\,;\alpha] = K'_m[x;y;\eta\,;\alpha]\, \bar{I}'[x;y;\eta\,;\alpha]\, \left(M_m[x;y;\eta\,;\alpha]\right)^{-1}, \quad (4.20)$$

$m = 1, \dots, n$

The last claim results from the commutativity of diagram (3.20) and formulae (4.3).

Remark 4.3. The numbers $\mu_{\mathfrak{l},\mathrm{m}}$ are related to the transformation properties of the elliptic weight functions:

$$Q_m W_{\mathfrak{l}} = \mu_{\mathfrak{l},\mathrm{m}} \prod_{1\le j\le m} (x_j/y_j)^{\ell}\, W_{\mathfrak{l}}\,, \quad (4.21)$$

$$Q_m W'_{\mathfrak{l}} = \mu^{-1}_{\mathfrak{l},\mathrm{m}} \prod_{1\le j\le m} (x_j/y_j)^{-\ell}\, W'_{\mathfrak{l}}\,. \quad (4.22)$$

Remark 4.4. Consider the system of difference equations

$$Q_m \Psi(x;y) = K_m(x;y)\,\Psi(x;y)\,, \qquad m = 1,\dots,n\,, \tag{4.23}$$

Its solutions have the form $\Psi = \bar{I}\,Y$ where $Y \in \mathrm{End}\,(V^*)$ solves the system of difference equations

$$Q_m Y(x;y) = M_m^{-1}\,Y(x;y)\,, \qquad m = 1,\dots,n\,.$$

Notice that the operators $M_1,\dots,M_n$ are invariant with respect to the shift operators $Q_1^h,\dots,Q_n^h$ for any nonzero h, so the last system of difference equations effectively has constant coefficients. The factor Y plays the role of an adjusting map in [30].

In Section [5.] we show that the system (4.23) can be viewed as the periodic section equation for a certain cohomological bundle with the discrete Gauss-Manin connection, see Theorem [5.1]. Solutions of the system (4.23) are described in Theorem [5.2] via invariant sections of the elliptic hypergeometric bundle. Slightly modified, the last construction results into the hypergeometric map, so Theorem [4.2] appears to be nearly the same as Theorem [5.2], see formulae (5.23), (5.24).

In Section [7.] we identify the system (4.23) with the qKZ equation with values in a weight subspace of a tensor product of $U_q(\mathfrak{sl}_2)$ Verma modules, see (7.11), (7.12). We also identify the system of difference equations

$$Q_m \Psi'(x;y) = K'_m(x;y)\,\Psi'(x;y)\,, \qquad m = 1,\dots,n\,, \tag{4.24}$$

with the dual qKZ equation (7.3) for a functionΨ' taking values in the dual space of the weight subspace.

4.4 Asymptotics of the hypergeometric maps

Let $\mathbb{A}$ be the following asymptotic zone of the parameters $x_1,\dots,x_n$, $y_1,\dots,y_n$:

$$\mathbb{A} = \left\{ \begin{array}{ll} |x_m/x_{m+1}| \ll 1\,, & m = 1,\dots,n-1 \\ |x_m/y_m| \simeq 1\,, & m = 1,\dots,n \end{array} \right\}. \tag{4.25}$$

We say that $(x;y)$ tends to limit in $\mathbb{A}$ and write $(x;y) \rightrightarrows \mathbb{A}$ if

$$x_m/x_{m+1} \to 0\,, \qquad m = 1,\dots,n-1\,,$$

and the ratios x_m/y_m and y_m/x_m remain bounded for any $m = 1,\dots,n$. If a function$f(x;y)$ has a finite limit as $(x;y) \rightrightarrows \mathbb{A}$, we denote this limit by $\lim_{\mathbb{A}} f$. Notice that the limit $\lim_{\mathbb{A}} f$ can depend on $x_1,\dots,x_n,y_1,\dots,y_n$, but it is invariant with respect to the shift operators $Q_1^h,\dots,Q_n^h$ for any nonzero h.

The hypergeometric pairings I and I' have finite limits in the asymptotic zone, $(x;y) \rightrightarrows \mathbb{A}$, and the limits can be represented by triangular matrices. Namely, define the functions $I_{\mathfrak{l}\,\mathfrak{m}}(x;y)$ and $I'_{\mathfrak{l}\,\mathfrak{m}}(x;y)$ by the formulae:

$$\begin{aligned} I_{\mathfrak{l}\,\mathfrak{m}}(x;y) &= I\,[x;y;\alpha]\big(w_{\mathfrak{l}}(\cdot;x;y), W_{\mathfrak{m}}(\cdot;x;y;\alpha)\big)\,, \\ I'_{\mathfrak{l}\,\mathfrak{m}}(x;y) &= I'[x;y;\alpha]\big(w'_{\mathfrak{l}}(\cdot;x;y), W'_{\mathfrak{m}}(\cdot;x;y;\alpha)\big)\,. \end{aligned}$$

Proposition 4.4. *[30] For any $\mathfrak{l},\mathfrak{m} \in \mathcal{Z}^{\mathfrak{n}}_{\ell}$ the hypergeometric integral $I_{\mathfrak{l}\,\mathfrak{m}}(x;y)$ has a finite limit as $(x;y) \rightrightarrows \mathbb{A}$. Moreover $\lim_{\mathbb{A}} I_{\mathfrak{l}\,\mathfrak{m}} = 0$ unless $\mathfrak{l} \gg \mathfrak{m}$ or $\mathfrak{l} = \mathfrak{m}$, cf. (3.5), and*

$$\lim\nolimits_{\mathbb{A}} I_{\mathfrak{l}\,\mathfrak{l}} = \prod_{m=1}^{n} \prod_{s=0}^{\mathfrak{l}_m-1} \frac{\eta^{-s}\,(\eta^{-1})_\infty\,(\eta^s\alpha^{-1}_{\mathfrak{l},m})_\infty\,(p\,\eta^{1-s-\mathfrak{l}_m}\alpha_{\mathfrak{l},m}x_m/y_m)_\infty}{(\eta^{-s-1})_\infty\,(\eta^{-s}x_m/y_m)_\infty\,(p)_\infty}\,.$$

Recall that $\alpha_{\mathfrak{l},m} = \alpha\prod_{1\le j<m}\eta^{-2\mathfrak{l}_j}x_j/y_j$.

Proposition 4.5. *For any $\mathfrak{l},\mathfrak{m} \in \mathcal{Z}^{\mathfrak{n}}_{\ell}$ the hypergeometric integral $I'_{\mathfrak{l}\,\mathfrak{m}}(x;y)$ has a finite limit as $(x;y) \rightrightarrows \mathbb{A}$. Moreover $\lim_{\mathbb{A}} I'_{\mathfrak{l}\,\mathfrak{m}} = 0$ unless $\mathfrak{l} \ll \mathfrak{m}$ or $\mathfrak{l} = \mathfrak{m}$, cf. (3.5), and*

$$\lim\nolimits_{\mathbb{A}} I'_{\mathfrak{l}\,\mathfrak{l}} = \prod_{m=1}^{n} \prod_{s=0}^{\mathfrak{l}_m-1} \frac{-\eta^{-s}\alpha_{\mathfrak{l},m}\,(\eta)_\infty\,(p\,\eta^{-s}\alpha_{\mathfrak{l},m})_\infty\,(\eta^{s-1+\mathfrak{l}_m}\alpha^{-1}_{\mathfrak{l},m}y_m/x_m)_\infty}{(\eta^{s+1})_\infty\,(\eta^s y_m/x_m)_\infty\,(p)_\infty}\,.$$

In Section [6.] we prove Propositions [4.4] and [4.5] in the one-dimensional case to show the main idea of the proof. Proposition [4.4] in the general case is equivalent to Theorem 6.2 in [30]. Proposition [4.5] can be proved in a similar way.

Corollary 4.4. *For any α,η the hypergeometric maps $\bar I\,[x;y;\eta\,;\alpha]$ and $\bar I'[x;y;\eta\,;\alpha]$ have finite limits as $(x;y) \rightrightarrows \mathbb{A}$. Moreover,*

$$\lim\nolimits_{\mathbb{A}} \bar I\, v^*_{\mathfrak{l}} = v_{\mathfrak{l}}\,\lim\nolimits_{\mathbb{A}} I_{\mathfrak{l}\,\mathfrak{l}} + \sum_{\mathfrak{m}\gg\mathfrak{l}} v_{\mathfrak{m}}\,\lim\nolimits_{\mathbb{A}} I_{\mathfrak{m}\,\mathfrak{l}}$$

and

$$\lim\nolimits_{\mathbb{A}} \bar I' v^*_{\mathfrak{l}} = v_{\mathfrak{l}}\,\lim\nolimits_{\mathbb{A}} I'_{\mathfrak{l}\,\mathfrak{l}} + \sum_{\mathfrak{m}\ll\mathfrak{l}} v_{\mathfrak{m}}\,\lim\nolimits_{\mathbb{A}} I'_{\mathfrak{m}\,\mathfrak{l}}\,.$$

Remark 4.5. It is easy to check that the operators $K_1,\dots,K_n$, $K'_1,\dots,K'_n$, cf. (4.14), have finite limits as $(x;y) \rightrightarrows \mathbb{A}$, and the limits are respectively lower and upper triangular with respect to the basis $\{\,v_{\mathfrak{l}}\,\}_{\mathfrak{l}\in\mathcal{Z}^{\mathfrak{n}}_{\ell}}$ and ordering (3.5). The diagonal parts of the limits $\lim_{\mathbb{A}} K_m^{-1}$ and $\lim_{\mathbb{A}} K'_m$ are equal to M^*_m, cf. (4.16):

$$\lim\nolimits_{\mathbb{A}} K_m v_{\mathfrak{l}} = \mu^{-1}_{\mathfrak{l},m}\,v_{\mathfrak{l}} + \sum_{\mathfrak{n}\gg\mathfrak{l}} v_{\mathfrak{n}}\,\langle v^*_{\mathfrak{n}}\,, \lim\nolimits_{\mathbb{A}} K_m v_{\mathfrak{l}}\rangle\,,$$

$$\lim\nolimits_{\mathbb{A}} K'_m v_{\mathfrak{l}} = \mu_{\mathfrak{l},m}\,v_{\mathfrak{l}} + \sum_{\mathfrak{n}\ll\mathfrak{l}} v_{\mathfrak{n}}\,\langle v^*_{\mathfrak{n}}\,, \lim\nolimits_{\mathbb{A}} K'_m v_{\mathfrak{l}}\rangle\,.$$

Furthermore, it follows from (4.19) that for any $\mathfrak{l} \in \mathcal{Z}_\ell^{\mathfrak{n}}$ the vector $\lim_{\mathbb{A}} \bar{I}\, v_{\mathfrak{l}}^*$ is a common eigenvector of the operators $\lim_{\mathbb{A}} K_1, \dots, \lim_{\mathbb{A}} K_n$ with eigenvalues $\mu_{\mathfrak{l},1}^{-1}, \dots, \mu_{\mathfrak{l},\mathfrak{n}}^{-1}$, respectively. Similarly, (4.20) implies that for any $\mathfrak{l} \in \mathcal{Z}_\ell^{\mathfrak{n}}$ the vector $\lim_{\mathbb{A}} \bar{I}' v_{\mathfrak{l}}^*$ is a common eigenvector of the operators $\lim_{\mathbb{A}} K_1', \dots, \lim_{\mathbb{A}} K_n'$ with eigenvalues $\mu_{\mathfrak{l},1}, \dots, \mu_{\mathfrak{l},\mathfrak{n}}$, respectively.

4.5 Proof of the hypergeometric Riemann identity

Now we are going to prove the bilinear identity for the hypergeometric integrals. Two equivalent forms of the identity are given by Theorems [3.1] and [4.1]. We will prove the latter theorem. Proof of Theorem [4.1]. Consider the functions

$$G_{\mathfrak{l}\,\mathfrak{m}}(x;y;\eta\,;\alpha) = \big(\bar{I}'[x;y;\eta\,;\alpha]\, v_{\mathfrak{l}}^*, \bar{I}\,[x;y;\eta\,;\alpha]\, v_{\mathfrak{m}}^*\big)[x;y;\eta]\,.$$

We have to prove that for any $\mathfrak{l}, \mathfrak{m} \in \mathcal{Z}_\ell^{\mathfrak{n}}$

$$G_{\mathfrak{l}\,\mathfrak{m}}(x;y;\eta\,;\alpha) \;=\; (\!(v_{\mathfrak{l}}^*, v_{\mathfrak{m}}^*)\!)[x;y;\eta\,;\alpha]\,. \tag{4.26}$$

Since both sides of the above equality are analytic functions of α, we can assume that α is generic. In particular, we will use the following statement.

Lemma 4.2. *Let α be generic. Let $\mu_{\mathfrak{l},\mathfrak{k}}$ be defined by (4.16). If $\mathfrak{l}, \mathfrak{m} \in \mathcal{Z}_\ell^{\mathfrak{n}}$ are such that $\mu_{\mathfrak{l},\mathfrak{k}} = \mu_{\mathfrak{m},\mathfrak{k}}$ for any $k = 1, \dots, n$, then $\mathfrak{l} = \mathfrak{m}$.*

In the rest of the proof we do not write explicitly arguments α and η for all the involved functions.

Proof. By the definition of the Shapovalov pairing $S[x;y]$ it is easy to see that

$$S[x;y]\big(w_{{}^k\mathfrak{l}}(\cdot\,;{}^k x;{}^k y), w'_{{}^k\mathfrak{m}}(\cdot\,;{}^k x;{}^k y)\big) \;=\; S[x;y]\big(w_{\mathfrak{l}}(\cdot\,;x;y), w'_{\mathfrak{m}}(\cdot\,;x;y)\big)\,,$$

cf. (4.13), that is $S\,(L_k\, w_{\mathfrak{l}}, L_k'\, w_{\mathfrak{m}}') \;=\; S\,(w_{\mathfrak{l}}, w_{\mathfrak{m}}'),\;\; k = 1, \dots, n$, cf. (4.12). Therefore, for any $u, v \in V$ we have

$$\big(K_m[x;y]\, u, K_m'[x;y]\, v\big)[x;y] \;=\; (u,v)[x;y]\,, \qquad\qquad m = 1, \dots, n\,.$$

Since the hypergeometric maps $\bar{I}[x;y]$ and $\bar{I}'[x;y]$ satisfy the systems of difference equations (4.19) and (4.20), respectively, the function$G_{\mathfrak{l}\,\mathfrak{m}}(x;y)$ satisfies a system of difference equations

$$Q_k\, G_{\mathfrak{l}\,\mathfrak{m}}(x;y) \;=\; \mu_{\mathfrak{l},\mathfrak{k}}\, \mu_{\mathfrak{m},\mathfrak{k}}^{-1}\, G_{\mathfrak{l}\,\mathfrak{m}}(x;y)\,, \qquad\qquad k = 1, \dots, n\,, \tag{4.27}$$

where $\mu_{\mathfrak{l},\mathfrak{k}}$ are given by (4.16).

On the other hand, Corollary [4.4] shows that the function$G_{\mathfrak{l}\,\mathfrak{m}}(x;y)$ has a finite limit as $(x;y) \rightrightarrows \mathbb{A}$. Taking the limit in equations (4.27) we obtain that

$$\lim_{\mathbb{A}} G_{\mathfrak{l}\,\mathfrak{m}} = \mu_{\mathfrak{l},\mathfrak{k}}\,\mu_{\mathfrak{m},\mathfrak{k}}^{-1} \lim_{\mathbb{A}} G_{\mathfrak{l}\,\mathfrak{m}}\,, \qquad k=1,\dots,n\,.$$

Therefore, $\lim_{\mathbb{A}} G_{\mathfrak{l}\,\mathfrak{m}} = 0$ for $\mathfrak{l} \neq \mathfrak{m}$ by Lemma [4.2].

Observe now that $G_{\mathfrak{l}\,\mathfrak{m}}$ is an analytic functionof x, y, η, α. So it suffices to prove the claim

$$G_{\mathfrak{l}\,\mathfrak{m}}(x;y) = 0 \qquad \text{for} \qquad \mathfrak{l} \neq \mathfrak{m}$$

under the assumptions

$$|\alpha| = 1\,, \qquad |\eta| = 1\,, \qquad |x_m/y_m| = 1\,, \qquad m = 1,\dots,n\,.$$

Under these assumptions $|\mu_{\mathfrak{l},\mathfrak{m}}| = 1$ for any $\mathfrak{l} \in \mathcal{Z}_\ell^{\mathfrak{n}}$ and $m = 1,\dots,n$, and by (4.27) we have

$$|\,Q_k\, G_{\mathfrak{l}\,\mathfrak{m}}(x;y)| = |\,G_{\mathfrak{l}\,\mathfrak{m}}(x;y)|\,, \qquad k=1,\dots,n\,.$$

Therefore, $|\,G_{\mathfrak{l}\,\mathfrak{m}}(x;y)| = |\lim_{\mathbb{A}} G_{\mathfrak{l}\,\mathfrak{m}}| = 0$, i.e. $G_{\mathfrak{l}\,\mathfrak{m}}(x;y) = 0$. For the function$G_{\mathfrak{l}\,\mathfrak{l}}(x;y)$ the system (4.27) reads

$$Q_k\, G_{\mathfrak{l}\,\mathfrak{l}}(x;y) = G_{\mathfrak{l}\,\mathfrak{l}}(x;y)\,, \qquad k=1,\dots,n\,,$$

which gives $G_{\mathfrak{l}\,\mathfrak{l}}(x;y) = \lim_{\mathbb{A}} G_{\mathfrak{l}\,\mathfrak{l}}$. In particular, this shows that all the functions $G_{\mathfrak{l}\,\mathfrak{l}}(x;y)$ are invariant with respect to the shift operators $Q_1^h,\dots,Q_n^h$ for any nonzero h.

Obviously, the right hand side of formula (4.26) enjoys the same properties:

$$(\!(v_{\mathfrak{l}}^*, v_{\mathfrak{m}}^*)\!)[x;y] = 0 \qquad \text{for} \quad \mathfrak{l} \neq \mathfrak{m}\,,$$

and $(\!(v_{\mathfrak{l}}^*, v_{\mathfrak{l}}^*)\!)[x;y]$ is invariant with respect to $Q_1^h,\dots,Q_n^h$ for any nonzero h. Hence, it remains to verify that

$$\lim_{\mathbb{A}} G_{\mathfrak{l}\,\mathfrak{l}} = (\!(v_{\mathfrak{l}}^*, v_{\mathfrak{l}}^*)\!)\,, \qquad \mathfrak{l} \in \mathcal{Z}_\ell^{\mathfrak{n}}\,,$$

or, in other words,

$$\lim_{\mathbb{A}} I_{\mathfrak{l}\,\mathfrak{l}} \cdot \lim_{\mathbb{A}} I'_{\mathfrak{l}\,\mathfrak{l}} \cdot (v_{\mathfrak{l}}, v_{\mathfrak{l}}) = (\!(v_{\mathfrak{l}}^*, v_{\mathfrak{l}}^*)\!)\,,$$

cf. (4.7), (4.8) and Corollary [4.4]. This is a straightforward calculation using Propositions [4.4], [4.5] and formulae (4.9), (4.10). Theorem [4.1] is proved.

5. Discrete local systems and the discrete Gauss-Manin connection

In this section we explain the geometric origin of the systems of difference equations (4.19), (4.20) satisfied by the hypergeometric maps. We show that these systems are essentially the periodic section equations for the discrete Gauss-Manin connection assosiated with a suitable discrete local system. The notions of a discrete local system and the discrete Gauss-Manin connection were introduced in [30], [31].

5.1 Discrete flat connections and discrete local systems

Consider a vector space $\mathbb{C}^m$ called the *base space*. The lattice $\mathbb{Z}^m$ acts on the base space by dilations:

$$\mathfrak{l} : (\mathfrak{z}_1, \ldots, \mathfrak{z}_m) \mapsto (\mathfrak{p}^{\mathfrak{l}_1}\mathfrak{z}_1, \ldots, \mathfrak{p}^{\mathfrak{l}_m}\mathfrak{z}_m), \qquad \mathfrak{l} \in \mathbb{Z}^m.$$

Let $\mathbb{B} \subset \mathbb{C}^m$ be an invariant subset of the base space. Say that there is a *bundle with a discrete connection* over $\mathbb{B}$ if for any $z \in \mathbb{B}$ there are a vector space $V(z)$ and linear isomorphisms

$$A_k(z_1, \ldots, z_m) : V(z_1, \ldots, pz_k, \ldots, z_m) \to V(z_1, \ldots, z_m), \qquad k = 1, \ldots, m.$$

The connection is called *flat* if the isomorphisms $A_1, \ldots, A_m$ commute:

$$A_j(z_1, \ldots, z_m)\, A_k(z_1, \ldots, pz_j, \ldots, z_m) = \tag{5.1}$$
$$= A_k(z_1, \ldots, z_m)\, A_j(z_1, \ldots, pz_k, \ldots, z_m).$$

Say that a *discrete subbundle* in $\mathbb{B}$ is given if a subspace in every fiber is distinguished and the family of subspaces is invariant with respect to the connection.

A section $s : z \mapsto s(z)$ is called *periodic* if its values are invariant with respect to the connection:

$$A_k(z_1, \ldots, z_m)\, s(z_1, \ldots, pz_k, \ldots, z_m) = s(z_1, \ldots, z_m), \qquad k = 1, \ldots, m.$$

A function $f(z_1, \ldots, z_m)$ is called a *quasiconstant* if

$$f(z_1, \ldots, pz_k, \ldots, z_m) = f(z_1, \ldots, z_m), \qquad k = 1, \ldots, m.$$

Periodic sections form a module over the ring of quasiconstants.

The *dual bundle* with the *dual connection* has fibers $V^*(z)$ and isomorphisms

$$A_k^*(z_1, \ldots, z_m) : V^*(z_1, \ldots, z_m) \to V^*(z_1, \ldots, pz_k, \ldots, z_m). \tag{5.2}$$

Let $s_1, \ldots, s_N$ be a basis of sections of the initial bundle. Then the isomorphisms A_k of the connection are given by matrices $\mathrm{A}^{(k)}$:

$$A_k(z_1, \ldots, z_m)\, s_a(z_1, \ldots, pz_k, \ldots, z_m) = \sum_{b=1}^{N} \mathrm{A}^{(k)}_{ab}(z_1, \ldots, z_m)\, s_b(z_1, \ldots, z_m).$$

For a section $\psi : z \mapsto \psi(z)$ of the dual bundle denote by $\Psi : z \mapsto \Psi(z)$ its coordinate vector,

$$\Psi_a(z) = \langle \psi(z), s_a(z) \rangle.$$

The section ψ is periodic if and only if its coordinate vector satisfies the system of difference equations

$$\Psi(z_1, \ldots, pz_k, \ldots, z_m) = \mathrm{A}^{(k)}(z_1, \ldots, z_m)\, \Psi(z_1, \ldots, z_m),$$

$k = 1, \dots, m$. This system of difference equations is called the *periodic section equation.*

Say that functions $\varphi_1, \dots, \varphi_m$ in variables $z_1, \dots, z_m$ form a *system of connection coefficients* if

$$\varphi_j(z_1, \dots, z_m)\, \varphi_k(z_1, \dots, pz_j, \dots, z_m) = \\ = \varphi_k(z_1, \dots, z_m)\, \varphi_j(z_1, \dots, pz_k, \dots, z_m)$$

for all $j, k = 1, \dots, m$. These functions define a discrete flat connection on the trivial complex one-dimensional vector bundle, cf. (5.1). A periodic section $\bar{\Phi}$ of the dual bundle is called a *phase function of system of connection coefficients*:

$$\bar{\Phi}(z_1, \dots, pz_k, \dots, z_m) = \varphi_k(z_1, \dots, z_m)\, \bar{\Phi}(z_1, \dots, z_m)\,, \qquad k = 1, \dots, m\,.$$

The system of connection coefficients $\varphi_1, \dots, \varphi_m$ defines the twisted shift operators $Z_1, \dots, Z_m$ acting on a function$f(z_1, \dots, z_m)$ by the rule:

$$Z_k f(z_1, \dots, z_m) = \varphi_k(z_1, \dots, z_m)\, f(z_1, \dots, pz_k, \dots, z_m)\,,$$

$k = 1, \dots, m$. The operators $Z_1, \dots, Z_m$ commute with each other.

Let F be a vector space of functions of $z_1, \dots, z_m$ such that the operators $Z_1, \dots, Z_m$ induce linear isomorphisms of F:

$$Z_k : \mathrm{F} \to \mathrm{F}\,, \qquad k = 1, \dots, m\,.$$

We say that the space F and the connection coefficients $\varphi_1, \dots, \varphi_m$ form a *one-dimensional discrete local system.* F is called the *functional space* of the local system.

The *de Rham complex*, the *cohomology* and *homology groups of* $\mathbb{C}^{\times m}$ *with coefficients in the discrete local system* are defined in [30], [31]. In particular, the top cohomology group H^m is canonically isomorphic to the quotient space $\mathrm{F}/D\mathrm{F}$ where

$$D\mathrm{F} = \{ \sum_{k=1}^{m} (Z_k f_k - f_k) \mid f_k \in \mathrm{F}\,, \quad k = 1, \dots, m \}\,.$$

The elements of the subspace $D\mathrm{F}$ are called the *twisted total differences.*

In this paper we take the equality $H^m = \mathrm{F}/D\mathrm{F}$ for the definition of the space H^m. The dual space $H_m = (H^m)^*$ is the top homology group. It can be considered as a subspace of the space F^* spanned by linear functionals annihilating the twisted total differences.

5.2 Discrete Gauss-Manin connection

There is a geometric construction of bundles with discrete flat connections, a discrete version of the Gauss-Manin connection construction, see [30], [31].

Let $\pi : \mathbb{C}^{\ell+m} \to \mathbb{C}^m$ be an affine projection onto the base with fiber $\mathbb{C}^\ell$. $\mathbb{C}^{\ell+m}$ is called the *total space*. Let $z_1, \ldots, z_m$ be coordinates on the base, $t_1, \ldots, t_\ell$ coordinates on the fiber, so that $t_1, \ldots, t_\ell, z_1, \ldots, z_m$ are coordinates on the total space.

Let F, $\varphi_1, \ldots, \varphi_\ell,\ \chi_1, \ldots, \chi_m$ be a local system on $\mathbb{C}^{\times(\ell+m)}$. We refer to this local system as the *total local system*, so that F is the *total functional space*. We denote the corresponding twisted shift operators by $T_1, \ldots, T_\ell,\ Z_1, \ldots, Z_m$.

For a point $z \in \mathbb{C}^{\times m}$ we define a local system $\mathrm{F}|_z$, $\varphi_1(\cdot\,; z), \ldots, \varphi_\ell(\cdot\,; z)$ on the fiber over z restricting all the functions to the fiber:

$$\mathrm{F}|_z = \{ f|_{\pi^{-1}(z)} \mid f \in \mathrm{F} \}, \qquad \varphi_a(\cdot\,; z) = \varphi_a|_{\pi^{-1}(z)} \,.$$

We denote by $H^\ell|_z$ and $H_\ell|_z$ the top cohomology and homology spaces of the fiber.

This construction provides a bundle over $\mathbb{C}^{\times m}$ with a discrete flat connection with fibers $\mathrm{F}|_z$ and isomorphisms

$$Z_k|_{(z_1,\ldots,z_m)} : \mathrm{F}|_{(z_1,\ldots,pz_k,\ldots,z_m)} \to \mathrm{F}|_{(z_1,\ldots,z_m)} \,, \qquad k = 1, \ldots, m \,, \tag{5.3}$$

$$Z_k|_{(z_1,\ldots,z_m)}\, f = \varphi_k(\cdot\,; z)\, f \,,$$

induced by the operators $Z_1, \ldots, Z_m$ acting on the total functional space F. On the other hand, the operators $T_1, \ldots, T_\ell$ can be naturally restricted to fibers $\mathrm{F}|_z$ and their action on the fibers is consistent with the isomorphisms (5.3):

$$T_a|_{(z_1,\ldots,z_m)}\, Z_k|_{(z_1,\ldots,z_m)} = Z_k|_{(z_1,\ldots,z_m)}\, T_a|_{(z_1,\ldots,pz_k,\ldots,z_m)} \,,$$

which follows from the commutativity of the operators T_a and Z_k acting on the total functional space. Hence the isomorphisms (5.3) induce isomorphisms

$$Z_k|_{(z_1,\ldots,z_m)} : H^\ell|_{(z_1,\ldots,pz_k,\ldots,z_m)} \to H^\ell|_{(z_1,\ldots,z_m)} \,, \qquad k = 1, \ldots, m \,,$$

denoted by the same letters, so that the vector spaces $H^\ell|_z$ form a bundle over $\mathbb{C}^{\times m}$ with a discrete flat connection. This connection is called the *discrete Gauss-Manin connection.*

The Gauss-Manin connection on the cohomological bundle induces the dual flat connection on the homological bundle:

$$Z_k^*|_{(z_1,\ldots,z_m)} : H_\ell|_{(z_1,\ldots,z_m)} \to H_\ell|_{(z_1,\ldots,pz_k,\ldots,z_m)} \,.$$

In what follows we usually do not distinguish explicitly in notations between the twisted shift operators acting on the total functional space and the induced isomorphisms of the functional or cohomological spaces of fibers, if it causes no confusion.

5.3 Discrete local system associated with the hypergeometric integrals

In this section we describe a certain discrete local system. The local system is closely related to the trigonometric $\hat{\mathfrak{sl}}_2$- type local system studied in [30].

Take a nonzero complex number η such that $\eta^{r+1} \notin p^{\mathbb{Z}}$. Let $\mathbb{B} \subset \mathbb{C}^{2n}$ be the subset of generic points:

$$\mathbb{B} = \{ (x;y) \in \mathbb{C}^{2n} \mid x_1,\dots,x_n,\ y_1,\dots,y_n \text{ obey } (3.1) \}$$

The set $\mathbb{B}$ is clearly invariant with respect to the dilations by p. From now on we deal with discrete bundles over $\mathbb{B}$.

Consider an affine projection $\pi : \mathbb{C}^{\ell+2n} \to \mathbb{C}^{2n}$ of the total space $\mathbb{C}^{\ell+2n}$ onto the base with fiber $\mathbb{C}^\ell$. Let $x_1,\dots,x_n,\ y_1,\dots,y_n$ be coordinates on the base, $t_1,\dots,t_\ell$ – coordinates on the fiber, so that $t_1,\dots,t_\ell,\ x_1,\dots,x_n,\ y_1,\dots,y_n$ are coordinates on the total space. Consider the space $\widehat{\mathcal{F}}[\eta]$ of rational function on the total space with at most simple poles at the following hyperplanes

$$t_a = p^{-s}x_m\,, \qquad t_a = p^{s+1}y_m\,, \qquad t_a = p^{s+1}\eta\, t_b\,, \qquad t_b = p^s\eta\, t_a\,, \tag{5.4}$$

$1 \le a < b \le \ell$, $m = 1,\dots,n$, $s \in \mathbb{Z}_{\ge 0}$, and any poles at the coordinate hyperplanes

$$t_a = 0\,, \qquad x_m = 0\,, \qquad y_m = 0\,,$$

$a = 1,\dots,\ell$, $m = 1,\dots,n$. Let $\varphi_1,\dots,\varphi_\ell,\ \chi_1,\dots,\chi_n,\ \phi_1,\dots,\phi_n$ be the following system of connection coefficients:

$$\varphi_a(t;x;y;\eta;\alpha) = \alpha\,\eta^{1-\ell}\prod_{m=1}^n x_m/y_m \prod_{m=1}^n \frac{t_a-y_m}{t_a-x_m}\ \times \tag{5.5}$$

$$\times \prod_{a<b\le\ell}\frac{t_a-\eta\, t_b}{\eta\, t_a - t_b}\prod_{1\le b<a}\frac{p\, t_a-\eta\, t_b}{p\eta\, t_a-t_b}\,, a=1,\dots,\ell\,,$$

$$\chi_m(t;x;y) = \prod_{a=1}^\ell (t_a - p\, x_m)\,, \tag{5.6}$$

$$m = 1,\dots,n\,.$$

$$\phi_m(t;x;y) = \prod_{a=1}^\ell (t_a - p\, y_m)^{-1}\,,$$

Denote by $T_1[\eta;\alpha],\dots,T_\ell[\eta;\alpha],\ X_1,\dots,X_m,\ Y_1,\dots,Y_m$ the corresponding twisted shift operators. It is easy to check that they induce linear isomorphism of the space $\widehat{\mathcal{F}}[\eta]$. Hence, we have a discrete local system with the functional space $\widehat{\mathcal{F}}[\eta]$ and the system of connection coefficients (5.5), (5.6).

A phase function of the system of connection coefficients (5.5), (5.6) has the form $g(t;x;y)\,\Phi(t;x;y;\eta)$ where

$$\Phi(t;x;y;\eta) = \prod_{m=1}^{n}\prod_{a=1}^{\ell}\frac{(t_a/x_m)_\infty}{(t_a/y_m)_\infty}\prod_{1\le a<b\le\ell}\frac{(\eta\,t_a/t_b)_\infty}{(\eta^{-1}t_a/t_b)_\infty}\,, \tag{5.7}$$

cf. (3.17), and $g(t;x;y)$ is an arbitrary functionsuch that

$$g(t_1,\dots,pt_a,\dots,t_\ell;x;y) = \alpha\,\eta^{2-2a}\,g(t_1,\dots,t_\ell;x;y)\,, \tag{5.8}$$

$$g(t;x_1,\dots,px_m,\dots,x_n;y) = (-p\,x_m)^\ell\,g(t;x_1,\dots,x_n;y)\,, \tag{5.9}$$

$$g(t;x;y_1,\dots,py_m,\dots,y_n) = (-p\,y_m)^{-\ell}\,g(t;x;y_1,\dots,y_n)\,.$$

We take the introduced local system as the total local system and consider the induced local systems on fibers. The functional space $\widehat{\mathcal{F}}[x;y;\eta]$ of the local system on a fiber over a point $(x;y)\in\mathbb{C}^{2n}$ is the space of rational functions of the variables $t_1,\dots,t_\ell$ with at most simple poles at the hyperplanes (5.4) and any poles at the coordinate hyperplanes $t_a=0$, $a=1,\dots,\ell$. The system of connection coefficients of the local system on the fiber $\varphi_1(\cdot;x;y),\dots,\varphi_\ell(\cdot;x;y)$ is given by formulae (5.5). We denote by $\widehat{\mathcal{H}}[x;y;\eta\,;\alpha]$ the corresponding top cohomology space:

$$\widehat{\mathcal{H}}[x;y;\eta\,;\alpha] = \widehat{\mathcal{F}}[x;y;\eta]\,/D\widehat{\mathcal{F}}[x;y;\eta\,;\alpha]$$

where

$$D\widehat{\mathcal{F}}[x;y;\eta\,;\alpha] = \tag{5.10}$$

$$= \{\sum_{a=1}^{\ell}(T_a[\eta\,;\alpha]|_{(x,y)}f_a - f_a)\ \mid\ f_a\in\widehat{\mathcal{F}}[x;y;\eta]\,,a=1,\dots,\ell\}\,.$$

There is an action of the symmetric group $\mathbf{S}_\ell$ on the total functional space $\widehat{\mathcal{F}}[\eta]$ given by (3.2):

$$[\,f\,]_\sigma(t_1,\dots,t_\ell;x;y) = f(t_{\sigma_1},\dots,t_{\sigma_\ell};x;y)\prod_{\substack{1\le a<b\le\ell\\ \sigma_a>\sigma_b}}\frac{t_{\sigma_b}-\eta\,t_{\sigma_a}}{\eta\,t_{\sigma_b}-t_{\sigma_a}}\,,$$

and a similar $\hat{}\ \mathbf{S}_\ell$-action on the functional space of a fiber $\widehat{\mathcal{F}}[x;y;\eta]$. We denote by $\widehat{\mathcal{F}}_\Sigma[\eta]$ and $\widehat{\mathcal{F}}_\Sigma[x;y;\eta]$ the subspaces of $\hat{}\ \mathbf{S}_\ell$-invariant functions and by $\widehat{\mathcal{H}}_\Sigma[x;y;\eta\,;\alpha]$ the subspace in $\widehat{\mathcal{H}}[x;y;\eta\,;\alpha]$ generated by $\widehat{\mathcal{F}}_\Sigma[x;y;\eta]$. The $\hat{}\ \mathbf{S}_\ell$-action on the total functional space commutes with the operators $X_1,\dots,X_n$, $Y_1,\dots,Y_n$ and naturally transforms the operators $T_1[\eta;\alpha],\dots,T_\ell[\eta;\alpha]$, that is for any $f\in\widehat{\mathcal{F}}[\eta]$ and $\sigma\in\mathbf{S}_\ell$ we have

$$[X_m f\,]_\sigma = X_m[\,f\,]_\sigma\,,\qquad [Y_m f\,]_\sigma = Y_m[\,f\,]_\sigma\,,\qquad [T_a f\,]_\sigma = T_{\sigma_a}[\,f\,]_\sigma\,.$$

This means that the spaces $\widehat{\mathcal{H}}_{\Sigma}[x;y;\eta;\alpha]$ form a discrete subbundle of the cohomological bundle with the discrete Gauss-Manin connection.

Let $\mathcal{F}[\eta] \subset \widehat{\mathcal{F}}_{\Sigma}[\eta]$ be the space of rational functions $f(t;x;y)$ which are homogeneous functions of degree zero: $f(ht;hx;hy) = f(t;x;y)$ for any $h \in \mathbb{C}^{\times}$, and such that the product

$$f(t;x;y) \prod_{a=1}^{\ell} t_a^{-1} \prod_{m=1}^{n} \prod_{a=1}^{\ell} (t_a - x_m) \prod_{1 \le a < b \le \ell} \frac{\eta t_a - t_b}{t_a - t_b}$$

is a symmetric polynomial of degree less than n in each of the variables $t_1, \ldots, t_\ell$. Similarly, let $\mathcal{F}_{\circ}[\eta] \subset \widehat{\mathcal{F}}_{\Sigma}[\eta]$ be the space of rational functions $f(t;x;y)$ which are homogeneous functions of degree zero and such that the product

$$f(t;x;y) \prod_{m=1}^{n} \prod_{a=1}^{\ell} (t_a - x_m) \prod_{1 \le a < b \le \ell} \frac{\eta t_a - t_b}{t_a - t_b}$$

is a symmetric polynomial of degree less than n in each of the variables $t_1, \ldots, t_\ell$. We call the spaces $\mathcal{F}[\eta]$ and $\mathcal{F}_{\circ}[\eta]$ the *total trigonometric hypergeometric spaces.* The trigonometric hypergeometric spaces introduced in Section [3.], see (3.12), (3.13), can be obtained from the total trigonometric hypergeometric spaces by the same specialization to a fiber that produces the functional space of the fiber from the total functional space:

$$\mathcal{F}[x;\eta] = \{ f(\cdot;x;y) \mid f \in \mathcal{F}[\eta] \},$$

$$\mathcal{F}'[x;\eta^{-1}] = \mathcal{F}_{\circ}[x;\eta] = \{ f(\cdot;x;y) \mid f \in \mathcal{F}_{\circ}[\eta] \}.$$

Notice that a particular choice of y is irrelevant for the above relations.

The *hypergeometric cohomology spaces* $\mathcal{H}[x;y;\eta;\alpha]$ and $\mathcal{H}_{\circ}[x;y;\eta;\alpha]$ are subspaces of $\widehat{\mathcal{H}}_{\Sigma}[x;y;\eta;\alpha]$ generated by $\mathcal{F}[x;\eta]$ and $\mathcal{F}_{\circ}[x;\eta]$, respectively.

It is not clear in advance that the hypergeometric cohomology spaces form a discrete subbundle with respect to the Gauss-Manin connection . But they do form the subbundle after a suitable reduction of the connection.

Proposition 5.1. *[33] The collections of spaces* $\{\mathcal{H}[x;y;\eta;\alpha]\}_{(x;y)\in\mathbb{B}}$ *and* $\{\mathcal{H}_{\circ}[x;y;\eta;\alpha]\}_{(x;y)\in\mathbb{B}}$ *are invariant with respect to the operators* $X_1Y_1, \ldots, X_nY_n$.

The proposition follows from Lemma [5.1] and Proposition [5.2].

Fix nonzero complex numbers $\nu_1, \ldots, \nu_n$ and consider the following subspace of the base space

$$\mathbb{C}^n_\nu = \{ (x;y) \in \mathbb{C}^{2n} \mid y_m = \nu_m x_m, \quad m = 1, \ldots, n \},$$

$x_1, \ldots, x_n$ being coordinates on $\mathbb{C}^n_\nu$. We take $\mathbb{C}^n_\nu$ as a new base space and $\mathbb{C}^\ell \oplus \mathbb{C}^n_\nu$ as a new total space. In other words we allow only simultaneous dilations of the coordinates x_m and y_m, thus preserving the ratios x_m/y_m, $m = 1, \ldots, n$. Introduce a discrete local system on $\mathbb{C}^\ell \oplus \mathbb{C}^n_\nu$ taking

$$\widehat{\mathcal{F}}_\nu[\eta] = \{\, f|_{\mathbb{C}^\ell \oplus \mathbb{C}^n_\nu} \mid f \in \widehat{\mathcal{F}}[\eta] \,\}$$

as the total functional space of the local system and the restrictions of functions $\varphi_1, \ldots, \varphi_\ell$, $\chi_1 \phi_1, \ldots, \chi_n \phi_n$ as the system of connection coefficients, so that the corresponding twisted shift operators are $T_1, \ldots, T_\ell$, $X_1 Y_1, \ldots, X_n Y_n$. The local systems on fibers are clearly kept intact under the described reduction of the base space, so the obtained cohomological bundle over $\mathbb{C}^{\times n}_\nu = \mathbb{C}^n_\nu \cap \mathbb{C}^{\times 2n}$ is a natural reduction of the original cohomological bundle over $\mathbb{C}^{\times 2n}$.

Corollary 5.1. *The hypergeometric cohomological spaces form discrete subbundles of the cohomological bundle over $\mathbb{C}^{\times n}_\nu$ with respect to to the discrete Gauss-Manin connection.*

The statement is equivalent to Proposition [5.1].

Let $w_\mathfrak{l}$, $\mathfrak{l} \in \mathcal{Z}^n_\ell$, be the trigonometric weight functions, see (4.1), and let

$$w^\circ_\mathfrak{l}(t; x; y; \eta) = \prod_{m=1}^{n} x_m^{\mathfrak{l}_m} \prod_{a=1}^{\ell} t_a^{-1}\, w_\mathfrak{l}(t; x; y; \eta)\,,$$

so that $w^\circ_\mathfrak{l}(t; x; y; \eta) = w'_\mathfrak{l}(t; y; x; \eta^{-1})$, cf. (4.3). Then $w_\mathfrak{l} \in \mathcal{F}[\eta]$ and $w^\circ_\mathfrak{l} \in \mathcal{F}_\circ[\eta]$ for any $\mathfrak{l} \in \mathcal{Z}^n_\ell$, and the sets

$$\{\, w_\mathfrak{l}(\cdot\,; x; y; \eta) \mid \mathfrak{l} \in \mathcal{Z}^n_\ell \,\}\,, \qquad \{\, w^\circ_\mathfrak{l}(\cdot\,; x; y; \eta) \mid \mathfrak{l} \in \mathcal{Z}^n_\ell \,\}$$

are bases of the spaces $\mathcal{F}[x;\eta]$, $\mathcal{F}_\circ[x;\eta]$, respectively, see Proposition [4.1] and Corollary [4.1]. Let

$${}^k w_\mathfrak{l}(t; x; y; \eta) = w_{{}^k\mathfrak{l}}(t; {}^k x; {}^k y; \eta)\,, \qquad {}^k w^\circ_\mathfrak{l}(t; x; y; \eta) = w^\circ_{{}^k\mathfrak{l}}(t; {}^k x; {}^k y; \eta)\,,$$

where

$$\begin{aligned} {}^k\mathfrak{l} &= (\mathfrak{l}_{\mathfrak{k}+1}, \ldots, \mathfrak{l}_n, \mathfrak{l}_1, \ldots, \mathfrak{l}_\mathfrak{k})\,, \\ {}^k x &= (x_{k+1}, \ldots, x_n, x_1, \ldots, x_k)\,, \\ {}^k y &= (y_{k+1}, \ldots, y_n, y_1, \ldots, y_k)\,, \end{aligned}$$

$k = 0, \ldots, n$, cf. (4.13).

Lemma 5.1. *For any $k = 1, \ldots, n$ the sets $\{\, {}^k w_\mathfrak{l}(\cdot\,; x; y; \eta) \mid \mathfrak{l} \in \mathcal{Z}^n_\ell \,\}$, $\{\, {}^k w^\circ_\mathfrak{l}(\cdot\,; x; y; \eta) \mid \mathfrak{l} \in \mathcal{Z}^n_\ell \,\}$ are bases of the spaces $\mathcal{F}[x;\eta]$, $\mathcal{F}_\circ[x;\eta]$, respectively.*

The statement is clear, since $\mathcal{F}[x;\eta] = \mathcal{F}[{}^k x;\eta]$ and $\mathcal{F}_\circ[x;\eta] = \mathcal{F}_\circ[{}^k x;\eta]$.

Set $Z_k = X_1 Y_1 \ldots X_k Y_k$, $k = 1, \ldots, n$. Let

$$\beta_{\mathfrak{l},\mathfrak{k}}[\eta\,;\alpha] = \Bigl(\alpha\,\eta^{1-\ell} \prod_{m=1}^{n} x_m/y_m\Bigr)^{\mathfrak{l}^\mathfrak{k}}, \qquad \mathfrak{l}^\mathfrak{k} = \mathfrak{l}_1 <, \ldots, +\mathfrak{l}_\mathfrak{k}\,.$$

Proposition 5.2. *The trigonometric weight functions have the following properties:*

$$Z_k w_{\mathfrak{l}} = \beta_{\mathfrak{l},\mathfrak{e}}[\eta\,;\alpha]\,{}^k w_{\mathfrak{l}} + \sum_{a=1}^{\ell} \bigl(T_a[\eta\,;\alpha]\, f_{\mathfrak{l},a} - f_{\mathfrak{l},a}\bigr)\,, \tag{5.11}$$

$$Z_k w_{\mathfrak{l}}^{\circ} = \beta_{\mathfrak{l},\mathfrak{e}}[\eta\,;\alpha]\,{}^k w_{\mathfrak{l}}^{\circ} + \sum_{a=1}^{\ell} \bigl(T_a[\eta\,;\alpha]\, g_{\mathfrak{l},a} - g_{\mathfrak{l},a}\bigr)\,, \tag{5.12}$$

$k = 1,\dots,n$, where $f_{\mathfrak{l},a}$, $g_{\mathfrak{l},a}$, $a = 1,\dots,\ell$, are certain functions belonging to $\widehat{\mathcal{F}}[\eta]$.

The first equality is proved in [32]. The proof of the second one is similar. To show the idea of the proof we consider below the one-dimensional example.

Proof. Proof of Proposition [5.2] for $\ell = 1$. Let $\mathfrak{e}(\mathfrak{m}) = (0,\dots,1_{m\text{-th}},\dots,0)$. Notice that ${}^k\mathfrak{e}(\mathfrak{m}) = \mathfrak{e}(\mathfrak{m}-\mathfrak{k})$ for $1 \le k < m$ and ${}^k\mathfrak{e}(\mathfrak{m}) = \mathfrak{e}(\mathfrak{m}-\mathfrak{k}+\mathfrak{n})$ for $m \le k \le n$. Recall that

$$w_{\mathfrak{e}(\mathfrak{m})}(t;x;y) = \tfrac{t}{t-x_m} \prod\nolimits_{1\le l<m} \tfrac{t-y_l}{t-x_l}\,,$$

$$T_a[\alpha]\, f(t;x;y) = \alpha\, f(pt;x;y) \prod_{m=1}^{n} x_m/y_m \prod_{m=1}^{n} \frac{t-y_m}{t-x_m}\,.$$

$$Z_k f(t;x;y) =$$

$$= f(t;px_1,\dots,px_m,x_{m+1},\dots,x_n;py_1,\dots,py_m,y_{m+1},\dots,y_n)\times$$

$$\times \prod_{1\le j\le k} \frac{t-px_j}{t-py_j}\,.$$

These formulae taken together imply

$$Z_k w_{\mathfrak{e}(\mathfrak{m})} = {}^k w_{\mathfrak{e}(\mathfrak{m})} \qquad \text{for} \quad k<m\,,$$

$$T_a[\alpha]\, Z_k w_{\mathfrak{e}(\mathfrak{m})} = \alpha \prod_{m=1}^{n} x_m/y_m\, {}^k w_{\mathfrak{e}(\mathfrak{m})} \qquad \text{for} \quad k \ge m\,.$$

Since $\beta_{\mathfrak{e}(\mathfrak{m}),\mathfrak{e}} = 1$ for $k < m$ and $\beta_{\mathfrak{e}(\mathfrak{m}),\mathfrak{e}} = \alpha \prod_{m=1}^{n} x_m/y_m$ for $k \ge m$, this gives (5.11). The proof of (5.12) is similar.

Consider a section ψ of the homological bundle, that is for any $(x;y) \in \mathbb{B}$ we have a linear functional $\psi(x;y)$ on $\widehat{\mathcal{F}}[x;y;\eta]$ vanishing on $D\widehat{\mathcal{F}}[x;y;\eta\,;\alpha]$. For any function$f \in \widehat{\mathcal{F}}[\eta]$ the section ψ defines a function $\langle\psi, f\rangle$ on $\mathbb{B}$ by the rule

$$\langle\psi, f\rangle(x;y) = \bigl\langle \psi(x;y), f(\cdot\,;x;y)\bigr\rangle\,.$$

The function$\langle\psi, f\rangle$ is identically zero if $f = \sum_{a=1}^{\ell} \bigl(T_a[\eta\,;\alpha]\, f_a - f_a\bigr)$ for some $f_1,\dots,f_\ell \in \widehat{\mathcal{F}}[\eta]$. The tensor coordinates on the trigonometric hypergeometric space, see (4.7), translate the section ψ into a function Ψ taking values in the vector space $V = \bigoplus_{\mathfrak{m} \in \mathcal{Z}_\ell^n} \mathbb{C}\, v_{\mathfrak{m}}$:

$$\Psi(x;y;\eta) = \sum_{\mathfrak{m} \in \mathcal{Z}_\ell^n} \langle \psi(x;y), w_{\mathfrak{m}}(\cdot;x;y;\eta) \rangle\, v_{\mathfrak{m}}\,. \tag{5.13}$$

The function Ψ is similar to a coordinate vector of the section ψ.

Recall that the operators $Q_1, \dots, Q_n$ act on functions of x, y as follows:

$$(Q_m f)(x_1, \dots, x_n; y_1, \dots, y_n) =$$
$$= f(px_1, \dots, px_m, x_{m+1}, \dots, x_n; py_1, \dots, py_m, y_{m+1}, \dots, y_n)\,,$$

cf. (4.17), and for any $f \in \widehat{\mathcal{F}}[\eta]$ we have

$$Q_m \langle \psi, f \rangle = \langle Z_m^* \psi, Z_m f \rangle\,, \qquad m = 1, \dots, n\,, \tag{5.14}$$

by definition of the operators $Z_1^*, \dots, Z_n^*$.

Theorem 5.1. *Let ψ be a section of the homological bundle, invariant with respect to the operators $Z_1^*, \dots, Z_n^*$. Then the function Ψ, cf. (5.13), satisfies a system of difference equations*

$$Q_m \Psi(x;y) = K_m(x;y)\, \Psi(x;y)\,, \qquad m = 1, \dots, n\,, \tag{5.15}$$

where the operators $K_1, \dots, K_n$ are given by (4.14). The system coincides with (4.23).

Proof. Applying the section ψ to both sides of the equality (5.11) we obtain

$$Q_k \Psi(x;y;\eta) = \sum_{\mathfrak{l} \in \mathcal{Z}_\ell^n} \Bigl(\alpha\, \eta^{1-\ell} \prod_{m=1}^{n} x_m/y_m\Bigr)^{\mathfrak{l}^k} \langle \psi(x;y), {}^k w_{\mathfrak{l}}(\cdot;x;y;\eta) \rangle\, v_{\mathfrak{l}}\,,$$

Here relations (5.14) and the invariance of ψ have been used to transform the left hand side. Taking into account the definition of the operators $K_1, \dots, K_n$, see (4.12) and (4.14), we prove the statement.

The last theorem shows that the system of difference equations satisfied by the hypergeometric map is nearly the same as the periodic section equation for a certain discrete bundle with the discrete Gauss-Manin connection. In the next section we will construct periodic sections of the homological bundle via the hypergeometric integral.

Remark 5.1. It is proved in [30] that for generic x, y, η, α the trigonometric hypergeometric space $\mathcal{F}[x;\eta]$ is isomorphic to the hypergeometric cohomology space $\mathcal{H}[x;y;\eta;\alpha]$ via the canonical projection. In other words, the space $\mathcal{F}[x;\eta]$ contains no functions which are the total twisted differences. This assertion results from Corollary [A.1]. The same is true for the spaces $\mathcal{F}_\circ[x;\eta]$ and $\mathcal{H}_\circ[x;y;\eta;\alpha]$. Therefore,

$$\dim \mathcal{H}[x;y;\eta\,;\alpha] \;=\; \dim \mathcal{H}_{\circ}[x;y;\eta\,;\alpha] \;=\; (\,n\,) + \ell - 1n - 1\,.$$

Furthermore, it is possible to show explicitly that the hypergeometric cohomology spaces $\mathcal{H}[x;y;\eta\,;\alpha]$ and $\mathcal{H}_{\circ}[x;y;\eta\,;\alpha]$ are the same, which implies that the functionΨ in addition to difference equations (5.15) satisfies a difference equation with respect to the transformation $\alpha \mapsto p\,\alpha$. The corresponding linear isomorphism of the total functional space $\widehat{\mathcal{F}}[\eta\,]$ is

$$f(t;x;y) \;\mapsto\; t_1 \ldots t_\ell\, f(t;x;y)$$

and it maps bijectively the trigonometric hypergeometric space $\mathcal{F}_{\circ}[x;\eta\,]$ to $\mathcal{F}[x;\eta\,]$.

Moreover, Aomoto implicitly showed in [1] that for almost all x, y, η, α

$$\dim \widehat{\mathcal{H}}_{\Sigma}[x;y;\eta\,;\alpha] \;=\; (\,n\,) + \ell - 1n - 1\,.$$

Thereby, the trigonometric hypergeometric space $\mathcal{F}[x;\eta\,]$ is a model for $\widehat{\mathcal{H}}_{\Sigma}[x;y;\eta\,;\alpha]$, and the function$\Psi$ defined by a periodic section ψ of the homological bundle solves difference equations corresponding to all the twisted shift operators $X_1, \ldots, X_n, Y_1, \ldots, Y_n$, not only to their products $X_1Y_1, \ldots, X_nY_n$. Unfortunately, an explicit form of the additional difference equations is not known yet.

5.4 Periodic sections of the homological bundle via the hypergeometric integral

Consider the elliptic hypergeometric space $\mathcal{F}_{ell}[x;\eta\,;\alpha]$ introduced in Section [3.], cf. (3.15), (3.16). Let us extend the hypergeometric pairing $I\,[x;y;\eta\,;\alpha]$, cf. (3.18), to the pairing $\widehat{\mathcal{F}}[x;y;\eta\,] \otimes \mathcal{F}_{ell}[x;\eta\,;\alpha] \to \mathbb{C}$ in such a way that it will vanish on $D\widehat{\mathcal{F}}[x;y;\eta;\alpha] \otimes \mathcal{F}_{ell}[x;\eta\,;\alpha]$, thus inducing a map from the elliptic hypergeometric space $\mathcal{F}_{ell}[x;\eta\,;\alpha]$ to the homology space $\widehat{\mathcal{H}}^*[x;y;\eta;\alpha]$. We describe this extension explicitly below.

Let $\Phi(t;x;y;\eta)$ be the phase function (3.17) and $Int\,[x;y;\eta\,]$ be the hypergeometric integral (3.11). Take functions $f \in \widehat{\mathcal{F}}[x;y;\eta\,]$ and $g \in \mathcal{F}_{ell}[x;\eta\,;\alpha]$. The product $fg\,\Phi(\cdot\,;x;y;\eta)$ is eligible for substituting as an integrand into $Int\,[p^{-s}x;p^{s}y;p^{s}\eta\,]$ if s is a large positive integer, and the result does not depend on s. So, for given f, g we define

$$I\,[x;y;\eta\,;\alpha](f,g) \;=\; Int\,[p^{-s}x;p^{-s}y;p^{s}\eta\,](fg\,\Phi) \tag{5.16}$$

taking a sufficiently large s, and extend the hypergeometric pairing as follows:

$$\hat{I}\,[x;y;\eta\,;\alpha] \;:\; \widehat{\mathcal{F}}[x;y;\eta\,] \otimes \mathcal{F}_{ell}[x;\eta\,;\alpha] \;\to\; \mathbb{C}, \tag{5.17}$$

$$f \otimes g \;\mapsto\; I\,[x;y;\eta\,;\alpha](f,g)\,.$$

In the one-dimensional case definitions (5.16), (5.17) are equivalent to (2.8), (2.9). The restriction of $\hat{I}\,[x;y;\eta\,;\alpha]$ on $\mathcal{F}[x;\eta\,] \otimes \mathcal{F}_{ell}[x;\eta\,;\alpha]$ clearly coincides with $I\,[x;y;\eta\,;\alpha]$.

Proposition 5.3. *[30] For any $f \in D\widehat{\mathcal{F}}[x;y;\eta;\alpha]$ and $g \in \mathcal{F}_{ell}[x;\eta;\alpha]$ we have $I[x;y;\eta;\alpha](f,g) = 0$.*

Proof (idea).

Proof. Denote for a while by U_a the shift operator with respect to t_a:

$$U_a f(t_1,\dots,t_\ell) = f(t_1,\dots,pt_a,\dots,t_\ell).$$

For any eligible function$F(t)$ the hypergeometric integral $Int[x;y;\eta]$ has the property:

$$Int[x;y;\eta](U_a F) = Int[x;y;\eta](F), \qquad a = 1,\dots,\ell, \tag{5.18}$$

Observe that the product $g\,\Phi(\cdot;x;y;\eta)$ is a phase function of the system of connection coefficients of the local system of a fiber:

$$U_a\left(g\,\Phi(\cdot;x;y;\eta)\right) = \varphi_a(\cdot;x;y;\eta;\alpha)\,g\,\Phi(\cdot;x;y;\eta)$$

cf. (5.5), (5.7), (5.8). Hence, we have

$$(T_a[\eta;\alpha]\,h)\,g\,\Phi(\cdot;x;y;\eta) = U_a\left(h\,g\,\Phi(\cdot;x;y;\eta)\right)$$

for any $h \in \widehat{\mathcal{F}}[x;y;\eta]$. Therefore, formulae (5.16) and (5.18) show that

$$I[x;y;\eta;\alpha](T_a[\eta;\alpha]\,h - h, g) = 0, \qquad a = 1,\dots,\ell,$$

which implies the required statement, cf. (5.10).

The last proposition means that $\hat{I}[x;y;\eta;\alpha]$ maps the elliptic hypergeometric space $\mathcal{F}_{ell}[x;\eta;\alpha]$ into the homology space $\widehat{\mathcal{H}}^*[x;y;\eta;\alpha]$.

Notice that the hypergeometric integral $Int[x;y;\eta]$ is invariant with respect to the transformations $x_m \mapsto p\,x_m$ and $y_m \mapsto p\,y_m$, that is

$$\begin{gathered} Int[x_1,\dots,px_m,\dots,x_n;y;\eta](F) = \\ Int[x;y_1,\dots,py_m,\dots,y_n;\eta](F) = Int[x;y;\eta](F) \end{gathered} \tag{5.19}$$

whenever the hypergeometric integrals are meaningful.

Consider a bundle over $\mathbb{B}$ with a fiber $\mathcal{F}_{ell}[x;\eta;\alpha]$ over a point $(x;y) \in \mathbb{B}$. To have a section g of this bundle means for any $(x;y) \in \mathbb{B}$ to have a functiong$(\cdot;x;y) \in \mathcal{F}_{ell}[x;\eta;\alpha]$. Notice, that for any point in the $\hat{}\,\mathbb{Z}^{2n}$-orbit of $(x;y)$ the fibers are the same and equal to $\mathcal{F}_{ell}[x;\eta;\alpha]$. Introduce a dual discrete flat connection on the bundle defining commuting operators $\widetilde{X}_1,\dots,\widetilde{X}_n, \widetilde{Y}_1,\dots,\widetilde{Y}_n$ acting on sections:

$$\widetilde{X}_m\,g(\cdot;x_1,\dots,px_m,\dots,x_n;y) = (-p\,x_m)^\ell\,g(\cdot;x_1,\dots,x_n;y), \tag{5.20}$$

$$\widetilde{Y}_m\,g(\cdot;x;y_1,\dots,py_m,\dots,y_n) = (-p\,y_m)^{-\ell}\,g(\cdot;x;y_1,\dots,y_n),$$

$m = 1, \ldots, n$, and taking these operators for the twisted shift operators. We call the obtained bundle with the discrete flat connection the *elliptic hypergeometric bundle.*

For a section g of the elliptic hypergeometric bundle denote by $\hat{I}\,[\eta\,;\alpha]\,g$ the corresponding section of the homological bundle:

$$(\hat{I}\,[\eta\,;\alpha]\,g)(x;y) \;=\; \hat{I}\,[x;y;\eta\,;\alpha]\,g(\cdot;x;y)\,.$$

Lemma 5.2. *The map $\hat{I}\,[\eta\,;\alpha]$ transforms the discrete connection (5.20) on the elliptic hypergeometric bundle to the dual Gauss-Manin connection on the homological bundle. That is*

$$\hat{I}\,[\eta\,;\alpha]\,\widetilde{X}_m \;=\; X^*_m\,\hat{I}\,[\eta\,;\alpha], \qquad \hat{I}\,[\eta\,;\alpha]\,\widetilde{Y}_m \;=\; Y^*_m\,\hat{I}\,[\eta\,;\alpha], \qquad m = 1, \ldots, n\,,$$

where the operators $X^*_1, \ldots, X^*_n, Y^*_1, \ldots, Y^*_n$ act on sections of the homological bundle according to the dual Gauss-Manin connection.

Proof. Take a function $f \in \widehat{\mathcal{F}}[\eta\,]$ and a section g of the elliptic hypergeometric bundle. Let

$$F_m(x;y) \;=\; \hat{I}\,[x;y;\eta\,;\alpha]\big(f(\cdot;x;y), \widetilde{X}_m\, g(\cdot;x;y)\big)\,.$$

Denote for a while by U_m the shift operator with respect to x_m:

$$U_m\, h(x_1, \ldots, x_n) \;=\; h(x_1, \ldots, p x_m, \ldots, x_n)\,.$$

To prove the first relation we have to show that for any $m = 1, \ldots, n$

$$U_m F_m(x;y) \;=\; \hat{I}\,[x;y;\eta\,;\alpha]\big(X_m f(\cdot;x;y), g(\cdot;x;y)\big)$$

where X_m is the twisted shift operator. Formulae (5.6), (5.7), (5.9) and (5.20) imply that $X_m f\, g\, \Phi = U_m\big(f\, \widetilde{X}_m g\, \Phi\big)$. So the claim follows from the invariance (5.19) of the hypergeometric integral.

The proof of the second relation is similar.

Given a section g of the elliptic hypergeometric bundle the map $\hat{I}\,[\eta\,;\alpha]$ combined with with the tensor coordinates on the trigonometric hypergeometric space produces a function $\Psi_g(x;y;\eta\,;\alpha)$ with values in the vector space V by the rule

$$\Psi_g(x;y;\eta\,;\alpha) \;=\; \sum_{\mathfrak{m} \in \mathcal{Z}^n_\ell} I\,[x;y;\eta\,;\alpha]\big(w_{\mathfrak{m}}(\cdot;x;y;\eta); g(\cdot;x;y)\big)\, v_{\mathfrak{m}}\,, \qquad (5.21)$$

cf. (5.13).

Theorem 5.2. *Let the section g be invariant with respect to the operators $\widetilde{X}_1\widetilde{Y}_1, \ldots, \widetilde{X}_n\widetilde{Y}_n$, that is*

$$g(\cdot;x_1, \ldots, p x_m, \ldots, x_n; y_1, \ldots, p y_m, \ldots, y_n) \;=\; (x_m/y_m)^\ell\, g(\cdot;x;y)\,, \quad (5.22)$$

cf. (5.9). Then the function Ψ_g satisfies the difference equations (5.15). Moreover, for generic α all solutions of the system (5.15) can be obtained in this way.

Proof. Proof (sketch). The first part of the statement follows from Lemma [5.2] and Theorem [5.1]. The second part follows from Proposition [3.2] and the nondegeneracy of the hypergeometric pairing, cf. Proposition [3.3].

The elliptic weight function $W_{\mathfrak{l}}(t;x;y;\eta;\alpha)$, cf. (4.2), defines a section $W_{\mathfrak{l}}[\eta;\alpha]$ of the elliptic hypergeometric bundle in a natural way:

$$W_{\mathfrak{l}}[\eta;\alpha](x;y) = W_{\mathfrak{l}}(\cdot;x;y;\eta;\alpha).$$

Its transformation properties differ a little from (5.22), containing one more multiplier

$$Q_m W_{\mathfrak{l}} = \mu_{\mathfrak{l},\mathfrak{m}} \prod_{1\le j\le m} (x_j/y_j)^{\ell}\, W_{\mathfrak{l}},$$

$$\mu_{\mathfrak{l},\mathfrak{m}} = \Big(\alpha\,\eta^{1-\mathfrak{l}^{\mathfrak{m}}} \prod_{1\le j\le m} x_j/y_j\Big)^{-\mathfrak{l}^{\mathfrak{m}}}, \qquad \mathfrak{l}^{\mathfrak{m}} = \mathfrak{l}_1 <,\dots,+\mathfrak{l}_{\mathfrak{m}},$$

cf. (4.21). This results in a slight modification of the difference equations for the function$\Psi_{W_{\mathfrak{l}}[\eta;\alpha]}$ compared with (5.15):

$$Q_m \Psi_{W_{\mathfrak{l}}[\eta;\alpha]}(x;y) = \mu_{\mathfrak{l},\mathfrak{m}}\, K_m(x;y)\, \Psi_{W_{\mathfrak{l}}[\eta;\alpha]}(x;y), \tag{5.23}$$

$m = 1,\dots,n$. The last system of difference equations is nothing else but the system of difference equations (4.19) satisfied by the hypergeometric map $\bar{I}\,[x;y;\eta;\alpha]$ because

$$\Psi_{W_{\mathfrak{l}}[\eta;\alpha]}(x;y) = \bar{I}\,[x;y;\eta;\alpha]\, v_{\mathfrak{l}}^{*} \tag{5.24}$$

by the definition of the tensor coordinates on the elliptic hypergeometric space.

6. Asymptotics of the hypergeometric maps

In this section we prove Propositions [4.4] and [4.5] for the one-dimensional case to illustrate the idea of the proof. The proof for the multidimensional case is given in [30].

All over this section we assume that $\ell = 1$. We use notations introduced in Section [2.] for the one-dimensional case. In particular, the labels $\mathfrak{e}(1),\dots,\mathfrak{e}(\mathfrak{n})$, which should be used for the weight functions according to notations in the general case, are replaced by labels $1,\dots,n$, the ordering (3.5) being translated as follows: $\mathfrak{e}(\mathfrak{l}) \ll \mathfrak{e}(\mathfrak{m})$ for $l > m$.

We consider an asymptotic zone $\mathbb{A}$ of the parameters $x_1,\dots,x_n,y_1,\dots,y_n$:

$$\mathbb{A} = \left\{ \begin{array}{ll} |x_m/x_{m+1}| \ll 1, & m = 1,\dots,n-1 \\ |x_m/y_m| \simeq 1, & m = 1,\dots,n \end{array} \right\},$$

writing $(x;y) \rightrightarrows \mathbb{A}$ if $x_m/x_{m+1} \to 0$ for any $m = 1,\dots,n-1$ and the ratios x_m/y_m and y_m/x_m remain bounded for any $m = 1,\dots,n$, cf. (4.25). We consider the functions

$$I_{lm}(x;y) = I[x;y;\alpha]\bigl(w_l(\cdot;x;y), W_m(\cdot;x;y;\alpha)\bigr), \qquad (6.1)$$
$$I'_{lm}(x;y) = I'[x;y;\alpha]\bigl(w'_l(\cdot;x;y), W'_m(\cdot;x;y;\alpha)\bigr)$$

where the functions w_l, w'_l, W_m, W'_m are given by (2.18). Propositions [4.4] and [4.5] are now equivalent to the following statement.

Proposition 6.1. *For any $l,m = 1,\dots,n$ the hypergeometric integrals $I_{lm}(x;y)$ and $I'_{lm}(x;y)$ have finite limits as $(x;y) \rightrightarrows \mathbb{A}$. Moreover $\lim_{\mathbb{A}} I_{lm} = 0$ unless $l \le m$, $\lim_{\mathbb{A}} I'_{lm} = 0$ unless $l \ge m$, and*

$$\lim_{\mathbb{A}} I_{mm} = \frac{(\alpha_m^{-1})_\infty\,(p\,\alpha_m x_m/y_m)_\infty}{(x_m/y_m)_\infty\,(p)_\infty},$$

$$\lim_{\mathbb{A}} I'_{mm} = \frac{-\alpha_m\,(p\,\alpha_m)_\infty\,(\alpha_m^{-1}y_m/x_m)_\infty}{(y_m/x_m)_\infty\,(p)_\infty}.$$

Here $\alpha_m = \alpha\prod_{1\le k<m} x_k/y_k$.

Proof. We give a proof for the integrals $I_{lm}(x;y)$. The proof for the integrals $I'_{lm}(x;y)$ is similar.

The function$I_{lm}(x;y)$ is given by

$$I_{lm}(x;y) = \frac{1}{2\pi i}\int_{\widetilde{\mathbb{T}}_r[x;y]} w_l(t;x;y)\,W_m(t;x;y;\alpha)\,\Phi(t;x;y)\,\frac{dt}{t}, \qquad (6.2)$$

cf. (2.8) and (2.4). Using formulae (2.6) and (2.18) we obtain that the integrand has the form

$$\frac{\theta(\alpha_m^{-1}t/x_m)}{(t-x_l)\,(p\,x_m/t)_\infty\,(t/y_m)_\infty\,(p)_\infty}\cdot$$
$$\cdot\prod_{1\le i<l}\frac{t-y_i}{t-x_i}\prod_{1\le j<m}\frac{(p\,y_j/t)_\infty}{(p\,x_j/t)_\infty}\prod_{m<k\le n}\frac{(t/x_k)_\infty}{(t/y_k)_\infty},$$

so it has singularities at most at the following points:

$$t=0, \qquad t=p^s x_j, \qquad t=p^{-s}y_k, \qquad 1\le j\le m\le k\le n, \qquad s\in\mathbb{Z}_{\ge 0}.$$

It is easy to see that the hypergeometric integral $\mathit{Int}[x;y](w_l\,W_m\,\Phi)$ does not change if we make simultaneous dilations of $t, x_1,\dots,x_n, y_1,\dots,y_n$ by the same multiple. So without loss of generality we can assume that as $(x;y) \rightrightarrows \mathbb{A}$ the parameters x_m, y_m remain bounded and separated from zero, while $x_j, y_j \to 0$ for $1\le j<m$ and $x_k, y_k \to \infty$ for $m<k\le n$. Therefore, we can replace in (6.2) the integration contour $\widetilde{\mathbb{T}}_r[x;y]$ by

$$\widetilde{\mathbb{T}}_r^{(N)}[x_m, y_m] = \mathbb{T}_r + \sum_{\substack{s \in \mathbb{Z}_{<N} \\ |p^s x_m| > r}} C[p^s x_m] - \sum_{\substack{s \in \mathbb{Z}_{>-N} \\ |p^s y_m| < r}} C[p^s y_m]$$

for a sufficiently large positive integer N, since all other small circles $C[p^s x_i]$, $C[p^s y_i]$ appearing in (2.3) contain no poles of the integrand inside and, thereby, do not contribute to the integral.

Furthermore, the integrand have a finite limit on $\widetilde{\mathbb{T}}_r^{(N)}[x_m, y_m]$ as $x_j, y_j \to 0$ for $1 \le j < m$ and $x_k, y_k \to \infty$ for $m < k \le n$, while x_m, y_m remain fixed. The limit equals

$$\begin{array}{lll} \dfrac{\theta(\alpha_m^{-1} t/x_m)}{t\,(p\,x_m/t)_\infty\,(t/y_m)_\infty\,(p)_\infty} & \text{for} & l < m\,, \\ \dfrac{\theta(\alpha_m^{-1} t/x_m)}{t\,(x_m/t)_\infty\,(t/y_m)_\infty\,(p)_\infty} & \text{for} & l = m\,, \\ 0 & \text{for} & l > m\,. \end{array}$$

The remaining integrals

$$\frac{1}{2\pi i} \int_{\widetilde{\mathbb{T}}_r^{(N)}[x_m, y_m]} \frac{\theta(\alpha_m^{-1} t/x_m)}{(p\,x_m/t)_\infty\,(t/y_m)_\infty\,(p)_\infty}\,\frac{dt}{t}$$

$$\text{and} \quad \frac{1}{2\pi i} \int_{\widetilde{\mathbb{T}}_r^{(N)}[x_m, y_m]} \frac{\theta(\alpha_m^{-1} t/x_m)}{(x_m/t)_\infty\,(t/y_m)_\infty\,(p)_\infty}\,\frac{dt}{t}$$

can be calculated using the formula

$$\frac{1}{2\pi i} \int_C \frac{\theta(c\,t)}{(a/t)_\infty (b\,t)_\infty}\,\frac{dt}{t} = \frac{(ac)_\infty\,(p\,b/c)_\infty}{(ab)_\infty}$$

where C is an anticlockwise oriented contour around the origin $t = 0$ separating the sets $\{\,p^s a \mid s \in \mathbb{Z}_{\le 0}\}$ and $\{\,p^s/b \mid s \in \mathbb{Z}_{\ge 0}\}$.

7. The quantum loop algebra $U_q'(\widetilde{\mathfrak{gl}_2})$ and the qKZ equation

In Sections 7. and 8. we show in what way the hypergeometric spaces and the tensor coordinates arise in the representation theory of the quantum loop algebra $U_q'(\widetilde{\mathfrak{gl}_2})$ and the elliptic quantum group $E_{\rho,\gamma}(\mathfrak{sl}_2)$. We identify the difference equations (4.23) and (5.14) with the trigonometric qKZ equation, cf. (7.11), (7.12), and the difference equation (4.24) with the dual qKZ equation (7.3). Further, we interpret the hypergeometric map as a map from a module over the elliptic quantum group to the corresponding module over the quantum loop algebra.

7.1 Highest weight $U_q(\mathfrak{sl}_2)$-modules

Let q be a nonzero complex number which is not a root of unity. Consider the quantum group $U_q(\mathfrak{sl}_2)$ with generators $E, F, q^{\pm H}$ and relations:

$$q^H q^{-H} = q^{-H} q^H = 1\,,$$
$$q^H E = q\,E\,q^H\,, \qquad q^H F = q^{-1} F\,q^H\,,$$
$$[E,F] = \tfrac{q^{2H}-q^{-2H}}{q-q^{-1}}\,.$$

Let the coproduct $\Delta : U_q(\mathfrak{sl}_2) \to \mathfrak{U}_q(\mathfrak{sl}_2) \otimes \mathfrak{U}_q(\mathfrak{sl}_2)$ be given by

$$\begin{aligned} \Delta(q^H) &= q^H \otimes q^H, & \Delta(q^{-H}) &= q^{-H} \otimes q^{-H}, \\ \Delta(E) &= E \otimes 1 + q^{2H} \otimes E\,, & \Delta(F) &= F \otimes q^{-2H} + 1 \otimes F\,. \end{aligned} \tag{7.1}$$

The coproduct defines a $U_q(\mathfrak{sl}_2)$-modulestructure on a tensor product of $U_q(\mathfrak{sl}_2)$-modules.

Remark 7.1. We take here the coproduct for $U_q(\mathfrak{sl}_2)$ which differs slightly from the coproduct used in [30]. This results also in a modification of the embedding $U_q(\mathfrak{sl}_2) \hookrightarrow \mathfrak{U}'_q(\widetilde{\mathfrak{gl}_2})$ and the evaluation homomorphism $\epsilon : U'_q(\widetilde{\mathfrak{gl}_2}) \to U_q(\mathfrak{sl}_2)$, cf. (7.4) and (7.5).

Let V be the $U_q(\mathfrak{sl}_2)$-modulewith highest weight q^Λ. Let $V = \bigoplus\limits_{l=0}^{\infty} V_{\Lambda-l}$ be its weight decomposition. For any nonzero complex number u we define an operator $u^{\Lambda-H} \in \operatorname{End}(V)$ by

$$u^{\Lambda-H} v = u^l v \qquad \text{for any} \quad v \in V_{\Lambda-l}\,.$$

7.2 The quantum loop algebra $U'_q(\widetilde{\mathfrak{gl}_2})$

The *quantum loop algebra* $U_q(\widetilde{\mathfrak{gl}_2})$ is a unital associative algebra with generators $L_{ij}^{(+0)}$, $L_{ji}^{(-0)}$, $1 \le j \le i \le 2$, and $L_{ij}^{(s)}$, $i,j = 1,2$, $s = \pm 1, \pm 2, \ldots$, subject to relations (7.2), (7.3).

Let e_{ij} be the 2×2 matrix with the only nonzero entry 1 at the intersection of the ˆi-th row and ˆj-th column. Set

$$\begin{aligned} R(u) = \quad & (uq - q^{-1})\,(e_{11} \otimes e_{11} + e_{22} \otimes e_{22}) + \\ & + (u-1)\,(e_{11} \otimes e_{22} + e_{22} \otimes e_{11}) + \\ & + u(q-q^{-1})\,e_{12} \otimes e_{21} + (q-q^{-1})\,e_{21} \otimes e_{12}\,. \end{aligned}$$

Introduce the generating series $L_{ij}^{\pm}(u) = L_{ij}^{(\pm 0)} + \sum\limits_{s=1}^{\infty} L_{ij}^{(\pm s)} u^{\pm s}$. The relations in the quantum loop algebra $U_q(\widetilde{\mathfrak{gl}_2})$ have the form

$$L_{ii}^{(+0)} L_{ii}^{(-0)} = 1\,, \qquad L_{ii}^{(-0)} L_{ii}^{(+0)} = 1\,, \qquad i = 1,2\,, \tag{7.2}$$

$$R(x/y)\, L_{(1)}^{+}(x)\, L_{(2)}^{+}(y) = L_{(2)}^{+}(y)\, L_{(1)}^{+}(x)\, R(x/y)\,,$$

$$R(x/y)\, L_{(1)}^{+}(x)\, L_{(2)}^{-}(y) = L_{(2)}^{-}(y)\, L_{(1)}^{+}(x)\, R(x/y)\,,$$

$$R(x/y)\, L_{(1)}^{-}(x)\, L_{(2)}^{-}(y) = L_{(2)}^{-}(y)\, L_{(1)}^{-}(x)\, R(x/y)\,,$$

where $L_{(1)}^{\pm}(u) = \sum_{ij} e_{ij} \otimes 1 \otimes L_{ij}^{\pm}(u)$ and $L_{(2)}^{\pm}(u) = \sum_{ij} 1 \otimes e_{ij} \otimes L_{ij}^{\pm}(u)$, see [24], [7]. The elements $L_{11}^{(+0)} L_{22}^{(+0)}$, $L_{22}^{(+0)} L_{11}^{(+0)}$, $L_{11}^{(-0)} L_{22}^{(-0)}$, $L_{22}^{(-0)} L_{11}^{(-0)}$ are central in $U_q(\widetilde{\mathfrak{gl}_2})$. We impose the following relations:

$$\begin{aligned} L_{11}^{(+0)} L_{22}^{(+0)} &= 1\,, & L_{22}^{(+0)} L_{11}^{(+0)} &= 1\,, \\ L_{11}^{(-0)} L_{22}^{(-0)} &= 1\,, & L_{22}^{(-0)} L_{11}^{(-0)} &= 1\,, \end{aligned} \tag{7.3}$$

in addition to relations (7.2) and denote the corresponding quotient algebra by $U'_q(\widetilde{\mathfrak{gl}_2})$.

The quantum loop algebra $U'_q(\widetilde{\mathfrak{gl}_2})$ is a Hopf algebra with a coproduct $\Delta : U'_q(\widetilde{\mathfrak{gl}_2}) \to U'_q(\widetilde{\mathfrak{gl}_2}) \otimes U'_q(\widetilde{\mathfrak{gl}_2})$:

$$\Delta : L_{ij}^{\pm}(u) \mapsto \sum_k L_{ik}^{\pm}(u) \otimes L_{kj}^{\pm}(u)\,.$$

There is an important one-parametric family of automorphisms $\rho_x : U'_q(\widetilde{\mathfrak{gl}_2}) \to U'_q(\widetilde{\mathfrak{gl}_2})$:

$$\rho_x : L_{ij}^{\pm}(u) \mapsto L_{ij}^{\pm}(u/x)\,,$$

The quantum loop algebra $U'_q(\widetilde{\mathfrak{gl}_2})$ contains $U_q(\mathfrak{sl}_2)$ as a Hopf subalgebra, the embedding being given by

$$q^H \mapsto L_{11}^{(-0)} q^{-H}\, E \mapsto -L_{21}^{(+0)}/(q-q^{-1}) F\, q^H \mapsto L_{12}^{(-0)}/(q-q^{-1}). \tag{7.4}$$

There is also an *evaluation homomorphism* $\epsilon : U'_q(\widetilde{\mathfrak{gl}_2}) \to U_q(\mathfrak{sl}_2)$:

$$L^{\pm}(u) \mapsto \mp\, u^{\vartheta_\pm} \begin{pmatrix} q^H - q^{-H} u^{-1} & F\, q^H (q-q^{-1}) \\ q^{-H} E\, (q-q^{-1})\, u^{-1} & q^{-H} - q^H u^{-1} \end{pmatrix} \tag{7.5}$$

where $\vartheta_+ = 1$, $\vartheta_- = 0$ and we arrange the generating series $L_{ij}^{\pm}(u)$ into a 2×2 matrix $L^{\pm}(u)$. Both the automorphisms ρ_x and the homomorphism ϵ restricted to the subalgebra $U_q(\mathfrak{sl}_2)$ are the identity maps.

For any $U_q(\mathfrak{sl}_2)$-module/ V denote by $V(x)$ the $\hat{}\,U'_q(\widetilde{\mathfrak{gl}_2})$- module which is obtained from the module V via the homomorphism $\epsilon \circ \rho_x$. The module $V(x)$ is called the *evaluation module.*

Let V_1, V_2 be Verma modules over $U_q(\mathfrak{sl}_2)$ with generating vectors v_1, v_2, respectively. For generic complex numbers x, y the $\hat{}\,U_q'(\widetilde{\mathfrak{gl}_2})$- modules $V_1(x) \otimes V_2(y)$ and $V_2(y) \otimes V_1(x)$ are isomorphic. The modules are generated by the vectors $v_1 \otimes v_2$ and $v_2 \otimes v_1$. The intertwiner is unique up to normalization and maps the subspace $\mathbb{C}\,(v_1 \otimes v_2)$ to $\mathbb{C}\,(v_2 \otimes v_1)$. The normalized intertwiner has the form $P_{V_1V_2}\,R_{V_1V_2}\,(x/y)$ where $P_{V_1V_2}\, : V_1 \otimes V_2 \to V_2 \otimes V_1$ is the permutation map and $R_{V_1V_2}\,(z)$ is a functionwith values in End $(V_1 \otimes V_2)$ having the properties:

$$R_{V_1V_2}(z)\,(F \otimes q^{-2H} + 1 \otimes F) \,=\, (F \otimes 1 + q^{-2H} \otimes F)\,R_{V_1V_2}(z)\,, \quad (7.6)$$

$$R_{V_1V_2}(z)\,(F \otimes 1 + z\,q^{-2H} \otimes F) \,=\, (F \otimes q^{-2H} + z\,1 \otimes F)\,R_{V_1V_2}(z)\,.$$

$$R_{V_1V_2}(z)\,v_1 \otimes v_2 \,=\, v_1 \otimes v_2\,,$$

Such a function$R_{V_1V_2}(z)$ exists and is uniquely determined. It is called the $\mathfrak{sl}_2$ *trigonometric R-matrix* for the tensor product $V_1 \otimes V_2$. The trigonometric R-matrix also satifies the following relations:

$$R_{V_1V_2}(z)\,q^H \otimes q^H \,=\, q^H \otimes q^H R_{V_1V_2}(z)\,. \quad (7.7)$$

$$R_{V_1V_2}(z)\,(E \otimes 1 + q^{2H} \otimes E) \,=\, (E \otimes q^{2H} + 1 \otimes E)\,R_{V_1V_2}(z)\,,$$

$$R_{V_1V_2}(z)\,(z\,E \otimes q^{2H} + 1 \otimes E) \,=\, (z\,E \otimes 1 + q^{2H} \otimes E)\,R_{V_1V_2}(z)\,.$$

In particular, $R_{V_1V_2}(z)$ respects the weight decomposition of $V_1 \otimes V_2$.

Proposition 7.1. *Let V_1, V_2 be Verma modules over $U_q(\mathfrak{sl}_2)$ with highest weights q^{Λ_1}, q^{Λ_2}, respectively. Let $(V_1 \otimes V_2)_\ell$ be the weight subspace of weight $q^{\Lambda_1+\Lambda_2-\ell}$. Then the restriction of $R_{V_1V_2}(z)$ to $(V_1 \otimes V_2)_\ell$ is a rational function of z which is regular for $z \neq q^{2\Lambda_1+2\Lambda_2-2r}$, $r = 0, \dots, \ell-1$. Moreover, $R_{V_1V_2}(z)$ is nondegenerate/ for $z \neq q^{2r-2\Lambda_1-2\Lambda_2}$, $r = 0, \dots, \ell-1$.*

The trigonometric R-matrix satisfies the inversion relation

$$P_{V_1V_2}\;R_{V_1V_2}(z) \,=\, \bigl(R_{V_2V_1}\,(z^{-1})\bigr)^{-1}\,P_{V_1V_2}\,. \quad (7.8)$$

For $U_q(\mathfrak{sl}_2)$ Verma modules V_1, V_2, V_3 the corresponding R-matrices satisfy the Yang-Baxter equation:

$$R_{V_1V_2}\,(x/y)\,R_{V_1V_3}\,(x)\,R_{V_2V_3}\,(y) \,=\, R_{V_2V_3}\,(y)\,R_{V_1V_3}\,(x)\,R_{V_1V_2}\,(x/y)\,. \quad (7.9)$$

All these properties of trigonometric R-matrix are well known (cf. [5], [14], [29], [3]).

7.3 The trigonometric qKZ equation

Let $V_1, \ldots, V_n$ be Verma modules over $U_q(\mathfrak{sl}_2)$ with highest weights $q^{\Lambda_1}, \ldots, q^{\Lambda_n}$, respectively. Let $R_{V_i V_j}(u)$ be the corresponding trigonometric $\hat{}R$-matrices. Let $R_{ij}(x) \in \mathrm{End}\,(V_1 \otimes \ldots \otimes V_n)$ be defined in a standard way:

$$R_{ij}(u) \;=\; \sum \mathrm{id} \otimes \ldots \otimes \% \overset{i-th}{r(u)} \otimes \ldots \otimes \% \overset{j-th}{r'(u)} \otimes \ldots \otimes \mathrm{id}$$

provided that $R_{V_i V_j}(u) = \sum r(u) \otimes r'(u) \in \mathrm{End}\,(V_i \otimes V_j)$. For any $X \in U_q(\mathfrak{sl}_2)$ set

$$X_m \;=\; \mathrm{id} \otimes \ldots \otimes \% \overset{m-th}{X} \otimes \ldots \otimes \mathrm{id}\,.$$

Let p, κ be complex numbers. For any $m = 1, \ldots, n$ set

$$\begin{aligned} A_m(z_1, \ldots, z_n) \quad &= R_{m,m-1}(p\,z_m/z_{m-1}) \ldots R_{m,1}(p\,z_m/z_1) \times \\ &\times \kappa^{\Lambda_m - H_m} R_{m,n}(z_m/z_n) \ldots R_{m,m+1}(z_m/z_{m+1})\,. \end{aligned} \tag{7.10}$$

Theorem 7.1. *[9] The linear maps $A_1(z), \ldots, A_n(z)$ obey the flatness conditions*

$$\begin{aligned} A_l(z_1, \ldots, p z_m, \ldots, z_n)\, A_m(z_1, \ldots, z_n) \;=& \\ =\; A_m(z_1, \ldots, p\, z_l, \ldots, z_n)\, A_l(z_1, \ldots, z_n)\,,& \end{aligned}$$

$l, m = 1, \ldots, n.$

The statement follows from the Yang-Baxter equation (7.9) and the inversion relation (7.8).

The maps $A_1(z), \ldots, A_n(z)$ define a discrete flat connection on the trivial bundle over $\mathbb{C}^{\times n}$ with fiber $V_1 \otimes \ldots \otimes V_n$. This connection is called the *qKZ connection.*

By (7.7) the operators $A_1(z), \ldots, A_n(z)$ commute with the action of q^H in $V_1 \otimes \ldots \otimes V_n$:

$$[A_m(z_1, \ldots, z_n)\,,\, q^H\,] = 0\,, \qquad m = 1, \ldots, n\,,$$

and, therefore, respect the weight decomposition of $V_1 \otimes \ldots \otimes V_n$.

Lemma 7.1. *Let $(V_1 \otimes \ldots \otimes V_n)_\ell$ be the weight subspace of weight $q^{\sum \Lambda_m - \ell}$. For any $z \in \mathbb{C}^{\times n}$ such that $q^{2\Lambda_l + 2\Lambda_m - 2r} z_l/z_m \neq p^s$ for all $r = 0, \ldots, \ell - 1$, and $l, m = 1, \ldots, n$, $l \neq m$, $s \in \mathbb{Z}$, the linear maps $A_1(z), \ldots, A_n(z)$ define isomorphisms of $(V_1 \otimes \ldots \otimes V_n)_\ell$.*

The statement follows from (7.10) and Proposition [7.1].

The *qKZ equation* for a function $\Psi(z_1, \ldots, z_n)$ with values in $V_1 \otimes \ldots \otimes V_n$ is the following system of difference equations [9] :

$$\Psi(z_1, \ldots, p z_m, \ldots, z_n) \;=\; A_m(z_1, \ldots, z_n)\, \Psi(z_1, \ldots, z_n)\,, \tag{7.11}$$

$m = 1, \dots, n$. The *qKZ* equation is a remarkable system of difference equations with applications in representation theory and quantum integrable models, see [9], [26], [15], [21], [19].

To make easier the future comparing of the *qKZ* equation and the difference equations for the hypergeometric maps we rewrite the *qKZ* equation in the following way:

$$\Psi(p\, z_1, \dots, p\, z_m, z_{m+1}, \dots, z_n) = A^{\{m\}}(z_1, \dots, z_n)\, \Psi(z_1, \dots, z_n), \qquad (7.12)$$

$m = 1, \dots, n$, where

$$A^{\{m\}}(z_1, \dots, z_n) = \prod_{l=1}^{m} \kappa^{\Lambda_l - H_l}\, \mathbf{R_m}(\mathbf{z_1}, \dots, \mathbf{z_n}), \qquad (7.13)$$

$$\begin{aligned} \mathbf{R_m}(\mathbf{z_1}, \dots, \mathbf{z_n}) = \quad & R_{1,n}(z_1/z_n) \dots R_{1,m+1}(z_1/z_{m+1}) \times \\ & \times R_{2,n}(z_2/z_n) \dots R_{2,m+1}(z_2/z_{m+1}) \times \\ & \times \dots\dots\dots\dots\dots\dots\dots \times \\ & \times R_{m,n}(z_m/z_n) \dots R_{m,m+1}(z_m/z_{m+1}). \end{aligned} \qquad (7.14)$$

Let $\mathbf{P_m} : \mathbf{V_1} \otimes \dots \otimes \mathbf{V_n} \to \mathbf{V_{m+1}} \otimes \dots \otimes \mathbf{V_n} \otimes \mathbf{V_1} \otimes \dots \otimes \mathbf{V_m}$ be the rotation map:

$$\mathbf{P_m} : \mathbf{v^{(1)}} \otimes \dots \otimes \mathbf{v^{(n)}} \mapsto \mathbf{v^{(m+1)}} \otimes \dots \otimes \mathbf{v^{(n)}} \otimes \mathbf{v^{(1)}} \otimes \dots \otimes \mathbf{v^{(m)}}.$$

Then the product $\mathbf{P_m R_m}(\mathbf{z_1}, \dots, \mathbf{z_n})$ is the normalized intertwiner of the tensor products of $U'_q(\widetilde{\mathfrak{gl}_2})$ evaluation modules:

$$\begin{aligned} \mathbf{P_m R_m}(\mathbf{z_1}, \dots, \mathbf{z_n}) : \mathbf{V_1}(\mathbf{z_1}) \otimes \dots \otimes \mathbf{V_n}(\mathbf{z_n}) \to \\ \to V_{m+1}(z_{m+1}) \otimes \dots \otimes V_n(z_n) \otimes V_1(z_1) \otimes \dots \otimes V_m(z_m). \end{aligned} \qquad (7.15)$$

$$\mathbf{P_m R_m}(\mathbf{z_1}, \dots, \mathbf{z_n}) : \mathbf{v_1} \otimes \dots \otimes \mathbf{v_n} \mapsto \mathbf{v_{m+1}} \otimes \dots \otimes \mathbf{v_n} \otimes \mathbf{v_1} \otimes \dots \otimes \mathbf{v_m}.$$

Here $v_1, \dots, v_n$ are generating vectors of the modules $V_1, \dots, V_n$, respectively.

The dual *qKZ* equation for a functionΨ' with values in $(V_1 \otimes \dots \otimes V_n)^*$ is the following system of difference equations:

$$\begin{aligned} \Psi'(p\, z_1, \dots, p\, z_m, z_{m+1}, \dots, z_n) = \\ = \left(A^{\{m\}}(z_1, \dots, z_n)\right)^{*-1} \Psi'(z_1, \dots, z_n) m = 1, \dots, n. \end{aligned} \qquad (7.16)$$

7.4 Tensor coordinates on the trigonometric hypergeometric spaces

Let $V_1, \ldots, V_n$ be Verma modules over $U_q(\mathfrak{sl}_2)$ with highest weights $q^{\Lambda_1}, \ldots, q^{\Lambda_n}$ and generating vectors $v_1, \ldots, v_n$, respectively. Let

$$V_1 \otimes \ldots \otimes V_n = \bigoplus_{k=0}^{\infty} (V_1 \otimes \ldots \otimes V_n)_k$$

be the weight decomposition of the tensor product, the weight subspace $(V_1 \otimes \ldots \otimes V_n)_k$ having weight $q^{\sum \Lambda_m - k}$. Denote by $(v_1 \otimes \ldots \otimes v_n)^*$ a linear functional on $V_1 \otimes \ldots \otimes V_n$ such that

$$\langle (v_1 \otimes \ldots \otimes v_n)^*, v_1 \otimes \ldots \otimes v_n \rangle = 1,$$

$$\langle (v_1 \otimes \ldots \otimes v_n)^*, v \rangle = 0 \qquad \text{for any} \quad v \in \bigoplus_{k=1}^{\infty} (V_1 \otimes \ldots \otimes V_n)_k .$$

Fix a nonnegative integer ℓ and consider the vector spaces $V = \bigoplus_{\mathfrak{m} \in \mathcal{Z}^n_\ell} \mathbb{C}\, v_{\mathfrak{m}}$ and $V' = \bigoplus_{\mathfrak{m} \in \mathcal{Z}^n_\ell} \mathbb{C}\, v'_{\mathfrak{m}}$ like in Section 4.. Assume that

$$q^{4\Lambda_m - 2r} \neq 1, \qquad r = 0, \ldots, \ell - 1.$$

We identify the space V with the weight subspace $(V_1 \otimes \ldots \otimes V_n)_\ell$ and the space V' with the dual space $(V_1 \otimes \ldots \otimes V_n)^*_\ell$ by the rule:

$$\begin{aligned} v_{\mathfrak{l}} &= \textstyle\prod_{1 \le j < k \le n} q^{-2\mathfrak{l}_j \Lambda_k} F^{\mathfrak{l}_1} v_1 \otimes \ldots \otimes F^{\mathfrak{l}_n} v_n , \\ v'_{\mathfrak{l}} &= \textstyle\prod_{1 \le j < k \le n} q^{2\mathfrak{l}_j \Lambda_k} \left(E_1^{\mathfrak{l}_1} \otimes \ldots \otimes E_n^{\mathfrak{l}_n} \right)^* (v_1 \otimes \ldots \otimes v_n)^* . \end{aligned} \tag{7.17}$$

Then

$$\langle v'_{\mathfrak{l}}, v_{\mathfrak{m}} \rangle = \delta_{\mathfrak{l}\,\mathfrak{m}} \prod_{m=1}^{n} \prod_{s=0}^{\mathfrak{l}_m - 1} \frac{(1 - q^{2s+2})\,(q^{4\Lambda_m - 2s} - 1)}{1 - q^2} . \tag{7.18}$$

Furthermore, we identify the parameters which are related to the hypergeometric integrals with the parameters related to representations of the quantum loop algebra and the qKZ equation as follows:

$$\begin{aligned} &p = p, \qquad && \eta = q^2, \qquad && \alpha = \kappa\, q^{2\ell - 2 - 2\sum \Lambda_m}, \\ &x_m = q^{2\Lambda_m} z_m , \qquad && y_m = q^{-2\Lambda_m} z_m , \qquad && m = 1, \ldots, n . \end{aligned} \tag{7.19}$$

Consider the dual spaces $V^* = \bigoplus_{\mathfrak{m} \in \mathcal{Z}^n_\ell} \mathbb{C}\, v^*_{\mathfrak{m}}$ and $V'^* = \bigoplus_{\mathfrak{m} \in \mathcal{Z}^n_\ell} \mathbb{C}\, v'^*_{\mathfrak{m}}$ with the dual bases:

$$\langle v^*_{\mathfrak{l}}, v_{\mathfrak{m}} \rangle = \delta_{\mathfrak{l}\,\mathfrak{m}} , \qquad \langle v'^*_{\mathfrak{l}}, v'_{\mathfrak{m}} \rangle = \delta_{\mathfrak{l}\,\mathfrak{m}} .$$

Recall that the tensor coordinates on the trigonometric hypergeometric spaces $\mathcal{F}[x;\eta]$ and $\mathcal{F}'[y;\eta]$ are the following linear maps:

$$\begin{aligned} B[x;y;\eta]: V^* &\to \mathcal{F}[x;\eta], & B'[x;y;\eta]: V'^* &\to \mathcal{F}'[y;\eta], \\ v^*_{\mathrm{m}} &\mapsto w_{\mathrm{m}}(t;x;y;\eta), & v'^*_{\mathrm{m}} &\mapsto w'_{\mathrm{m}}(t;x;y;\eta), \end{aligned} \tag{7.20}$$

cf. (4.7), where w_{m} and w'_{m} are the trigonometric weight functions, cf. (4.1) and (4.3). Notice that, unlike in Section [4.], we distinguish here the vector spaces associated with the tensor coordinates of different types. In particular, now $(\ ,\ )[x;y;\eta]$, see (4.8), is a pairing of V and V', not a bilinear form on V. Comparing formulae (7.18) and (4.9) we immediately find that the pairing $(\ ,\ )[x;y;\eta]$ of V and V' actually coincides with the canonical pairing of the weight subspace $(V_1 \otimes \ldots \otimes V_n)_\ell$ and its dual space $(V_1 \otimes \ldots \otimes V_n)^*_\ell$. The same statement in different words is

$$\big(B[x;y;\eta]\big)^* S\,[x;y;\eta]\,B'[x;y;\eta] = \mathrm{id} \in \mathrm{End}\,\big((V_1 \otimes \ldots \otimes V_n)_\ell\,\big).$$

Here $S\,[x;y;\eta] : \mathcal{F}'[y;\eta] \to \big(\mathcal{F}[x;\eta]\big)^*$ is the Shapovalov pairing of the trigonometric hypergeometric spaces.

Theorem 7.2. *[17], [30] Let $v_{\mathfrak{l}}, v'_{\mathfrak{l}}$, $\mathfrak{l} \in \mathcal{Z}^n_\ell$, be given by formulae (7.17). Then*

$$\begin{aligned} L^-_{12}(t_1)\ldots L^-_{12}(t_\ell)\, v_1 \otimes \ldots \otimes v_n &= (q-q^{-1})^\ell\, q^{\ell \sum \Lambda_m - \ell(\ell-1)/2} \times \\ \times \prod_{a=1}^{\ell} \prod_{m=1}^{n} (1 - x_m/t_a) \prod_{1 \le a < b \le \ell} \tfrac{\eta\, t_a - t_b}{t_a - t_b} & \sum_{\mathrm{m} \in \mathcal{Z}^n_\ell} w_{\mathrm{m}}(t;x;y;\eta)\, v_{\mathrm{m}}\,, \end{aligned} \tag{7.21}$$

$$\begin{aligned} \big(L^-_{21}(t_1)\ldots L^-_{21}(t_\ell)\big)^*(v_1 \otimes \ldots \otimes v_n)^* &= q^{\ell(\ell-1)/2 - \ell \sum \Lambda_m} \times \\ \times \prod_{a=1}^{\ell} \prod_{m=1}^{n} (1 - y_m/t_a) \prod_{1 \le a < b \le \ell} \tfrac{\eta^{-1} t_a - t_b}{t_a - t_b} & \sum_{\mathrm{m} \in \mathcal{Z}^n_\ell} w'_{\mathrm{m}}(t;x;y;\eta)\, v'_{\mathrm{m}}\,. \end{aligned} \tag{7.22}$$

Proof. Proof (idea). Consider the formula (7.21) as a definition of the functions $w_{\mathrm{m}}(t;x;y;\eta)$. Then using the coproduct and the commutation relations for the quantum loop algebra one can show that these functions belong to the trigonometric hypergeometric space $\mathcal{F}[x;\eta]$ and have the double triangularity properties (4.6). Hence, by Lemma [4.1] they are linear combinations of the trigonometric weight functions which also have the double triangularity properties (4.6). This means that the connection matrix is diagonal, and one has only to show that its diagonal entries equal one, which can be done by comparing either residues at points $x \triangleright \mathrm{m}[\eta]$ or values at points $y \triangleright \mathrm{m}[\eta^{-1}]$.

The proof of formula (7.22) is similar.

Formulae (7.21) and (7.22) are the key points in connecting the trigonometric hypergeometric spaces and the tensor coordinates with the representations of the quantum loop algebra.

Corollary 7.1. *[30], [31] For any $v \in (V_1 \otimes \ldots \otimes V_n)_\ell$ and $v^* \in (V_1 \otimes \ldots \otimes V_n)^*_\ell$ we have*

$$\langle v^* , L^-_{12}(t_1) \dots L^-_{12}(t_\ell)\, v_1 \otimes \dots \otimes v_n \rangle = (B[x;y;\eta]\, v^*)(t_1,\dots,t_\ell) \times$$

$$\times\ (q-q^{-1})^\ell\, q^{\ell \sum \Lambda_m - \ell(\ell-1)/2} \prod_{a=1}^{\ell} \prod_{m=1}^{n} (1 - x_m/t_a) \prod_{1\le a<b\le \ell} \frac{\eta\, t_a - t_b}{t_a - t_b}\,,$$

$$\langle (v_1 \otimes \dots \otimes v_n)^* , L^-_{21}(t_1) \dots L^-_{21}(t_\ell)\, v \rangle = (B'[x;y;\eta]\, v)(t_1,\dots,t_\ell) \times$$

$$,\times\, q^{\ell(\ell-1)/2 - \ell \sum \Lambda_m} \prod_{a=1}^{\ell} \prod_{m=1}^{n} (1 - y_m/t_a) \prod_{1\le a<b\le \ell} \frac{\eta^{-1} t_a - t_b}{t_a - t_b}\,.$$

The statement follows from (7.20) and formulae (7.21), (7.22).

Formulae (7.15) and Corollary [7.1] imply that after a suitable identification of notations, see (7.17), (7.19), the operators $A\#1,\dots,A\#n$, $(A\#1)^{*-1},\dots,(A\#n)^{*-1}$ given by (7.13) almost coincide with the operators $K_1,\dots,K_n$, $K'_1,\dots,K'_n$ given by (4.14):

$$A^{\{m\}} = q^{-2\ell \sum_{j=1}^{m} \Lambda_j} K_m\,,\ (A^{\{m\}})^{*-1} = q^{2\ell \sum_{j=1}^{m} \Lambda_j} K'_m\,. \tag{7.23}$$

That is the system of difference equations (4.23) nearly coincides with the qKZ equation (7.12), the functionΨ taking values in the weight subspace $(V_1\otimes \dots\otimes V_n)_\ell$, while the system of difference equations (4.24) nearly coincides with the dual qKZ equation (7.3), the functionΨ' taking values in $(V_1 \otimes\dots\otimes V_n)^*_\ell$.

Let $\mathcal{F}_{ell}[x;\eta\,;\alpha]$ and $\mathcal{F}'_{ell}[y;\eta\,;\alpha]$ be the elliptic hypergeometric spaces. Let g and g' be sections of the elliptic hypergeometric bundles with fibers $\mathcal{F}_{ell}[x;\eta\,;\alpha]$ and $\mathcal{F}'_{ell}[y;\eta\,;\alpha]$, respectively, that is for any point $(x;y)$ functions $g(\cdot;x;y) \in \mathcal{F}_{ell}[x;\eta\,;\alpha]$ and $g'(\cdot;x;y) \in \mathcal{F}'_{ell}[y;\eta\,;\alpha]$ are given. Let $I\,[x;y;\eta\,;\alpha]$ and $I'[x;y;\eta\,;\alpha]$ be the hypergeometric pairings (3.18). The sections g and g' define functions Ψ_g and $\Psi'_{g'}$ taking values in $(V_1\otimes\dots\otimes V_n)_\ell$ and $(V_1\otimes\dots\otimes V_n)^*_\ell$, respectively:

$$\Psi_g(x;y;\eta\,;\alpha) = \sum_{\mathrm{m}\in \mathcal{Z}^n_\ell} I\,[x;y;\eta\,;\alpha]\big(w_{\mathrm{m}}(\cdot;x;y;\eta); g(\cdot;x;y)\big)\, v_{\mathrm{m}}\,,$$

$$\Psi'_{g'}(x;y;\eta\,;\alpha) = \sum_{\mathrm{m}\in \mathcal{Z}^n_\ell} I'[x;y;\eta\,;\alpha]\big(w'_{\mathrm{m}}(\cdot;x;y;\eta); g'(\cdot;x;y)\big)\, v'_{\mathrm{m}}\,,$$

cf. (5.21).

Recall that the parameters $x_1,\dots,x_n$, $y_1,\dots,y_n$ and $z_1,\dots,z_n$, $\Lambda_1,\dots,\Lambda_n$ are related via (7.19).

Theorem 7.3. *Assume that the section g has the property*

$$g(\cdot;x_1,\dots,px_m,\dots,x_n;y_1,\dots,py_m,\dots,y_n) = q^{2\ell\Lambda_m}\, g(\cdot;x;y)\,,$$

$m=1,\dots,n$. *Then the function*Ψ_g *is a solution of the* qKZ *equation (7.12).*

Proof. The claim follows from Theorem [5.2] and formulae (7.23).

Theorem 7.4. *Assume that the section* g' *has the property*

$$g'(\cdot;x_1,\dots,px_m,\dots,x_n;y_1,\dots,py_m,\dots,y_n) = q^{-2\ell\Lambda_m}\, g'(\cdot;x;y)\,,$$

$m=1,\dots,n$. *Then the function*$\Psi'_{g'}$ *is a solution of the dual* qKZ *equation (7.3).*

The proof is similar to the proof of Theorem [7.3].

The pairing $\langle \Psi'_{g'}, \Psi_g\rangle$ of solutions of the qKZ and dual qKZ equations is clearly a ˆp-periodic functionof $z_1,\dots,z_n$, i.e. a quasiconstant. The hypergeometric Riemann identity expresses this quasiconstant in terms of the elliptic Shapovalov pairing of the sections g and g'.

Theorem 7.5.

$$\langle \Psi'_{g'}(x;y;\eta\,;\alpha), \Psi_g(x;y;\eta\,;\alpha)\rangle = \mathcal{S}_{ell}[x;y;\eta\,;\alpha]\big(g(\cdot;x;y), g'(\cdot;x;y)\big)\,.$$

8. The elliptic quantum group $E_{\rho,\gamma}(\mathfrak{sl}_2)$

8.1 Modules over the elliptic quantum group $E_{\rho,\gamma}(\mathfrak{sl}_2)$

In this section we recall the definitions concerning the elliptic quantum group $E_{\rho,\gamma}(\mathfrak{sl}_2)$ and the elliptic dynamical R-matrices associated with tensor products of ˆ$E_{\rho,\gamma}(\mathfrak{sl}_2)$- modules. For a more detailed exposition on the subject and proofs see [8], [12].

Fix two complex numbers ρ,γ such that $\operatorname{Im}\rho>0$. Set $p=e^{2\pi i\rho}$ and $\eta=e^{-4\pi i\gamma}$. Let

$$\alpha(x,\lambda) = \frac{\eta\,\theta(x)\,\theta(\lambda/\eta)}{\theta(\eta\,x)\,\theta(\lambda)}\,, \qquad \beta(x,\lambda) = \frac{\theta(\eta)\,\theta(x\lambda)}{\theta(\eta\,x)\theta(\lambda)}\,.$$

Let e_{ij} be the 2×2 matrix with the only nonzero entry 1 at the intersection of the ˆi-th row and ˆj-th column. Set

$$R(x,\lambda) = e_{11}\otimes e_{11} + e_{22}\otimes e_{22} + \alpha(x,\lambda)\, e_{11}\otimes e_{22} +$$
$$+\,\alpha(x,\lambda^{-1})\, e_{22}\otimes e_{11} + \beta(x,\lambda)\, e_{12}\otimes e_{21} + \beta(x,\lambda^{-1})\, e_{21}\otimes e_{12}\,.$$

Remark 8.1. In [12] the elliptic quantum group $E_{\rho,\gamma}(\mathfrak{sl}_2)$ is described in terms of the additive theta-function

$$\Theta(u) = -\sum_{m=-\infty}^{\infty} \exp\big(\pi i(m+1/2)^2\rho + 2\pi i(m+1/2)(u+1/2)\big)\,,$$

which is related to the multiplicative theta-function$\theta(x)$ by the equality

$$\Theta(u) = i\,\exp(\pi i\rho/4 - \pi i u)\,\theta(e^{2\pi i u})\,.$$

Let $\mathfrak{h}$ be the one-dimensional Lie algebra with the generator H. Let V be an $\hat{}\,\mathfrak{h}$- module. Say that V is diagonalizable if V is a direct sum of finite-dimensional eigenspaces of H:

$$V = \bigoplus_{\mu} V_{\mu}, \qquad Hv = \mu v \qquad \text{for } v \in V_{\mu}.$$

For a function$X(\mu)$ taking values in End (V) we set $X(H)\,v = X(\mu)\,v$ for any $v \in V_{\mu}$.

Let $V_1, \ldots, V_n$ be diagonalizable $\hat{}\,\mathfrak{h}$- modules. We have the decomposition

$$V_1 \otimes \ldots \otimes V_n = \bigoplus_{\mu_1,\ldots,\mu_n} (V_1)_{\mu_1} \otimes \ldots \otimes (V_n)_{\mu_n}.$$

Set $H_m = \mathrm{id} \otimes \ldots \otimes \% \overset{m-th}{H} \otimes \ldots \otimes \mathrm{id}$.

For a function$X(\mu_1, \ldots, \mu_n)$ taking values in End $(V_1 \otimes \ldots \otimes V_n)$ we set $X(H_1, \ldots, H_n)\,v = X(\mu_1, \ldots, \mu_n)\,v$ for any $v \in (V_1)_{\mu_1} \otimes \ldots \otimes (V_n)_{\mu_n}$.

By definition, see [12], a *module over the elliptic quantum group* $E_{\rho,\gamma}(\mathfrak{sl}_2)$ is a diagonalizable $\hat{}\,\mathfrak{h}$- module V together with four $\hat{}$ End (V)-valued functions $T_{ij}(u,\lambda)$, $i,j = 1,2$, which are meromorphic in $u, \lambda \in \mathbb{C}^{\times}$ and obey the following relations

$$[\,T_{ij}(u,\lambda)\,, H\,] = (j-i)\,H\,, \qquad i,j = 1,2\,,$$

$$R(x/y, \eta^{1\otimes 1\otimes 2H}\,\lambda)\,T_{(1)}(x,\lambda)\,T_{(2)}(y, \eta^{2H\otimes 1\otimes 1}\,\lambda) =$$

$$= T_{(2)}(y,\lambda)\,T_{(1)}(x, \eta^{1\otimes 2H\otimes 1}\,\lambda)\,R(x/y,\lambda)\,.$$

Here $T_{(1)}(u,\lambda) = \sum_{ij} e_{ij} \otimes \mathrm{id} \otimes T_{ij}(u,\lambda)$, $T_{(2)}(u,\lambda) = \sum_{ij} \mathrm{id} \otimes e_{ij} \otimes T_{ij}(u,\lambda)$, and H acts in $\mathbb{C}^2$ as $(e_{11} - e_{22})/2$.

Example. Fix a complex number Λ. Consider a diagonalizable $\hat{}\,\mathfrak{h}$- module $V^{\Lambda} = \bigoplus_{k\in\mathbb{Z}_{\geq 0}} \mathbb{C}v^{[k]}$ such that

$$Hv^{[k]} = (\Lambda - k)\,v^{[k]}\,.$$

Let z be a nonzero complex number. Set

$$T_{11}(u,\lambda)\,v^{[k]} = \frac{\theta(\eta^{\Lambda-k}u/z)\,\theta(\eta^{-k}\lambda)}{\theta(\eta^{\Lambda}u/z)\,\theta(\lambda)}\,\eta^k v^{[k]}\,,$$

$$T_{12}(u,\lambda)\,v^{[k]} = \frac{\theta(\eta^{\Lambda-k-1}\lambda\,u/z)\,\theta(\eta)}{\theta(\eta^{\Lambda}u/z)\,\theta(\lambda)}\,\eta^k v^{[k+1]}\,,$$

$$T_{21}(u,\lambda)\,v^{[k]} = \frac{\theta(\eta^{k-1-\Lambda}\lambda^{-1}u/z)\,\theta(\eta^{2\Lambda-k+1})\,\theta(\eta^k)}{\theta(\eta^{\Lambda}u/z)\,\theta(\lambda^{-1})\,\theta(\eta)}\,\eta^{1-k}v^{[k-1]}\,,$$

$$T_{22}(u,\lambda)\, v^{[k]} = \frac{\theta(\eta^{k-\Lambda} u/z)\,\theta(\eta^{2\Lambda-k}\lambda)}{\theta(\eta^{\Lambda} u/z)\,\theta(\lambda)}\, v^{[k]} .$$

These formulae make V^{Λ} into an $\hat{}E_{\rho,\gamma}(\mathfrak{sl}_2)$- module $V^{\Lambda}(z)$ which is called the *evaluation Verma module* with *highest weight* Λ and *evaluation point* z [12].

For any complex vector space V denote by $Fun(V)$ the space of $\hat{}V$-valued meromorphic functions on $\mathbb{C}^{\times}$. The space V is naturally embedded in $Fun(V)$ as the subspace of constant functions.

Let V_1, V_2 be complex vector spaces. Any function$\varphi \in Fun(\mathrm{Hom}\,(V_1, V_2))$ induces a linear map

$$^{Fun}\varphi : Fun(V_1) \to Fun(V_2) , \qquad {}^{Fun}\varphi : f(\lambda) \mapsto \varphi(\lambda)\, f(\lambda) .$$

For any $\hat{}E_{\rho,\gamma}(\mathfrak{sl}_2)$- module V we define the associated *operator algebra* acting on the space $Fun(V)$. The operator algebra is generated by meromorphic functions in λ, η^{H} acting pointwise,

$$\varphi(\lambda, \eta^{H}) : f(\lambda) \mapsto \varphi(\lambda, \eta^{H}) f(\lambda) ,$$

and by values and residues with respect to u of the operator-valued meromorphic functions $\widetilde{T}_{ij}(u)$, $i, j = 1, 2$, defined below:

$$\widetilde{T}_{i1}(u) : f(\lambda) \mapsto T_{i1}(u,\lambda)\, f(\eta\,\lambda) ,$$

$$\widetilde{T}_{i2}(u) : f(\lambda) \mapsto T_{i2}(u,\lambda)\, f(\eta^{-1}\lambda) .$$

The relations obeyed by the generators of the operator algebra are described in detail in [12].

Let V_1, V_2 be $\hat{}E_{\rho,\gamma}(\mathfrak{sl}_2)$- modules. An element $\varphi \in Fun(\mathrm{Hom}\,(V_1, V_2))$ such that the induced map $^{Fun}\varphi$ intertwines the actions of the respective operator algebras is called a *morphism* of $\hat{}E_{\rho,\gamma}(\mathfrak{sl}_2)$- modules V_1, V_2. A morphism φ is called an *isomorphism* if the linear map $\varphi(\lambda)$ is nondegenerate/ for generic λ.

Example. Evaluation Verma modules $V^{\Lambda}(x)$ and $V^{\mathrm{M}}(x)$ are isomorphic if $\eta^{\Lambda} = \eta^{\mathrm{M}}$ with the tautological isomorphism.

The elliptic quantum group $E_{\rho,\gamma}(\mathfrak{sl}_2)$ has the coproduct Δ^{ell}:

$$\Delta^{ell} : H \mapsto H \otimes 1 + 1 \otimes H ,$$

$$\Delta^{ell} : T_{ij}(u,\lambda) \mapsto \sum_{k} \bigl(1 \otimes T_{ik}(u, \eta^{2H\otimes 1}\lambda)\bigr)\, \bigl(T_{kj}(u,\lambda) \otimes 1\bigr) .$$

The precise meaning of the coproduct is that it defines an $\hat{}E_{\rho,\gamma}(\mathfrak{sl}_2)$-module structure on the tensor product $V_1 \otimes V_2$ of $\hat{}E_{\rho,\gamma}(\mathfrak{sl}_2)$- modules V_1, V_2. If V_1, V_2, V_3 are $\hat{}E_{\rho,\gamma}(\mathfrak{sl}_2)$- modules, then the modules $(V_1 \otimes V_2) \otimes V_3$ and $V_1 \otimes (V_2 \otimes V_3)$ are naturally isomorphic [8].

Remark 8.2. Notice that we take the coproduct Δ^{ell} which is opposite to the coproduct used in [8], [12], [10]. The coproduct Δ^{ell} is in a sence opposite to the coproduct Δ taken for the quantum loop algebra $U'_q(\widetilde{\mathfrak{gl}_2})$.

Theorem 8.1. *[12], [10] Let $V^\Lambda(x), V^M(y)$ be evaluation Verma modules. Then for any Λ, M and generic x, y there is a unique isomorphism $\widetilde{R}_{(x,y)}$ of $\hat{}E_{\rho,\gamma}(\mathfrak{sl}_2)$- modules $V^\Lambda(x) \otimes V^M(y)$, $V^M(y) \otimes V^\Lambda(x)$ s.t. $\widetilde{R}_{(x,y)}(\lambda) v^{[0]} \otimes v^{[0]} = v^{[0]} \otimes v^{[0]}$.*
Moreover, $\widetilde{R}_{(x,y)}$ has the form

$$\widetilde{R}_{(x,y)}(\lambda) = P_{V^\Lambda V^M} R^{ell}_{V^\Lambda V^M}(x/y, \lambda)$$

where

$$P_{V^\Lambda V^M} : V^\Lambda \otimes V^M \to V^M \otimes V^\Lambda$$

is the permutation map and $R^{ell}_{V^\Lambda V^M}(u, \lambda)$ is a meromorphic function of $u, \lambda \in \mathbb{C}^\times$ with values in End $(V^\Lambda \otimes V^M)$.

Corollary 8.1. *The function$R^{ell}_{V^\Lambda V^M}(x, \lambda)$ satisfies the inversion relation*

$$P_{V^\Lambda V^M} R^{ell}_{V^\Lambda V^M}(x, \lambda) = \left(R^{ell}_{V^M V^\Lambda}(x^{-1}, \lambda)\right)^{-1} P_{V^\Lambda V^M} .$$

The function$R^{ell}_{V^\Lambda V^M}(x, \lambda)$ is called the $\mathfrak{sl}_2$ *dynamical elliptic R-matrix* for the tensor product $V^\Lambda \otimes V^M$.

Theorem 8.2. *[12], [10] For any complex numbers Λ, M, N the corresponding elliptic R-matrices satisfy the dynamical Yang-Baxter equation:*

$$R^{ell}_{V^\Lambda V^M}(x/y, \eta^{1\otimes 1\otimes 2H} \lambda) R^{ell}_{V^\Lambda V^N}(x, \lambda) R^{ell}_{V^M V^N}(y, \eta^{2H\otimes 1\otimes 1} \lambda) =$$

$$= R^{ell}_{V^M V^N}(y, \lambda) R^{ell}_{V^\Lambda V^N}(x, \eta^{1\otimes 2H\otimes 1} \lambda) R^{ell}_{V^\Lambda V^M}(x/y, \lambda) .$$

8.2 Tensor coordinates on the elliptic hypergeometric spaces

Let $V_1^e(z_1), \ldots, V_n^e(z_n)$ be evaluation Verma modules over $E_{\rho,\gamma}(\mathfrak{sl}_2)$ with highest weights $\Lambda_1, \ldots, \Lambda_n$ and evaluation points $z_1, \ldots, z_n$, respectively. Let $V_1^e, \ldots, V_n^e$ be the corresponding $\hat{}\mathfrak{h}$- modules. Let

$$V_1^e \otimes \ldots \otimes V_n^e = \bigoplus_{\ell=0}^{\infty} (V_1^e \otimes \ldots \otimes V_n^e)_\ell$$

be the weight decomposition of the tensor product, the weight subspace $(V_1^e \otimes \ldots \otimes V_n^e)_\ell$ having weight $\sum \Lambda_m - \ell$ with respect to $\mathfrak{h}$. The weight subspace has a basis given by the monomials $v^{[\mathfrak{l}_1]} \otimes \ldots \otimes v^{[\mathfrak{l}_n]}$, $\mathfrak{l} \in \mathcal{Z}_\ell^n$. We denote by $(v^{[\mathfrak{l}_1]} \otimes \ldots \otimes v^{[\mathfrak{l}_n]})^*$, $\mathfrak{l} \in \mathcal{Z}_\ell^n$, the dual basis of $(V_1^e \otimes \ldots \otimes V_n^e)_\ell^*$.

Recall that the elliptic weight functions $W_{\mathrm{m}}(t;x;y;\eta;\alpha)$, $W'_{\mathrm{m}}(t;x;y;\eta;\alpha)$ are given by formulae (4.2), (4.3). We relate the parameters $x_1,\dots,x_n$, $y_1,\dots,y_n$ and $z_1,\dots,z_n$, $\Lambda_1,\dots,\Lambda_n$ as follows:

$$x_m = \eta^{\Lambda_m} z_m\,, \qquad y_m = \eta^{-\Lambda_m} z_m\,, \qquad m = 1,\dots,n\,, \tag{8.1}$$

The next theorem is an elliptic analogue of Theorem [7.2].

Theorem 8.3. *[10], [30]*

$$T^-_{12}(t_1,\lambda)\dots T^-_{12}(t_\ell,\eta^{1-\ell}\lambda)\, v^{[0]}\otimes\dots\otimes v^{[0]} = \tag{8.2}$$
$$= \eta^{\ell(\ell-1)/2} \prod_{s=0}^{\ell-1} \tfrac{\theta(\eta)}{\theta(\eta^{-s}\lambda)} \prod_{1\le a<b\le\ell} \tfrac{\theta(\eta^{-1}t_a/t_b)}{\theta(t_a/t_b)} \times$$
$$\times \sum_{\mathrm{m}\in\mathcal{Z}^n_\ell} W'_{\mathrm{m}}(t;x;y;\eta;\eta^{-1}\lambda)\, v^{[\mathrm{m}_1]}\otimes\dots\otimes v^{[\mathrm{m}_n]}\,,$$

$$\bigl(T^-_{21}(t_1,\eta^{1-\ell}\lambda)\dots T^-_{21}(t_\ell,\lambda)\bigr)^*(v^{[0]}\otimes\dots\otimes v^{[0]})^* = \tag{8.3}$$
$$\prod_{s=0}^{\ell-1} \frac{\theta(\eta^{1-s-\ell+2\sum\Lambda_m}\lambda)}{\theta(\eta)} \prod_{a=1}^{\ell}\prod_{m=1}^{n} \frac{\theta(t_a/x_m)}{\theta(t_a/y_m)} \prod_{1\le a<b\le\ell} \frac{\theta(\eta\, t_a/t_b)}{\theta(t_a/t_b)} \times$$
$$\times \sum_{\mathrm{m}\in\mathcal{Z}^n_\ell} W_{\mathrm{m}}(t;x;y;\eta;\lambda)\, N_{\mathrm{m}}(\lambda)\, (v^{[\mathrm{m}_1]}\otimes\dots\otimes v^{[\mathrm{m}_n]})^*$$

where

$$N_{\mathrm{l}}(\lambda) = \prod_{m=1}^{n}\prod_{s=0}^{\mathrm{l}_m-1} \frac{\theta(\eta^{s+1})\,\theta(\eta^{2\Lambda_m-s})}{\eta^s\,\theta(\eta^s\lambda^{-1}_{\mathrm{l},\mathrm{m}})\,\theta(\eta^{1-s-\mathrm{l}_m+2\Lambda_m}\lambda_{\mathrm{l},\mathrm{m}})} \tag{8.4}$$

and $\lambda_{\mathrm{l},\mathrm{m}} = \lambda\prod_{1\le j<m}\eta^{2\Lambda_j-2\mathrm{l}_j}$.

The proof is similar to the proof of Theorem [7.2].

Consider the vector spaces

$$\mathcal{V}_{ell} = \bigoplus_{\mathrm{m}\in\mathcal{Z}^n_\ell} \mathbb{C}\, \mathrm{v}\,\mathrm{m}\,, \qquad \mathcal{V}'_{ell} = \bigoplus_{\mathrm{m}\in\mathcal{Z}^n_\ell} \mathbb{C}\, \mathrm{v}\,\mathrm{m}'$$

and their dual spaces

$$\mathcal{V}^*_{ell} = \bigoplus_{\mathrm{m}\in\mathcal{Z}^n_\ell} \mathbb{C}\, \mathrm{v}\,\mathrm{m}^*\,, \qquad {\mathcal{V}'_{ell}}^* = \bigoplus_{\mathrm{m}\in\mathcal{Z}^n_\ell} \mathbb{C}\, \mathrm{v}\,\mathrm{m}'^*\,,$$

so that $\langle \mathrm{v}\,\mathrm{l}^*, \mathrm{v}\mathrm{m}\rangle = \delta_{\mathrm{l}\,\mathrm{m}}$, $\langle \mathrm{v}\,\mathrm{l}'^*, \mathrm{v}\mathrm{m}'\rangle = \delta_{\mathrm{l}\,\mathrm{m}}$. Motivated by formulae (8.2) and (8.3) we identify the spaces $\mathcal{V}'_{ell}$ and $\mathcal{V}^*_{ell}$ with the weight subspace $(V^e_1\otimes\dots\otimes V^e_n)_\ell$ by the rule

$$\begin{aligned} \mathrm{v}_{\mathfrak{l}}' &= v^{[\mathfrak{l}_1]} \otimes \ldots \otimes v^{[\mathfrak{l}_n]}, \\ \mathrm{v}_{\mathfrak{l}}^* &= \frac{\theta(\eta)^\ell}{(p)_\infty^{3\ell}\, N_{\mathfrak{l}}(\lambda)}\, v^{[\mathfrak{l}_1]} \otimes \ldots \otimes v^{[\mathfrak{l}_n]}. \end{aligned} \tag{8.5}$$

Notice that the right hand side of the second formula depends on λ.

We consider the tensor coordinates on the elliptic hypergeometric spaces $\mathcal{F}_{ell}[x;\eta;\lambda]$ and $\mathcal{F}'_{ell}[y;\eta;\lambda]$ as the following linear maps:

$$\begin{aligned} B_{ell}[x;y;\eta;\lambda] : \mathcal{V}^*_{ell} &\to \mathcal{F}_{ell}[x;\eta;\lambda], \\ \mathrm{v}_{\mathfrak{m}}^* &\mapsto \mathfrak{W}_{\mathfrak{m}}(t;\mathfrak{x};\mathfrak{y};\eta;\lambda), \end{aligned} \tag{8.6}$$

$$\begin{aligned} B'_{ell}[x;y;\eta;\lambda] : {\mathcal{V}'_{ell}}^* &\to \mathcal{F}'_{ell}[y;\eta;\lambda], \\ \mathrm{v}_{\mathfrak{m}}'^* &\mapsto \mathfrak{W}'_{\mathfrak{m}}(t;\mathfrak{x};\mathfrak{y};\eta;\lambda), \end{aligned}$$

cf. (4.7), (7.20), again distinguishing the vector spaces associated with the tensor coordinates of different types. Comparing formulae (8.4), (8.5) with (4.10) we obtain that the pairing $(\!(\, , \,)\!)$ of $\mathcal{V}^*_{ell}$ and ${\mathcal{V}'_{ell}}^*$, cf. (4.8), coincides with the canonical pairing of $(V_1^e \otimes \ldots \otimes V_n^e)_\ell$ and $(V_1^e \otimes \ldots \otimes V_n^e)^*_\ell$. In other words

$$(-1)^\ell \bigl(B_{ell}[x;y;\eta;\lambda]\bigr)^* \mathcal{S}_{ell}[x;y;\eta;\lambda]\, B'_{ell}[x;y;\eta;\lambda] = \\ = \mathrm{id} \in \mathrm{End}\bigl((V_1^e \otimes \ldots \otimes V_n^e)^*_\ell\bigr)$$

where $\mathcal{S}_{ell}[x;y;\eta;\lambda] : \mathcal{F}'_{ell}[y;\eta;\lambda] \to \bigl(\mathcal{F}_{ell}[x;\eta;\lambda]\bigr)^*$ is the Shapovalov pairing of the elliptic hypergeometric spaces.

Corollary 8.2. *For any* $v \in (V_1^e \otimes \ldots \otimes V_n^e)_\ell$ *and* $v^* \in (V_1^e \otimes \ldots \otimes V_n^e)^*_\ell$ *we have*

$$\begin{aligned} \bigl\langle v^*, T_{12}(t_1,\lambda) \ldots T_{12}(t_\ell,\eta^{1-\ell}\lambda)\, v^{[0]} \otimes \ldots \otimes v^{[0]} \bigr\rangle &= \\ = \bigl(B'_{ell}[x;y;\eta;\eta^{-1}\lambda]\, v^*\bigr)(t_1,\ldots,t_\ell) \,\times& \\ \times\, \eta^{\ell(\ell-1)/2} \prod_{s=0}^{\ell-1} \frac{\theta(\eta)}{\theta(\eta^{-s}\lambda)} \prod_{1\le a<b\le \ell} \frac{\theta(\eta^{-1}t_a/t_b)}{\theta(t_a/t_b)}&, \end{aligned} \tag{8.7}$$

$$\begin{aligned} \bigl\langle (v^{[0]} \otimes \ldots \otimes v^{[0]})^*, T_{21}(t_1,\eta^{1-\ell}\lambda) \ldots T_{21}(t_\ell,\lambda)\, v \bigr\rangle &= \\ = \bigl(B_{ell}[x;y;\eta;\lambda]\, v\bigr)(t_1,\ldots,t_\ell) \,\times& \\ \times \prod_{s=0}^{\ell-1} \frac{\theta(\eta^{1-s-\ell+2\sum \Lambda_m}\lambda)}{(p)^3_\infty} \prod_{a=1}^{\ell} \prod_{m=1}^{n} \frac{\theta(t_a/x_m)}{\theta(t_a/y_m)} \prod_{1\le a<b\le \ell} \frac{\theta(\eta\, t_a/t_b)}{\theta(t_a/t_b)}&. \end{aligned} \tag{8.8}$$

The statement follows from (8.6) and formulae (8.2), (8.3).

8.3 The hypergeometric maps

In Section [4.] we introduced the hypergeometric maps $\bar{I}$ and $\bar{I}'$, see (4.11). In notations of Sections [7.], [8.] they are the following linear maps

$$\bar{I}\,[x;y;\eta\,;\lambda]\,:\mathcal{V}_{ell}^{*}\to V\,,\qquad\qquad \bar{I}'[x;y;\eta\,;\lambda]\,:{\mathcal{V}_{ell}'}^{*}\to V'.$$

Taking into account the correspondence of the parameters, cf. (7.19), (8.1), and the identification of vector spaces, the hypergeometric maps induce the following linear maps:

$$\tilde{I}\,[z;\lambda]\,:V_1^e\otimes\ldots\otimes V_n^e\to V_1\otimes\ldots\otimes V_n\,,$$

$$\tilde{I}'[z;\lambda]\,:(V_1^e\otimes\ldots\otimes V_n^e)^*\to(V_1\otimes\ldots\otimes V_n)^*,$$

where $V_1,\ldots,V_n$ are Verma modules over $U_q(\mathfrak{sl}_2)$ with highest weights $q^{\Lambda_1},\ldots,q^{\Lambda_n}$, respectively. We describe below properties of the maps $\tilde{I}$ and $\tilde{I}'$ reformulating the corresponding properties of the hypergeometric maps.

Proposition 8.1. *For generic z and λ the maps $\tilde{I}[z;\lambda]$ and $\tilde{I}'[z;\lambda]$ are non-degenerate/.*

The statement follows from Propositions [3.3], [4.2] and Corollaries [3.1], [4.2].

Introduce operators $A^{\{m\}}(z_1,\ldots,z_n;\lambda)$, $m=1,\ldots,n$, acting in $V_1\otimes\ldots\otimes V_n$:

$$A^{\{m\}}(z_1,\ldots,z_n;\lambda)\;=\;\prod_{j=1}^{m}\,(\eta^{\,1\,+\,H}\lambda)^{\Lambda_j-H_j}\,\mathbf{R_m}(\mathbf{z_1},\ldots,\mathbf{z_n})\tag{8.9}$$

where $H=H_1<,\ldots,+H_n$ and $\mathbf{R_m}(\mathbf{z_1},\ldots,\mathbf{z_n})$ is given by (7.14). Introduce also operators $U\#1(\lambda),\ldots,U\#n(\lambda)$ acting in $V_1^e\otimes\ldots\otimes V_n^e$:

$$U^{\{m\}}(\lambda)\,v^{[\mathfrak{l}_1]}\otimes\ldots\otimes v^{[\mathfrak{l}_n]}\;=\tag{8.10}$$

$$=\lambda^{\mathfrak{l}^{m}}\eta^{\,\mathfrak{l}^{m}(2\Lambda^{\{m\}}-\mathfrak{l}^{m}+1)-\ell\sum\Lambda_{\mathfrak{k}}}\,v^{[\mathfrak{l}_1]}\otimes\ldots\otimes v^{[\mathfrak{l}_n]}\,,$$

$m=1,\ldots,n$, where $\mathfrak{l}^{m}=\mathfrak{l}_1<,\ldots,+\mathfrak{l}_m$, $\ell=\mathfrak{l}_1<,\ldots,+\mathfrak{l}_n$ and $\Lambda^{\{m\}}=\Lambda_1<,\ldots,+\Lambda_m$.

Theorem 8.4. *The maps $\tilde{I}\,[z;\lambda]$ and $\tilde{I}'[z;\lambda]$ obey the following systems of difference equations*

$$\tilde{I}\,[p\,z_1,\ldots,p\,z_m,z_{m+1},\ldots,z_n;\lambda]\;=\tag{8.11}$$

$$=\;A^{\{m\}}(z_1,\ldots,z_n;\lambda)\,\tilde{I}\,[z_1,\ldots,z_n;\lambda]\,{U^{\{m\}}(\lambda)}^{-1},$$

$$\tilde{I}'[p\,z_1,\ldots,p\,z_m,z_{m+1},\ldots,z_n;\lambda]\;=\tag{8.12}$$

$$=\;\bigl(A^{\{m\}}(z_1,\ldots,z_n;\lambda)\bigr)^{*-1}\,\tilde{I}'[z_1,\ldots,z_n;\lambda]\,\bigl(U^{\{m\}}(\lambda)\bigr)^{*}.$$

Proof. The statement follows from Theorem [4.2], Corollary [4.3] and formulae (7.23).

Remark 8.3. The system (8.11) is a modification of the qKZ equation (7.12) while the system (8.12) is a modification of the dual qKZ equation (7.3).

The next statement is equivalent to Theorem [4.1].

Theorem 8.5. *For any $v \in V_1^e \otimes \ldots \otimes V_n^e$ and $v^* \in (V_1^e \otimes \ldots \otimes V_n^e)^*$ we have*

$$\langle v^*, v\rangle = \left\langle \tilde I'[z;\lambda]\, v^*,\, \tilde I\,[z;\lambda]\, v\right\rangle .$$

Recall that for $\mathfrak{l}, \mathfrak{m} \in \mathbb{Z}^n_{\geq 0}$ we write $\mathfrak{l} \ll \mathfrak{m}$ if $\sum_{j=1}^{n} \mathfrak{l}_j = \sum_{j=1}^{n} \mathfrak{m}_j$ and $\sum_{j=1}^{k} \mathfrak{l}_j < \sum_{j=1}^{k} \mathfrak{m}_j$ for all $k = 1, \ldots, n-1$, see (3.5).

Theorem 8.6. *Assume that $z_m/z_{m+1} \to 0$ for all $m = 1, \ldots, n-1$. Then for any λ the maps $\tilde I\,[z;\lambda]$ and $\tilde I'[z;\lambda]$ have finite limits. Moreover,*

$$\lim \tilde I\, v^{[\mathfrak{l}_1]} \otimes \ldots \otimes v^{[\mathfrak{l}_n]} = F^{\mathfrak{l}_1} v_1 \otimes \ldots \otimes F^{\mathfrak{l}_n} v_n \lim \tilde I_{\mathfrak{l}\,\mathfrak{l}} + \\ + \sum_{\mathfrak{m} \gg \mathfrak{l}} F^{\mathfrak{m}_1} v_1 \otimes \ldots \otimes F^{\mathfrak{m}_n} v_n \lim \tilde I_{\mathfrak{m}\,\mathfrak{l}}\,,$$

$$\lim \tilde I'\, (v^{[\mathfrak{l}_1]} \otimes \ldots \otimes v^{[\mathfrak{l}_n]})^* = (F^{\mathfrak{l}_1} v_1 \otimes \ldots \otimes F^{\mathfrak{l}_n} v_n)^* \lim \tilde I'_{\mathfrak{l}\,\mathfrak{l}} + \\ + \sum_{\mathfrak{m} \ll \mathfrak{l}} (F^{\mathfrak{m}_1} v_1 \otimes \ldots \otimes F^{\mathfrak{m}_n} v_n)^* \lim \tilde I'_{\mathfrak{m}\,\mathfrak{l}}$$

where $\lim \tilde I_{\mathfrak{m}\,\mathfrak{l}}$, $\lim \tilde I'_{\mathfrak{m}\,\mathfrak{l}}$ are suitable functions of λ. In particular,

$$\lim \tilde I_{\mathfrak{l}\,\mathfrak{l}} = \frac{1}{\lim \tilde I'_{\mathfrak{l}\,\mathfrak{l}}} = \prod_{m=1}^{n} \prod_{s=0}^{\mathfrak{l}_m - 1} \frac{-\eta^{\,2s-2\Lambda_m} (p\eta^{s+1})_\infty\, (p\eta^{s-2\Lambda_m})_\infty\, (p)_\infty}{\lambda_{\mathfrak{l},\mathfrak{m}}\, (p\eta)_\infty\, (p\eta^{-s}\lambda_{\mathfrak{l},\mathfrak{m}})_\infty\, (\eta^{s-1+\mathfrak{l}_m - 2\Lambda_m}\lambda^{-1}_{\mathfrak{l},\mathfrak{m}})_\infty}$$

where $\lambda_{\mathfrak{l},\mathfrak{m}} = \lambda \prod_{1 \leq j < m} \eta^{\,2\Lambda_j - 2\mathfrak{l}_j}$.

The statement follows from to Propositions [4.4], [4.5].

A function$Y[z;\lambda]$ with values in $\mathrm{End}\,(V_1^e \otimes \ldots \otimes V_n^e)$ satisfying the difference equations

$$Y[p\,z_1, \ldots, p\,z_m, z_{m+1}, \ldots, z_n; \lambda] = U^{\{m\}}(\lambda)\, Y[z_1, \ldots, z_n; \lambda]\,,$$

$m = 1, \ldots, n$, is called an *adjusting map.* Given $v \in (V_1^e \otimes \ldots \otimes V_n^e)_\ell$, $v^* \in (V_1^e \otimes \ldots \otimes V_n^e)^*_\ell$ and an adjusting map $Y[z;\lambda]$, define functions

$$\Psi_v(z;\lambda) = \tilde I\,[z;\lambda]\, Y[z;\lambda]\, v\,, \qquad \Psi'_{v^*}(z;\lambda) = \tilde I'[z;\lambda]\, \left(Y[z;\lambda]\right)^{*^{-1}} v$$

taking values in $(V_1 \otimes \ldots \otimes V_n)_\ell$ and $(V_1 \otimes \ldots \otimes V_n)^*_\ell$, respectively.

Corollary 8.3. *Let $\lambda = \kappa\,\eta^{\,\ell - 1 - \sum \Lambda_m}$. Then*
i) The function$\Psi_v(z;\lambda)$ is a solution of the qKZ equation (7.12).

ii) The function $\Psi'_{v^*}(z;\lambda)$ *is a solution of the dual qKZ equation (7.3).*
iii) $\langle \Psi'_{v^*}(z;\lambda), \Psi_v(z;\lambda) \rangle = \langle v^*, v \rangle$.

The statement follows from Theorems [8.4], [8.5].

Corollaries [7.1] and [8.2] allow us to describe the maps $\tilde{I}\,[z;\lambda\,]$ and $\tilde{I}'[z;\lambda\,]$ entirely in terms of the actions of the quantum loop algebra $U'_q(\widetilde{\mathfrak{gl}_2})$ and the elliptic quantum group $E_{\rho,\gamma}(\mathfrak{sl}_2)$ in the corresponding modules $V_1(z_1) \otimes \ldots \otimes V_n(z_n)$ and $V_1^e(z_1) \otimes \ldots \otimes V_n^e(z_n)$. For instance, let $v \in (V_1^e \otimes \ldots \otimes V_n^e)_\ell$. Then we have

$$\tilde{I}\,[z;\lambda\,]\,v \,=\, \frac{q^{\,\ell(1-\ell)/2-\ell\sum \Lambda_m}}{(2\pi i)^\ell\,(q-q^{-1})^\ell\,\ell\,!}\,\prod_{s=0}^{\ell-1}\,\frac{(p)^3_\infty}{\theta(\eta^{1-s-\ell+2\sum\Lambda_m}\lambda)}\,\times \tag{8.13}$$

$$\times \int_{\widetilde{\mathbb{T}}^\ell[x;y;\eta\,]} L^-_{12}(t_1)\ldots L^-_{12}(t_\ell)\,v_1 \otimes \ldots \otimes v_n\,\times$$

$$\times\,\langle (v^{[0]} \otimes \ldots \otimes v^{[0]})^*\,,T_{21}(t_1,\eta^{1-\ell}\lambda)\ldots T_{21}(t_\ell,\lambda)\,v \rangle\,\times$$

$$\times\,\prod_{m=1}^{n}\prod_{a=1}^{\ell}\frac{(p\,y_m/t_a)_\infty}{(x_m/t_a)_\infty}\,\prod_{a,b=1 a\neq b}^{\ell}\frac{(t_a/t_b)_\infty}{(\eta^{-1}t_a/t_b)_\infty}\,(dt/t)^\ell$$

where the integration contour $\widetilde{\mathbb{T}}^\ell[x;y;\eta\,]$ is defined by (3.11) and we keep in mind relations (7.19), (8.1). The parameters $x_1,\ldots,x_n$, $y_1,\ldots,y_n$ in their turn, can be discovered via the quantum loop algebra action as well. Namely, for generic values of the parameters $V_1(z_1) \otimes \ldots \otimes V_n(z_n)$ is an irreducible $\hat{}\,U'_q(\widetilde{\mathfrak{gl}_2})$- module generated by the vector $v_1 \otimes \ldots \otimes v_n$ which obey relations

$$L^\pm_{21}(u)\,v_1 \otimes \ldots \otimes v_n \,=\, 0\,,$$

$$L^+_{11}(u)\,v_1 \otimes \ldots \otimes v_n \,=\, q^{-\sum\Lambda_m}\,\prod_{m=1}^{n}\,(1-u/y_m)\,v_1 \otimes \ldots \otimes v_n\,,$$

$$L^+_{22}(u)\,v_1 \otimes \ldots \otimes v_n \,=\, q^{\sum\Lambda_m}\,\prod_{m=1}^{n}\,(1-u/x_m)\,v_1 \otimes \ldots \otimes v_n\,,$$

$$L^-_{22}(u)\,v_1 \otimes \ldots \otimes v_n \,=\, q^{-\sum\Lambda_m}\,\prod_{m=1}^{n}\,(1-x_m/u)\,v_1 \otimes \ldots \otimes v_n\,.$$

Moreover, the $\hat{}\,U'_q(\widetilde{\mathfrak{gl}_2})$- module is uniquely determined by these properties up to isomorphism.

Similarly, for generic values of the parameters $V_1^e(z_1) \otimes \ldots \otimes V_n^e(z_n)$ is an irreducible $\hat{}\,E_{\rho,\gamma}(\mathfrak{sl}_2)$- module generated by the vector $v^{[0]} \otimes \ldots \otimes v^{[0]}$ which obey relations

$$T_{21}(u,\lambda)\,v^{[0]} \otimes \ldots \otimes v^{[0]} \,=\, 0\,,$$

$$T_{11}(u,\lambda)\, v^{[0]} \otimes \ldots \otimes v^{[0]} = v^{[0]} \otimes \ldots \otimes v^{[0]} ,$$

$$T_{22}(u,\lambda)\, v^{[0]} \otimes \ldots \otimes v^{[0]} = \frac{\theta(\lambda \prod x_m/y_m)}{\theta(\lambda)} \prod_{m=1}^{n} \frac{\theta(u/x_m)}{\theta(u/y_m)}\, v^{[0]} \otimes \ldots \otimes v^{[0]} ,$$

and the $\hat{}\, E_{\rho,\gamma}(\mathfrak{sl}_2)$- module is uniquely determined by these properties up to isomorphism. Here $\prod$ stands for $\prod_{m=1}^{n}$.

Therefore, $\tilde{I}\,[z;\lambda]$ naturally appears as a map from the module over the elliptic quantum group $E_{\rho,\gamma}(\mathfrak{sl}_2)$ to the corresponding module over the quantum loop algebra $U_q'(\widetilde{\mathfrak{gl}_2})$:

$$\tilde{I}\,[z;\lambda] \,:\, V_1^e(z_1) \otimes \ldots \otimes V_n^e(z_n) \,\to\, V_1(z_1) \otimes \ldots \otimes V_n(z_n)\,. \tag{8.14}$$

Similarly, $\tilde{I}'[z;\lambda]$ is a map of the right modules:

$$\tilde{I}'[z;\lambda] \,:\, (V_1^e(z_1) \otimes \ldots \otimes V_n^e(z_n))^* \,\to\, (V_1(z_1) \otimes \ldots \otimes V_n(z_n))^*.$$

Theorem [8.5] claims that these maps respect the canonical pairings.

The modules involved in (8.14) can be replaced by isomorphic modules. E.g. for any permutation τ, τ' one can consider a map

$$\tilde{I}_{\tau,\tau'}[z;\lambda] \,:\, V_{\tau_1'}^e(z_{\tau_1'}) \otimes \ldots \otimes V_{\tau_n'}^e(z_{\tau_n'}) \,\to\, V_{\tau_1}(z_{\tau_1}) \otimes \ldots \otimes V_{\tau_n}(z_{\tau_n})\,.$$

There are analogues of Theorems [8.4] and [8.6] for these maps, cf. [30]. In the next section we use this construction for τ being the identity permutation to describe asymptotic solutions of the qKZ equation.

9. Asymptotic solutions of the qKZ equation

One of the most important characteristics of a differential equation is the monodromy group of its solutions. For the differential KZ equation with values in a tensor product of representations of a simple Lie algebra its monodromy group is described in terms of the corresponding quantum group. This fact establishes a remarkable connection between representation theories of simple Lie algebras and their quantum groups, see [16], [6], [18], [27], [35].

The analogue of the monodromy group for difference equations is the set of transition functions between asymptotic solutions. For a difference equation one defines suitable asymptotic zones in the domain of the definition of the equation and then an asymptotic solution for every zone. Thus, for every pair of asymptotic zones one gets a transition function between the corresponding asymptotic solutions, cf. [30], [31].

In this section we describe asymptotic zones, asymptotic solutions, and their transition functions for the qKZ equation with values in a tensor product of $U_q(\mathfrak{sl}_2)$-modules. A remarkable fact is that the transition functions are described in terms of the elliptic R-matrices acting in the tensor product of the corresponding $\hat{}E_{\rho,\gamma}(\mathfrak{sl}_2)$- modules, cf. (9.3) – (9.5). This fact establishes a correspondence between representation theories of the quantum loop algebra and the elliptic quantum group [30].

Let $V_1, \ldots, V_n$ be Verma modules over $U_q(\mathfrak{sl}_2)$ with highest weights $q^{\Lambda_1}, \ldots, q^{\Lambda_n}$. Consider the qKZ equation for a functionΨ with values in $(V_1 \otimes \ldots \otimes V_n)_\ell$:

$$\Psi(z_1, \ldots, pz_m, \ldots, z_n) = A_m(z_1, \ldots, z_n)\, \Psi(z_1, \ldots, z_n)\,, \tag{9.1}$$

$m = 1, \ldots, n$ cf. (7.10), (7.11). For every permutation $\tau \in \mathbf{S_n}$ we consider an asymptotic zone $\mathbb{A}_\tau$ in $\mathbb{C}^{\times n}$ given by

$$\mathbb{A}_\tau = \{ z \in \mathbb{C}^{\times n} \mid |z_{\tau_m}/z_{\tau_{m+1}}| \ll 1, \quad m = 1, \ldots, n-1 \}.$$

Say that z tends to limit in the asymptotic zone, $z \rightrightarrows \mathbb{A}_\tau$, if $z_{\tau_m}/z_{\tau_{m+1}} \to 0$ for all $m = 1, \ldots, n-1$.

Say that a basis $\Psi_1, \ldots, \Psi_N$ of solutions of the qKZ equation (9.1) form an *asymptotic solution* in the asymptotic zone $\mathbb{A}_\tau$ if

$$\Psi_j(z) = h_j(z)\,\big(v_j + o(1)\big)\,,$$

where $h_1(z), \ldots, h_N(z)$ are meromorphic functions such that

$$h_j(z_1, \ldots, pz_m, \ldots, z_n) = a_{jm}\, h_j(z_1, \ldots, z_n)$$

for suitable numbers a_{jm}, $v_1, \ldots, v_N$ are constant vectors which form a basis in $(V_1 \otimes \ldots \otimes V_n)_\ell$, and $o(1)$ tends to 0 as $z \rightrightarrows \mathbb{A}_\tau$.

Remark 9.1. The operators $A_1, \ldots, A_n$ have finite limits in the asymptotic zone $\mathbb{A}_\tau$. Each vector v_j is their common eigenvector with eigenvalues $a_{j1}, \ldots, a_{jn}$, respectively.

Given for any permutation τ an asymptotic solution $\Psi_1^\tau, \ldots, \Psi_N^\tau$ of the qKZ equation in the asymptotic zone $\mathbb{A}_\tau$, one gets transition functions $\mathbf{C}_{\tau,\tau'}$ between the asymptotic solution:

$$\Psi_k^{\tau'}(z) = \sum_{j=1}^{N} \Psi_j^\tau(z)\,\big(\mathbf{C}_{\tau,\tau'}(\mathbf{z})\big)_{\mathbf{jk}}\,. \tag{9.2}$$

The transition functions $\mathbf{C}_{\tau,\tau'}(\mathbf{z})$ are $\hat{}p$ -periodic functions of each of the variables $z_1, \ldots, z_n$, i.e. quasiconstants, and they have the properties

$$\mathbf{C}_{\tau,\tau}(\mathbf{z}) = \mathrm{id}\,, \qquad \mathbf{C}_{\tau,\tau'}(\mathbf{z})\, \mathbf{C}_{\tau',\tau''}(\mathbf{z}) = \mathbf{C}_{\tau,\tau''}(\mathbf{z})\,.$$

Example. Let τ be the identity permutation. Let $\lambda = \kappa\, q^{2\ell-2-2\sum \Lambda_m}$ and let $Y[z;\lambda]$ be an adjusting map which is nondegenerate/ and diagonal in the monomial basis $\{ v^{[\mathfrak{l}_1]} \otimes \ldots \otimes v^{[\mathfrak{l}_n]} \mid \mathfrak{l} \in \mathcal{Z}_\ell^n \}$ of $(V_1^e \otimes \ldots \otimes V_n^e)_\ell$. Set

$$\Psi_{\mathfrak{l}}(z) = \tilde{I}[z;\lambda]\, Y[z;\lambda]\, v^{[\mathfrak{l}_1]} \otimes \ldots \otimes v^{[\mathfrak{l}_n]} .$$

Then $\{ \Psi_{\mathfrak{l}} \mid \mathfrak{l} \in \mathcal{Z}_\ell^n \}$, is an asymptotic solution of the qKZ equation (9.1) in the asymptotic zone $\mathbb{A}_{\mathrm{id}}$, see Corollary [8.3], Theorem [8.6] and Proposition [8.1].

An asymptotic solution of the qKZ equation in the asymptotic zone $\mathbb{A}_\tau$ for any permutation τ can be described in a similar way. Let $V_1^e(z_1), \ldots, V_n^e(z_n)$ be the evaluation Verma modules over the elliptic quantum group $E_{\rho,\gamma}(\mathfrak{sl}_2)$ with highest weights $\Lambda_1, \ldots, \Lambda_n$, respectively. Consider a map

$$\tilde{I}_\tau[z;\lambda] \,:\, V_{\tau_1}^e \otimes \ldots \otimes V_{\tau_n}^e \;\to\; V_1 \otimes \ldots \otimes V_n$$

analogous to $\tilde{I}[z;\lambda]$, for any $v \in (V_{\tau_1}^e \otimes \ldots \otimes V_{\tau_n}^e)_\ell$ set $\tilde{I}_\tau[z;\lambda]\, v$ to be equal to the right hand side of formula (8.13). Define operators $U_1^\tau(\lambda), \ldots, U_n^\tau(\lambda) \in \mathrm{End}\,(V_1^e \otimes \ldots \otimes V_n^e)$ by the rule

$$U_m^\tau(\lambda)\, v^{[\mathfrak{l}_{\tau_1}]} \otimes \ldots \otimes v^{[\mathfrak{l}_{\tau_n}]} =$$

$$= \lambda^{\mathfrak{l}_m}\, \eta^{(\mathfrak{l}_m - \ell)\Lambda_m - \mathfrak{l}_m(\mathfrak{l}_m - 1)} \prod_{1 \le \sigma_k \le \sigma_m} \eta^{2\Lambda_k \mathfrak{l}_m + 2\Lambda_m \mathfrak{l}_k - 2\mathfrak{l}_k \mathfrak{l}_m}\, v^{[\mathfrak{l}_{\tau_1}]} \otimes \ldots \otimes v^{[\mathfrak{l}_{\tau_n}]}$$

where $\ell = \mathfrak{l}_1 <, \ldots, +\mathfrak{l}_n$ and $\sigma = \tau^{-1}$. A solution $Y_\tau[z;\lambda]$ of the difference equations

$$Y_\tau[z_1, \ldots, p z_m, \ldots, z_n; \lambda] = U_m^\tau(\lambda)\, Y_\tau[z_1, \ldots, z_n; \lambda],$$

$m = 1, \ldots, n$, with values in $\mathrm{End}\,(V_1^e \otimes \ldots \otimes V_n^e)$ is called an adjusting map corresponding to the permutation τ.

Theorem 9.1. *[30] Let $\lambda = \kappa\, q^{2\ell - 2 - 2\sum \Lambda_m}$ and let $Y_\tau[z;\lambda]$ be an adjusting map which is nondegenerate/ and diagonal in the monomial basis $\{ v^{[\mathfrak{l}_1]} \otimes \ldots \otimes v^{[\mathfrak{l}_n]} \mid \mathfrak{l} \in \mathcal{Z}_\ell^n \}$ of $(V_{\tau_1}^e \otimes \ldots \otimes V_{\tau_n}^e)_\ell$. Set*

$$\Psi_{\mathfrak{l}}^\tau(z) = \tilde{I}_\tau[z;\lambda]\, Y_\tau[z;\lambda]\, v^{[\mathfrak{l}_{\tau_1}]} \otimes \ldots \otimes v^{[\mathfrak{l}_{\tau_n}]}, \qquad \mathfrak{l} \in \mathcal{Z}_\ell^n .$$

Then $\{ \Psi_{\mathfrak{l}}^\tau \mid \mathfrak{l} \in \mathcal{Z}_\ell^n \}$, is an asymptotic solution of the qKZ equation (9.1) in the asymptotic zone $\mathbb{A}_\tau$.

By construction, the ratio $\tilde{I}_\tau[z;\lambda]^{-1} \tilde{I}_{\tau'}[z;\lambda]$ equals the normalized intertwiner $\mathbf{P}_{\tau',\tau}\mathbf{R}^{\mathrm{ell}}_{\tau',\tau}(\mathbf{z};\lambda)$ of the $\hat{\ } E_{\rho,\gamma}(\mathfrak{sl}_2)$- modules:

$$\mathbf{P}_{\tau',\tau}\mathbf{R}^{\mathrm{ell}}_{\tau',\tau}(\mathbf{z};\lambda) \,:\, \mathbf{V}^{\mathbf{e}}_{\tau'_1}(\mathbf{z}_{\tau'_1}) \otimes \ldots \otimes \mathbf{V}^{\mathbf{e}}_{\tau'_n}(\mathbf{z}_{\tau'_n}) \to \tag{9.3}$$

$$\to V^e_{\tau_1}(z_{\tau_1}) \otimes \ldots \otimes V^e_{\tau_n}(z_{\tau_n}),$$

$$\mathbf{P}_{\tau',\tau}\mathbf{R}^{\mathrm{ell}}_{\tau',\tau}(\mathbf{z};\lambda)\, \mathbf{v}^{[0]} \otimes \ldots \otimes \mathbf{v}^{[0]} = \mathbf{v}^{[0]} \otimes \ldots \otimes \mathbf{v}^{[0]} .$$

Here $\mathbf{P}_{\tau',\tau}$ is the permutation map

$$\mathbf{P}_{\tau',\tau} \,:\, \mathbf{V}^{\mathbf{e}}_{\tau'_1} \otimes \ldots \otimes \mathbf{V}^{\mathbf{e}}_{\tau'_n} \,\to\, \mathbf{V}^{\mathbf{e}}_{\tau_1} \otimes \ldots \otimes \mathbf{V}^{\mathbf{e}}_{\tau_n} \,,$$

$$\mathbf{P}_{\tau',\tau}\, \mathbf{v}^{[\mathfrak{l}_{\tau'_1}]} \otimes \ldots \otimes \mathbf{v}^{[\mathfrak{l}_{\tau'_n}]} \,=\, \mathbf{v}^{[\mathfrak{l}_{\tau_1}]} \otimes \ldots \otimes \mathbf{v}^{[\mathfrak{l}_{\tau_n}]} \,,$$

and $\mathbf{R}^{\mathbf{ell}}_{\tau',\tau}(\mathbf{z};\lambda)$ is a suitable product of the dynamical elliptic R-matrices. For example, if $\tau' = \tau\cdot(m,m+1)$ where $(m,m+1)$ is a transposition, then

$$\mathbf{R}^{\mathbf{ell}}_{\tau\cdot(\mathbf{m},\mathbf{m}+\mathbf{1}),\tau}(\mathbf{z};\lambda) \,= \tag{9.4}$$

$$= R^{ell}_{V^e_{\tau_{m+1}} V^e_{\tau_m}}\bigl(z_{\tau_{m+1}}/z_{\tau_m}, \lambda\,\underbrace{\eta^{2H}\otimes\ldots\otimes\eta^{2H}}_{m-1}\otimes \mathrm{id}^{\otimes(n-m+1)}\bigr)\,.$$

By Theorem [9.1] for any permutation τ we have an asymptotic solution $\{\Psi^{\tau}_{\mathfrak{l}} \mid \mathfrak{l}\in\mathcal{Z}^{n}_{\ell}\}$ of the qKZ equation in the asymptotic zone $\mathbb{A}_{\tau}$. The corresponding transition functions $\mathbf{C}_{\tau,\tau'}$ between the asymptotic solutions, cf. (9.2), equal the intertwiners of the respective tensor products of evaluation Verma modules over the elliptic quantum group restricted to the weight subspace and twisted by the adjusting maps:

$$\mathbf{C}_{\tau,\tau'}(\mathbf{z}) \,:\, (\mathbf{V}^{\mathbf{e}}_{\tau'_1} \otimes \ldots \otimes \mathbf{V}^{\mathbf{e}}_{\tau'_n})_{\ell} \,\to\, (\mathbf{V}^{\mathbf{e}}_{\tau_1} \otimes \ldots \otimes \mathbf{V}^{\mathbf{e}}_{\tau_n})_{\ell} \,, \tag{9.5}$$

$$\mathbf{C}_{\tau,\tau'}(\mathbf{z}) \,=\, \mathbf{Y}_{\tau}[\mathbf{z};\lambda]^{-1}\mathbf{R}^{\mathbf{ell}}_{\tau',\tau}(\mathbf{z};\lambda)\,\mathbf{Y}_{\tau'}[\mathbf{z};\lambda]$$

where $\lambda = \kappa\, q^{2\ell-2-2\sum\Lambda_m}$.

A. Six determinant formulae

S We present here explicit formulae for determinants of the Shapovalov pairings S and $\mathcal{S}_{ell}$, and the hypergeometric pairing I.

Let $A = \alpha\,\eta^{1-\ell}\prod_{m=1}^{n} x_m$. Let $\mathcal{E}[A]$ be the space of holomorphic functions on $\mathbb{C}^\times$ such that $f(pu) = A(-u)^{-n} f(u)$. It is easy to see that $\dim \mathcal{E}[A] = n$, say by Laurent series.

Let $\omega = \exp(2\pi i/n)$. Fix complex numbers ξ and ζ such that $\xi^n = p$ and $\zeta^n = -A^{-1}$. Set

$$\vartheta_l(u) \;=\; u^{l-1} \prod_{m=1}^{n} \theta(-\zeta\,\xi^{l-1}\omega^m u)\,, \qquad l = 1, \dots, n\,.$$

Notice, that the functions $\vartheta_1, \dots, \vartheta_n$ do not depend on a particular choice of ξ and ζ.

Lemma A.1. *The functions $\vartheta_1, \dots, \vartheta_n$ form a basis in the space $\mathcal{E}[A]$.*

Proof. Clearly, $\vartheta_l \in \mathcal{E}[A]$ for any $l = 1, \dots, n$. Moreover,

$$\vartheta_l(\omega\, u) \;=\; \omega^{l-1}\vartheta_l(u)\,,$$

that is the functions $\vartheta_1, \dots, \vartheta_n$ are eigenfunctions of the translation operator with distinct eigenvalues. Hence, they are linearly independent.

For any $\mathfrak{l} \in \mathcal{Z}^{\mathfrak{n}}_{\ell}$ let $g_{\mathfrak{l}}(t; x; \eta)$ and $G_{\mathfrak{l}}(t; x; \eta\,; \alpha)$ be the following functions:

$$g_{\mathfrak{l}}(t; x; \eta) \;=\; \frac{1}{\mathfrak{l}_1! \dots \mathfrak{l}_{\mathfrak{n}}!} \;\times$$

$$\times \prod_{m=1}^{n} \prod_{a=1}^{\ell} \frac{1}{t_a - x_m} \prod_{1 \le a < b \le \ell} \frac{t_a - t_b}{\eta\, t_a - t_b} \sum_{\sigma \in \mathbf{S}_\ell} \prod_{m=1}^{n} \prod_{a \in \Gamma_m} t_{\sigma_a}^m\,,$$

$$G_{\mathfrak{l}}(t; x; \eta\,; \alpha) \;=\; \frac{1}{\mathfrak{l}_1! \dots \mathfrak{l}_{\mathfrak{n}}!} \;\times$$

$$\times \prod_{m=1}^{n} \prod_{a=1}^{\ell} \frac{1}{\theta(t_a/x_m)} \prod_{1 \le a < b \le \ell} \frac{\theta(t_a/t_b)}{\theta(\eta\, t_a/t_b)} \sum_{\sigma \in \mathbf{S}_\ell} \prod_{m=1}^{n} \prod_{a \in \Gamma_m} \vartheta_m(t_{\sigma_a})\,.$$

Here $\Gamma_m = \bigl\{\, 1 + \sum_{k=1}^{m-1} \mathfrak{l}^{\mathfrak{k}}, \dots, \sum_{\mathfrak{k}=1}^{\mathfrak{m}} \mathfrak{l}^{\mathfrak{k}} \,\bigr\}$, $m = 1, \dots, n$.

Lemma A.2. *The functions $g_{\mathfrak{l}}(t; x; \eta)$, $\mathfrak{l} \in \mathcal{Z}^{\mathfrak{n}}_{\ell}$, form a basis in the trigonometric hypergeometric space $\mathcal{F}[x; \eta\,; \ell\,]$. The functions $G_{\mathfrak{l}}(t; x; \eta\,; \alpha)$, $\mathfrak{l} \in \mathcal{Z}^{\mathfrak{n}}_{\ell}$, form a basis in the elliptic hypergeometric space $\mathcal{F}_{ell}[x; \eta\,; \alpha; \ell\,]$.*

Proof. We prove the second claim. The proof of the first one is similar.

The elliptic hypergeometric space $\mathcal{F}_{ell}[x;\eta\,;\alpha;\ell\,]$ is naturally isomorphic to the ^ℓ-th symmetric power of the space $\mathcal{E}[A]$ — the space of symmetric functions in $t_1,\dots,t_\ell$ which considered as functions of one variable t_a belong to $\mathcal{E}[A]$ for any $a=1,\dots,\ell$. The isomorphism reads as follows:

$$f(t_1,\dots,t_\ell)\;\mapsto\;f(t_1,\dots,t_\ell)\prod_{m=1}^{n}\prod_{a=1}^{\ell}\theta(t_a/x_m)\prod_{1\le a<b\le\ell}\frac{\theta(\eta\,t_a/t_b)}{\theta(t_a/t_b)}\,,$$

$f\in\mathcal{F}_{ell}[x;\eta\,;\alpha]$. Now the statement follows from Lemma [A.1].

Let $w_{\mathfrak{l}},W_{\mathfrak{l}}$, $\mathfrak{l}\in\mathcal{Z}^{\mathfrak{n}}_{\ell}$, be the weight functions (4.1), (4.2). Define matrices $Q(x;y;\eta)$ and $Q^{ell}(x;y;\eta\,;\alpha)$ by the rule:

$$w_{\mathfrak{l}}(t;x;y;\eta)\;=\;\sum_{\mathfrak{m}\in\mathcal{Z}^{\mathfrak{n}}_{\ell}}Q_{\mathfrak{l}\,\mathfrak{m}}(x;y;\eta)\,g_{\mathfrak{m}}(t;x;\eta)\,,$$

$$W_{\mathfrak{l}}(t;x;y;\eta\,;\alpha)\;=\;\sum_{\mathfrak{m}\in\mathcal{Z}^{\mathfrak{n}}_{\ell}}Q^{ell}_{\mathfrak{l}\,\mathfrak{m}}(x;y;\eta\,;\alpha)\,G_{\mathfrak{m}}(t;x;\eta\,;\alpha)\,.$$

Set

$$d(n,m,\ell,s)\;=\;\sum_{i,j\ge0\;\;i+j<\ell i-j=s}(\,m\,)-1+im-1\;(\,n\,)-m-1+jn-m-1\,. \tag{A.1}$$

Proposition A.1. *[29], [30, Theorem A.7]*

$$\det Q(x;y;\eta)\;=\;\prod_{s=0}^{\ell-1}\;\prod_{1\le l<m\le n}(\eta^s y_l-x_m)^{(\,n\,)+\ell-s-2n-1}\,.$$

Proposition A.2. *[30, Theorem B.8]*

$$\det Q^{ell}(x;y;\eta\,;\alpha)\;=\;\Xi\prod_{s=1-\ell}^{\ell-1}\prod_{m=1}^{n-1}\theta\big(\eta^{s+\ell-1}\alpha^{-1}\textstyle\prod_{1\le l\le m}y_l/x_l\big)^{d(n,m,\ell,s)}\;\times$$

$$\times\prod_{m=1}^{n}y_m^{(m-n)(\,n\,)+\ell-1n}\prod_{s=0}^{\ell-1}\;\prod_{1\le l<m\le n}\theta(\eta^s y_l/x_m)^{(\,n\,)+\ell-s-2n-1}$$

where

$$\Xi\;=\;\Big[\,(p)_\infty{}^{1-n^2}\prod_{m=1}^{n-1}\Big(\frac{\theta(\omega^m)}{\omega^m-1}\Big)^{n-m}\Big]^{(\,n\,)+\ell-1n}\,. \tag{A.2}$$

Consider a basis $g'_{\mathfrak{l}}(t)\;=\;t_1^{-1}\dots t_\ell^{-1}\,g_{\mathfrak{l}}(t;y;\eta^{-1})$, $\mathfrak{l}\in\mathcal{Z}^{\mathfrak{n}}_{\ell}$, of the trigonometric hypergeometric space $\mathcal{F}'[y;\eta\,;\ell\,]$ and a basis $G'_{\mathfrak{l}}(t)\;=\;G_{\mathfrak{l}}(t;y;\eta^{-1};\alpha^{-1})$, $\mathfrak{l}\in\mathcal{Z}^{\mathfrak{n}}_{\ell}$, of the elliptic hypergeometric space $\mathcal{F}'_{ell}[y;\eta\,;\alpha;\ell\,]$.

Proposition A.3.

$$\det\big[S(g_{\mathfrak{l}},g'_{\mathfrak{m}})\big]_{\mathfrak{l},\mathfrak{m}\in\mathcal{Z}^{n}_{\ell}} = (-1)^{n(n-1)/2\cdot(n)+\ell-1n}\,\eta^{n(n+1)/2\cdot(n)+\ell-1n+1}\times$$

$$\times\prod_{s=0}^{\ell-1}\left[\frac{(1-\eta)^n}{(1-\eta^{s+1})^n}\prod_{l,m=1}^{n}\frac{1}{x_l-\eta^s y_m}\right]^{(n)+\ell-s-2n-1},$$

$$\det\big[\mathcal{S}_{ell}(G_{\mathfrak{l}},G'_{\mathfrak{m}})\big]_{\mathfrak{l},\mathfrak{m}\in\mathcal{Z}^{n}_{\ell}} =$$

$$\Xi^{-2}\,(-1)^{n(n-1)/2\cdot(n)+\ell-1n}\,\eta^{n(3-n)/2\cdot(n)+\ell-1n+1}\prod_{m=1}^{n}x_m^{(n-1)(n)+\ell-1n}\times$$

$$\times\prod_{s=0}^{\ell-1}\left[\frac{\theta(\eta)^n\theta(\eta^s\alpha^{-1})\,\theta\big(\eta^{s+2-2\ell}\alpha\prod x_m/y_m\big)}{\theta'(1)^n\theta(\eta^{s+1})^n\prod\prod\theta(\eta^{-s}x_l/y_m)}\right]^{(n)+\ell-s-2n-1}.$$

Here Ξ is given by (A.2), $\prod$ stands for $\prod_{m=1}^{n}$ and $\prod\prod$ stands for $\prod_{l=1}^{n}\prod_{m=1}^{n}$.

Proof. We prove the second formula. The proof of the first one is similar. By Lemma [4.1] we have that

$$\det\big[\mathcal{S}_{ell}(G_{\mathfrak{l}},G'_{\mathfrak{m}})\big]_{\mathfrak{l},\mathfrak{m}\in\mathcal{Z}^{n}_{\ell}} = \big(\det Q(x;y;\eta;\alpha)\,\det Q(y;x;\eta^{-1};\alpha^{-1})\big)^{-1}\times$$

$$\times\prod_{\mathfrak{l}\in\mathcal{Z}^{n}_{\ell}}\prod_{m=1}^{n}\prod_{s=0}^{\mathfrak{l}_m-1}\frac{\eta^s\,\theta(\eta)\,\theta(\eta^s\alpha^{-1}_{\mathfrak{l},\mathfrak{m}})\,\theta(\eta^{1-s-\mathfrak{l}_m}\alpha_{\mathfrak{l},\mathfrak{m}}\,x_m/y_m)}{\theta'(1)\,\theta(\eta^{s+1})\,\theta(\eta^{-s}x_m/y_m)}.$$

To get the final answer we use Proposition [A.2] and simplify the triple product changing the order of the products and applying Lemma [A.3] several times.

Lemma A.3. *The following identity holds:*

$$\sum_{a=1}^{l}(j)+aj\,(j)+k+ak\,(l)+m-am = (j)+kk\,(j)+k+l+m+1j+k+m+1\,.$$

The statement can be proved by induction with respect to l and m.

Let $I=I\,[x;y;\eta;\alpha]$ be the hypergeometric pairing.

Proposition A.4. *[30, Theorem 5.9]*

$$\det\big[I(w_{\mathfrak{l}},W_{\mathfrak{m}})\big]_{\mathfrak{l},\mathfrak{m}\in\mathcal{Z}^{n}_{\ell}} =$$

$$=\eta^{-n(n)+\ell-1n+1}\prod_{s=1-\ell}^{\ell-1}\prod_{m=1}^{n-1}\theta\big(\eta^{s+\ell-1}\alpha^{-1}\textstyle\prod_{1\le l\le m}y_l/x_l\big)^{d(n,m,\ell,s)}\times$$

$$\times \prod_{s=0}^{\ell-1} \left[\frac{(\eta^{-1})_\infty^n\, (\eta^s \alpha^{-1})_\infty\, (p\,\eta^{s+2-2\ell} \alpha \prod x_m/y_m)_\infty}{(\eta^{-s-1})_\infty^n\, (p)_\infty^{2n-1} \prod (\eta^{-s} x_m/y_m)_\infty} \times \right.$$

$$\left. \times \prod_{1 \le l < m \le n} \frac{(\eta^s y_l/x_m)_\infty}{(\eta^{-s} x_l/y_m)_\infty} \right]^{(n)+\ell-s-2n-1}.$$

Here $\prod$ *stands for* $\prod_{m=1}^n$ *and the exponents* $d(n,m,\ell,s)$ *are given by (A.1).*

Corollary A.1. *[30]*

$$\det \left[I(g_{\mathfrak{l}}, G_{\mathfrak{m}}) \right]_{\mathfrak{l},\mathfrak{m} \in \mathcal{Z}_\ell^n} = \Xi^{-1}\, \eta^{-n(n+1)/2 \cdot (n)+\ell-1n+1} \times$$

$$\times \prod_{s=0}^{\ell-1} \left[\frac{(\eta^{-1})_\infty^n\, (\eta^s \alpha^{-1})_\infty\, (p\,\eta^{s+2-2\ell} \alpha \prod x_m/y_m)_\infty}{(\eta^{-s-1})_\infty^n\, (p)_\infty^{2n-1+n(n-1)/2} \prod\prod (\eta^{-s} x_l/y_m)_\infty} \right]^{(n)+\ell-s-2n-1}.$$

Here Ξ *is given by (A.2),* $\prod$ *stands for* $\prod_{m=1}^n$ *and* $\prod\prod$ *stands for* $\prod_{l=1}^n \prod_{m=1}^n$.

Proof. The statement follows from Propositions [A.1], [A.2] and [A.4].

B. The Jackson integrals via the hypergeometric integrals

Consider the hypergeometric integral $Int[x;y;\eta](f\,\widetilde{\Phi})$, cf. (3.11), for a function $f(t_1,\dots,t_\ell)$ of the form

$$f(t_1,\dots,t_\ell) = P(t_1,\dots,t_\ell)\, \Theta(t_1,\dots,t_\ell) \prod_{1 \le a < b \le \ell} (t_a/t_b)_\infty \tag{B.1}$$

where $P(t_1,\dots,t_\ell)$ is a symmetric polynomial of degree at most M in each of the variables $t_1,\dots,t_\ell$ and $\Theta(t_1,\dots,t_\ell)$ is a symmetric holomorphic function on $\mathbb{C}^{\times \ell}$ such that

$$\Theta(t_1,\dots,pt_a,\dots,t_\ell) = A\,(-t_a)^{-n}\, \Theta(t_1,\dots,t_\ell) \tag{B.2}$$

for some constant A. The hypergeometric integrals which appear in the definition of the hypergeometric pairings, see (3.18), fit this case for $M = n-1$ and A determined by $\alpha, \eta, x_1,\dots,x_n, y_1,\dots,y_n$.

For any $x \in \mathbb{C}^n$, $\mathfrak{l} \in \mathcal{Z}_\ell^{\mathfrak{n}}$, $\mathbf{s} \in \mathbb{Z}^\ell$, let $x \triangleright (\mathfrak{l},\mathbf{s})\,[\eta] \in \mathbb{C}^{\times \ell}$ be the following point:

$$x \triangleright (\mathfrak{l},\mathbf{s})\,[\eta] = (\mathbf{p}^{\mathbf{s_1}+\dots+\mathbf{s}_{\mathfrak{l}_1}} \eta^{\mathbf{1}-\mathfrak{l}_1} \mathbf{x_1},\, \mathbf{p}^{\mathbf{s_2}+\dots+\mathbf{s}_{\mathfrak{l}_1}} \eta^{\mathbf{2}-\mathfrak{l}_1} \mathbf{x_1}, \dots, \mathbf{p}^{\mathbf{s}_{\mathfrak{l}_1}} \mathbf{x_1},$$

$$p^{s_{\mathfrak{l}_1+1}+\dots+s_{\mathfrak{l}_1+\mathfrak{l}_2}} \eta^{1-\mathfrak{l}_2} x_2, \dots, p^{s_{\mathfrak{l}_1+\mathfrak{l}_2}} x_2, \dots, p^{s_{\ell-\mathfrak{l}_n+1}+s_\ell} \eta^{1-\mathfrak{l}_n} x_n, \dots, p^{s_\ell} x_n).$$

For instance, if $\mathbf{s} = (\mathbf{0},\dots,\mathbf{0})$, then $x \triangleright (\mathfrak{l},\mathbf{s})\,[\eta] = \mathbf{x} \triangleright \mathfrak{l}\,[\eta]$, cf. (3.6).

Proposition B.1. *Let parameters* x, y, η *be generic. Let a function* $f(t_1, \dots, t_\ell)$ *have the form (B.1), (B.2). Assume that*

$$|p^n A \prod_{m=1}^{n} x_m^{-1}| < \min(1, |\eta|^{1-\ell}).$$

Then the sum below is convergent and

$$Int[x; y; \eta](f \widetilde{\Phi}) = \sum_{\mathrm{m} \in \mathcal{Z}_\ell^n} \sum_{\mathrm{s} \in \mathbb{Z}_{\geq 0}^\ell} \operatorname{Res}\left(f(t)\, \widetilde{\Phi}(t; x; y; \eta)\, (dt/t)^\ell\right)\big|_{t = x \triangleright (\mathrm{m}, \mathrm{s})[\eta]}.$$

Similarly, if $|p^M A \prod_{m=1}^{n} y_m^{-1}| > \max(1, |\eta|^{\ell-1})$*, then the sum below is convergent and*

$$Int[x; y; \eta](f \widetilde{\Phi}) =$$

$$= (-1)^\ell \sum_{\mathrm{m} \in \mathcal{Z}_\ell^n} \sum_{\mathrm{s} \in \mathbb{Z}_{\leq 0}^\ell} \operatorname{Res}\left(f(t)\, \widetilde{\Phi}(t; x; y; \eta)\, (dt/t)^\ell\right)\big|_{t = y \triangleright (\mathrm{m}, \mathrm{s})[\eta^{-1}]}.$$

The proof is similar to the proof of Theorem F.1 in [30]. The sums in Proposition [B.1] coincide the symmetric A-type Jackson integrals, see for example [2].

Bibliography

1. K. Aomoto, *q-analogue of de Rham cohomology associated with Jackson integrals, I* Proceedings of Japan Acad. **66**, Ser A, 1990, 161–164
   and
   q-analogue of de Rham cohomology associated with Jackson integrals, II, Proceedings of Japan Acad. **66**, Ser A, 1990, 240–244
   and
   Finiteness of a cohomology associated with certain Jackson integrals, Tohoku Math. J. **43**, 1991, 75-101
2. K. Aomoto and Y. Kato *Gauss decomposition of connection matrices for symmetric A-type Jackson integrals*, Selecta Math., New Series **1**, 1995 issue 4, 623–666
3. V. Chari and A. Pressle, *A guide to quantum groups*, 1994, Cambridge University Press
4. K. Cho and K. Matsumoto *Intersection theory for twisted cohomologies and twisted Riemann's period relations I*, Nagoya Math. J. **139**, 1995 , 67–86
5. V. G. Drinfeld *Quantum groups* in Proceedings of the ICM, Berkley, 1986, A. M. Gleason ed. , 1987 , 798–820, AMS, Providence
6. V. G. Drinfeld *Quasi-Hopf algebras*, Leningrad Math. J. **1** , 1990 , 1419–1457
7. J. Ding and I. B. Frenkel *Isomorphism of two realizations of quantum affine algebra* $U_q\big(\widehat{\mathfrak{gl}}(n)\big)$, Comm. Math. Phys. **156** , 1993 , 277–300
8. G. Felder , *Conformal field theory and integrable systems associated to elliptic curves* in Proceedings of the ICM, Zürich, 1994, Birkhäuser , 1994 , 1247–1255
   and
   G. Felder *Elliptic quantum groups* in Proceedings of the ICMP, Paris 1994, D. Iagolnitzer ed., 1995, Intern. Press Boston , 211–218
9. I. B. Frenkel and N. Yu. Reshetikhin *Quantum affine algebras and holonomic difference equations*, Comm. Math. Phys. **146** , 1992 , 1–60
10. G. Felder, V. Tarasov / and A. Varchenko / *Solutions of elliptic qKZB equations and Bethe ansatz I* Amer. Math. Society Transl., Ser. 2 **180** , 1997 , 45–75
11. G. Felder, V. Tarasov / and A. Varchenko / *Monodromy of solutions of the elliptic quantum Knizhnik-Zamolodchikov-Bernard difference equations* Preprint , 1997 , 1–20
12. G. Felder and A. Varchenko / *On representations of the elliptic quantum group* $E_{\tau,\eta}(sl_2)$ Comm. Math. Phys. **181** , 1996 issue 3 , 741–761
13. G. Gasper and M. Rahman *Basic hypergeometric series* Encycl. Math. Appl., 1990, Cambridge University Press
14. M. Jimbo *Quantum R-matrix for the generalized Toda system* Comm. Math. Phys. **102** , 1986 , 537–547
15. M. Jimbo and T. Miwa *Algebraic analysis of solvable lattice models* CBMS Regional Conf. Series in Math. **85** , 1995
16. T. Kohno *Monodromy representations of braid groups and Yang-Baxter equations* Ann. Inst. Fourier **37** , 1987 , 139–160
    and
    T. Kohno *Linear representations of braid groups and classical Yang-Baxter equations* Contemp. Math. **78** , 1988 , 339–363
17. V. E. Korepin, N. M. Bogolyubov and A. G. Izergin *Quantum inverse scattering method and correlation functions*, 1993 Cambridge University Press
18. D. Kazhdan and G. Lusztig *Affine Lie algebras and quantum groups* Intern. Math. Research Notices **2**, 1991 , 21–29

19. D. Kazhdan and Ya. S. . Soibelman *Representation theory of quantum affine algebras* Selecta Math., New Series **3**, 1995 issue 3 , 537–595
20. M. Kita and M. Yoshida *Intersection theory fro twisted cycles I* Math. Nachrichten **166** , 1994 , 287-304
and
II Math. Nachrichten **168** , 1994 , 171–190
21. S. Lukyanov *Free field representation for massive integrable models* Comm. Math. Phys. **167** , 1995 , 183–226
22. E. Mukhin and A. Varchenko / *The quantized Knizhnik-Zamolodchikov equation in tensor products of irreducible* $\hat{\ }sl_2$*-modules* Preprint , 1997 , 1–32
23. N. Yu. Reshetikhin *Jackson-type integrals, Bethe vectors, and solutions to a difference analogue of the Knizhnik-Zamolodchikov system* Lett. Math. Phys. **26** , 1992 , 153–165
24. N. Yu. Reshetikhin and M. A. Semenov-Tian-Shansky *Central extensions of the quantum current groups* Lett. Math. Phys. **19** , 1990 , 133–142
25. F . A. Smirnov *On the deformation of Abelian integrals* Lett. Math. Phys. **36** , 1996 , 267–275
26. F . A. Smirnov *Form factors in completely integrable models of quantum field theory* Advanced Series in Math. Phys., vol. **14** , 1992, World Scientific, Singapore
27. V. V. Schechtman/ and A.N. Varchenko/ *Arrangements of hyperplanes and Lie algebras homology* Invent. Math. **106** , 1991 , 139–194
and
V. V. Schechtman/ and A.N. Varchenko/ *Quantum groups and homology of local systems* in Algebraic Geometry and Analytic Geometry, Proceedings of the ICM Satellite Conference, Tokyo, 1990 , 1991, Springer-Verlag , Berlin , 182–191
28. V. Tarasov , *Bilinear identity for q-hypergeometric integrals* Preprint , 1997 , 1–24 (to appear in Osaka J. Math.)
29. V. O. Tarasov *Irreducible monodromy matrices for the R-matrix of the XXZ-model and lattice local quantum Hamiltonians*, Theor. Math. Phys. **63**, 1985, 440–454
30. V. Tarasov and A. Varchenko , *Geometry of q-hypergeometric functions, quantum affine algebras and elliptic quantum groups*, Astérisque **246**, 1997, 1–135
31. V. Tarasov and A. Varchenko *Geometry of q-hypergeometric functions as a bridge between Yangians and quantum affine algebras* Invent. Math. , 1997 **128-3**, 501–588
32. V. Tarasov and A. Varchenko *Asymptotic solution to the quantized Knizhnik-Zamolodchikov equation and Bethe vectors* Amer. Math. Society Transl., Ser. 2 **174**, 1996, 235–273
33. V. O. Tarasov and A.N. Varchenko, *Jackson integral representations of solutions of the quantized Knizhnik-Zamolodchikov equation* St. Petersburg Math. J. **6-2**, 1995, 275–313
34. A. Varchenko / *Quantized Knizhnik-Zamolodchikov equations, quantum Yang-Baxter equation, and difference equations for q-hypergeometric functions*, Comm. Math. Phys. **162** , 1994 , 499–528
35. A. Varchenko /, *Multidimensional hypergeometric functions and representation theory of Lie algebras and quantum groups*, Advanced Series in Math. Phys., vol. **21** , 1995, World Scientific, Singapore
A. Varchenko / *Asymptotic solutions to the Knizhnik-Zamolodchikov equation and crystal base*, Comm. Math. Phys. **171** , 1995 , 99-137

Constructing symplectic invariants

Gang Tian

Department of Mathematics, Massachusetts Institute of Technology Cambridge, MA 02139

Introduction

These lectures notes have been written for a course given at the C.I.M.E. session "Quantum Cohomology", June 30-July 8, 1997, at Cetraro, Italy. Their aim is to first review construction of virtual fundamental classes for moduli problems in symplectic geoemtry. The construction here follows [LT3]. Then we will review construction of Gromov-Witten invariants for general symplectic manifolds and give some applications.

The Gromov-Witten invariants for a symplectic manifold X are multilinear maps

$$\psi^X_{A,g,k} : H^*(X;\mathbb{Q})^{\times k} \times H^*(\overline{\mathfrak{M}}_{g,k};\mathbb{Q}) \longrightarrow \mathbb{Q}, \tag{0.1}$$

where $A \in H_2(X,\mathbb{Z})$ is any homology class, k, g are two non-negative integers, and $\overline{\mathfrak{M}}_{g,k}$ is the Deligne-Mumford compactification of $\mathfrak{M}_{\mathfrak{g},\mathfrak{k}}$, the space of smooth k-pointed genus g curves. They are symplectic invariants and defined by enumerating holomorphic maps from Riemann surfaces into X. In [Gr], Gromov first studied symplectic manifolds by exploring invariant properties of the moduli space of pseudo-holomorphic maps from the Riemann sphere. Subsequent progress was made by Ruan [Ru]. He introduced certain symplectic invariants of genus zero and used them to distinguish symplectic manifolds. It is now known that Ruan's invariants are part of extensive invariants, refered to as Gromov-Witten invariants (abbriviated as GW-invariants). The first mathematical theory of GW-invariants and its applications to quantum cohomology was established by Ruan and Tian in late 1993 (cf. [RT1], [RT2]). They not only constructed the GW-invariants of any genus for semi-positive manifolds by using inhomogeneous Cauchy-Riemann equations, but also proved fundamental axioms for these invariants. All Fano-manifolds and Calabi-Yau manifolds are special examples of semi-positive symplectic manifolds. Also any symplectic manifold of complex dimension less than 4 is semi-positive. In 1995, the semi-positivity was removed by Li and Tian for algebraic manifolds by constructing virtual moduli cycles. An alternative construction of virtual moduli cycles was given later by Behrend and Fentachi. More recently, the construction of virtual moduli cycles was extended to general symplectic manifolds first by Fukaya and Ono, Li and Tian, followed by B. Siebert (cf. [FO], [LT2], [Si]). Another approach was proposed by Ruan [Ru], making use of a result in [Si].

I conclude this short introduction with my thanks to Paolo de Bartolomeis for giving me the oppotunity of presenting my lectures at Cetraro.

This work was partially supported by NSF grants.

1. Euler class of weakly Fredholm V-bundles

In this section, we will first introduce the notions of smoothly stratified orbispaces, pseudocycles and weakly Fredholm V-bundles, or for short, simply WFV-bundles, over any smooth stratified orbispaces. Then we will construct the Euler class for weakly Fredholm V-bundles. This construction is due to J. Li and myself [LT3] and refines the one in [LT2].

A smoothly stratified orbispace X is the generalization of orbifolds, which are those topological spaces with only quotient singularities by finite groups. As usual, a singularity of X is where X can not be made smooth locally. For the purpose of defining the GW-invariants, we suffice to consider the case that X has only singularities which are essentially quotients by finite groups. It is possible to extend this theory to the case of more complicated singularities, such as quotient singularities by compact Lie groups. For example, in the case of the Donaldson type invariants using Yang-Mills connections, one needs to construct the Euler class of more general bundles involving quotient singularities by compact Lie groups.

All topological spaces in this section will be assumed to be Hausdorff.

1.1 Smooth stratified orbispaces

In this subsection, we introduce a class of topological spaces which admit certain stratified structures.

Definition 1.1. *By a smoothly stratified space, we mean a topological space X which admits a locally finite [1] stratification $X = \bigcup_{\alpha\in I} X_\alpha$ by locally closed subsets X_i, called strata, such that each X_α is a smooth Banach manifold.*

Note that our definition doe not require any knowledge of the normal cone structure along each stratum X_α.

A *morphism* $f : X \mapsto Y$ between two smoothly stratified spaces is a continuous map $f : X \mapsto Y$ such that for each stratum X_α, the restriction $f_\alpha = f|_{X_\alpha}$ is a smooth map into a certain stratum Y_β of Y. If Y is a smooth manifold, then a morphism f above simply means that each restriction f_α is smooth.

In the following, we will call smoothly stratified space simply by stratified space if no confusion will occur.

Next let us introduce the notion of *smoothly stratified orbispaces* and their *local uniformization charts.* Let X be a topological space with a locally finite stratification $X = \bigcup_{\alpha\in I} X_\alpha$ by locally closed subsets X_α, called strata.

---

[1] By this, we mean for each $x \in X$ the set $\{\alpha \in I \mid a \in X_\alpha\}$ is finite.

Definition 1.2. *Let X be as above. A local uniformization chart of X consists of a stratified space $\tilde{\mathbf{U}}$, a finite group $G_{\tilde{\mathbf{U}}}$ acting effectively on $\tilde{\mathbf{U}}$ by morphisms and a $G_{\tilde{\mathbf{U}}}$-invariant continuous map $\pi_{\mathbf{U}} : \tilde{\mathbf{U}} \to X$ such that*
(1) $\mathbf{U} = \pi_{\mathbf{U}}(\tilde{\mathbf{U}})$ is an open subset of X;
(2) The strata of $\tilde{\mathbf{U}}$ are given by $\tilde{\mathbf{U}}_\alpha = \pi_{\tilde{\mathbf{U}}}^{-1}(\mathbf{U} \cap X_\alpha)$ and each stratum $\tilde{\mathbf{U}}_\alpha$ is invariant under $G_{\tilde{\mathbf{U}}}$;
(3) $\pi_{\mathbf{U}}$ induces a homeomorphism from $\tilde{\mathbf{U}}/G_{\tilde{\mathbf{U}}}$ onto $\mathbf{U}$.

If there is no confusion, we will call $(\tilde{\mathbf{U}}, G_{\tilde{\mathbf{U}}})$ a chart and $\mathbf{U}$ the quotient of the chart $(\tilde{\mathbf{U}}, G_{\tilde{\mathbf{U}}})$.

Let $\mathbf{V} \subset \mathbf{U}$ be an open subset, then $(\pi_{\tilde{\mathbf{U}}}^{-1}(\mathbf{V}), G_{\tilde{\mathbf{U}}})$ defines a new local uniformization chart of X, called the restriction of $(\tilde{\mathbf{U}}, G_{\tilde{\mathbf{U}}})$ to $\mathbf{V}$. For simplicity, we will often denote it by $(\tilde{\mathbf{U}}, G_{\tilde{\mathbf{U}}})|_{\mathbf{V}}$. Now let $(\tilde{\mathbf{U}}, G_{\tilde{\mathbf{U}}})$ and $(\tilde{\mathbf{V}}, G_{\tilde{\mathbf{V}}})$ be two charts of X. We say that the later is finer than the former if there is a homomorphism $G_{\tilde{\mathbf{V}}} \to G_{\tilde{\mathbf{U}}}$ and an equivariant continuous covering map

$$\varphi_{\tilde{\mathbf{U}}\tilde{\mathbf{V}}} : \pi_{\tilde{\mathbf{V}}}^{-1}(\mathbf{U} \cap \mathbf{V}) \longrightarrow \pi_{\tilde{\mathbf{U}}}^{-1}(\mathbf{U} \cap \mathbf{V}) \tag{1.1}$$

such that it is a morphism between stratified spaces and commutes with the projections $\pi_{\tilde{\mathbf{V}}}^{-1}(\mathbf{U} \cap \mathbf{V}) \to \mathbf{U} \cap \mathbf{V} \subset X$ and $\pi_{\tilde{\mathbf{U}}}^{-1}(\mathbf{U} \cap \mathbf{V}) \to \mathbf{U} \cap \mathbf{V}$.

Now we introduce the notion of smoothly stratified orbispaces.

Definition 1.3. *We say that X is a smoothly stratified orbispace if it is a topological space with a locally finite stratification $X = \bigcup_{\alpha \in I} X_\alpha$ by locally closed subsets X_α and can be covered by quotients of local uniformization charts as in Definition 1.2 such that for any finite collection of charts $\{(\tilde{\mathbf{U}}_i, G_{\tilde{\mathbf{U}}_i})\}_{1 \le i \le k}$ and $x \in \cap_{i=1}^k \mathbf{U}_\alpha$, there is a chart $(\tilde{\mathbf{V}}, G_{\tilde{\mathbf{V}}})$ such that x is contained in the quotient $\mathbf{V}$ of $(\tilde{\mathbf{V}}, G_{\tilde{\mathbf{V}}})$ and $(\tilde{\mathbf{V}}, G_{\tilde{\mathbf{V}}})$ is finer than all $(\tilde{\mathbf{U}}_i, G_{\tilde{\mathbf{U}}_i})$.*

We will call smoothly stratified orbispace simply stratified orbispace. Clearly, each stratified space is a stratified orbispace. A stratified orbispace is smooth if all its charts in the Definition 1.3 and corresponding morphisms defined in (1.1) are smooth. In particular, a finite dimensional smooth smoothly stratified orbispace is an orbifold. Also if X is a smoothly stratified orbispace, its strata X_α are Banach orbifolds, that is, they are the quotients of open Banach manifolds by finite groups.

If X and Y are two stratified orbispaces, by a map from X into Y, we mean a continuous map $f : X \mapsto Y$ such that for any $x \in X$, there is a chart $(\tilde{\mathbf{U}}, G_{\tilde{\mathbf{U}}})$ of X, a chart $(\tilde{\mathbf{V}}, G_{\tilde{\mathbf{V}}})$ of Y and a homomorphism $G_{\tilde{\mathbf{U}}} \mapsto G_{\tilde{\mathbf{V}}}$ and $G_{\tilde{\mathbf{U}}}$-equivariant map $\tilde{f} : \tilde{\mathbf{U}} \mapsto \tilde{\mathbf{V}}$ such that $x \in \mathbf{U}$, $f(x) \in \mathbf{V}$ and $\tilde{f}$ descends to $f|_{\mathbf{U}} : \mathbf{U} \mapsto \mathbf{V} \subset Y$. We will call such $\tilde{f}$ a local representative of f. We say that f is a morphism if all its representatives are morphisms. A morphism $f : X \mapsto Y$ is an isomorphism if it has an inverse morphism $f^{-1} : Y \mapsto X$.

1.2 Weakly pseudocycles

In this subsection, we will work with a continuous map $f: X \to Y$ between two topological spaces X and Y. All homology theories are with rational coefficients.[2]

First we recall the definition of manifolds with corner. A manifold M of dimension d with corner is a topological space which admits a covering by open subsets $\{U_\alpha\}$ such that (1) for each α, there is a homeomorphism ϕ_α from U_α onto an open subset in $L_\alpha = \{x_1 \geq 0, \cdots, x_k \geq 0\} \subset \mathbb{R}^d$, where $k = k(\alpha) \leq d$ and $x_1, \cdots, x_d$ are coordinates of $\mathbb{R}^d$; (2) For any α and β with $U_\alpha \cap U_\beta \neq \emptyset$, the transition function $\phi_\beta \circ \phi_\alpha^{-1}|_{\phi_\alpha(U_\alpha \cap U_\beta)}$ is a diffeomorphism. The boundary of such a manifold is defined to be

$$\partial M = \bigcup_{\alpha} \bigcup_{1 \leq i \leq k(\alpha)} \phi_\alpha^{-1}(L_\alpha \cap \{x_i = 0\}). \tag{1.2}$$

Clearly, ∂M is a manifold of dimension $d-1$ with corner.

Now we define pseudomanifolds. An oriented pseudomanifold S of dimension d in X is a pair (ρ, M), where M is an oriented, smooth manifold of dimension d with corner and $\rho: M \to X$ is a continuous map. For a pseudomanifold $S = (\rho, M)$, we define ∂S to be $(\rho|_{\partial M}, \partial M)$. We define $-S$ to be $(\rho, \overline{M})$, where $\overline{M}$ is M with the reversed orientation. We define the sum of two pseudomanifolds (ρ, M) and (ρ', N) to be $(\rho \cup \rho', M \cup N)$. We then can define the sum $\sum n_i S_i$, where $n_i \in \mathbb{Z}$, of pseudomanifolds inductively. For any (ρ, M), we say that it is 0 if there are two disjoint open subsets U_1 and U_2 of M so that the complement of their union in M is at most a $d-1$ dimensional set[3]. such that there is an orientation reversing isomorphism $\varphi: U_1 \to U_2$ so that $\rho|_{U_1} \equiv \rho \circ \varphi$. A rational pseudomanifold is a formal sum of oriented pseudomanifolds with rational coefficients. A rational pseudomanifold is zero if an integer multiple of it is integral and is zero.

Let $f: X \to Y$ be a continuous map between topological spaces. We now define weakly f-pseudocycles.

Definition 1.4. *A rational weakly d-dimensional f-pseudocycle is given by a triple (ρ, M, K), where (ρ, M) is a rational pseudomanifold in X, K is a closed subset in Y of homological dimension no more than $d-2$[4] such that the closure of $M - \rho^{-1}f^{-1}(K)$ is compact in M and the rational $(d-1)$-dimensional pseudomanifold $(\rho|_{\partial M}, \partial M - \rho^{-1}f^{-1}(K))$ is 0.*

Two such cycles (ρ_1, M_1, K_1) and (ρ_2, M_2, K_2) are quasi-cobordant if there is a $(d+1)$-dimensional pseudomanifold (ρ, M) in X, a closed subset $K \subset Y^{top}$ of homological dimension no more than $d-1$ such that $M - \rho^{-1}f^{-1}(K)$ is compact in M, K_1 and K_2 are contained in K and $(\rho_1, M_1) - (\rho_2, M_2) =$

---

[2] In fact, many discussions in the following hold for integers.

[3] By this we mean that it is the image of finitely many $d-1$ dimensional manifolds.

[4] By this we mean $\max\{k; H_k(K, \mathbb{Q}) \neq 0\} \leq d-2$.

$(\rho, \partial M)$. *Two such cycles* S *and* S' *are equivalent if there is a chain of weakly* f*-pseudocycles* $S_0, \cdots, S_l$ *so that* $S = S_0$, S_i *is quasi-cobordant to* S_{i+1} *and* $S_l = S'$.

The usefulness of this definition is illustrated by the following Proposition.

Proposition 1.1. *Let* (ρ, M, K) *be a weakly* d*-dimensional* f*-pseudocycle. Then* $f \circ \rho : M \to Y$ *defines a unique element in* $H_d(Y^{\text{top}}, K) = H_d(Y^{\text{top}})$. *Further, two equivalent such cycles define an identical element in* $H_d(Y)$.

Proof. First, the two homology groups are canonically isomorphic since K has homological dimension $\leq d-2$. Now we show that $f \circ \rho$ defines a cycle in $H_d(Y^{\text{top}}, K)$. Since M is smooth and the closure of $M - \rho^{-1} f^{-1}(K)$ is compact in M, we can give M a triangulation so that the boundary of $f \circ \rho(M)$ is contained in K, thus it defines an element in $H_d(Y, K)$. For the same reason, if two weakly f-pseudocycles S_1 and S_2 are quasi-cobordant, then they give rise to the same class in $H_d(Y^{\text{top}}, K)$. On the other hand, since K is of homological dimension $\leq d-1$, the natural homomorphism from $H_d(Y^{\text{top}})$ into $H_d(Y^{\text{top}}, K)$ is injective, so S_1 and S_2 define the same homology class in $H_d(Y^{\text{top}})$. It follows that equivalent weakly f-pseudocycles define identical elements in $H_d(Y)$.

1.3 Weakly Fredholm V-bundles

In this subsection, we introduce the notion of weakly Fredholm V-bundles over a smoothly stratified orbispace $X = \bigcup_{\alpha \in I} X_\alpha$. We assume throughout this and the next section that $0 \in I$ and the stratum X_0 is open and dense in X. Also, if $\tilde{\mathbf{U}}$ is a chart of X, then $\tilde{\mathbf{U}}$ has a stratification $\cup_{\alpha \in I} \tilde{\mathbf{U}}_\alpha$ so indexed that $\tilde{\mathbf{U}}_\alpha$ corresponds to the stratum $\mathbf{U} \cap X_\alpha$.

Definition 1.5. *Let* X *be a smoothly stratified orbispace. A* V*-bundle over* X *is a pair* $(\mathbf{E}, \mathbf{P})$, *where* $\mathbf{E}$ *is a smoothly stratified orbispace and* $\mathbf{P} : \mathbf{E} \to X$ *is a projection (of stratified orbispaces) such that* X *(resp.* $\mathbf{E}$*) is covered by a collection of charts* $(\tilde{\mathbf{U}}_i, G_{\tilde{\mathbf{U}}_i})$ *(resp.* $(\mathbf{E}_{\tilde{\mathbf{U}}_i}, G_{\tilde{\mathbf{U}}_i})$*) for which the following holds:*
(1) if $\tilde{\mathbf{U}}_i$ *is a chart over* $\mathbf{U}_i \subset X$ *then* $\mathbf{E}_{\tilde{\mathbf{U}}_i}$ *is a chart over* $\mathbf{P}^{-1}(\mathbf{U}_i)$ *and there is a projection* $\mathbf{P}_{\tilde{\mathbf{U}}_i} : \mathbf{E}_{\tilde{\mathbf{U}}_i} \to \tilde{\mathbf{U}}_i$ *representing* $\mathbf{P}$ *from* $\mathbf{P}^{-1}(\mathbf{U}_i)$ *onto* $\mathbf{U}_i$*;*
(2) Let $\tilde{\mathbf{U}}_\alpha$ *be any stratum of* $\tilde{\mathbf{U}}_i$*, then restricting to* $\tilde{\mathbf{U}}_\alpha$*, the bundle* $\mathbf{E}_{\tilde{\mathbf{U}}_i}|_{\tilde{\mathbf{U}}_\alpha}$ *is a smooth Banach vector bundle over* $\tilde{\mathbf{U}}_\alpha$ *with a* $G_{\tilde{\mathbf{U}}_i}$*-linear action. Furthermore, the strata of* $\mathbf{E}_{\tilde{\mathbf{U}}_i}$ *are given by* $\mathbf{E}_{\tilde{\mathbf{U}}_i}|_{\tilde{\mathbf{U}}_\alpha}$*, which are simply pull-backs of the strata of* $\tilde{\mathbf{U}}_i$ *by* $\mathbf{P}_{\tilde{\mathbf{U}}_i}$*.*
(3) For any two charts $(\tilde{\mathbf{U}}_i, G_{\tilde{\mathbf{U}}_i})$ *and* $(\tilde{\mathbf{U}}_j, G_{\tilde{\mathbf{U}}_j})$ *of* X *(over* $\mathbf{U}_i$ *and* $\mathbf{U}_j$ *respectively) and* $x \in \mathbf{U}_i \cap \mathbf{U}_j$*, there is a local uniformization chart* $(\tilde{\mathbf{U}}_k, G_{\tilde{\mathbf{U}}_k})$ *finer than both* $(\tilde{\mathbf{U}}_i, G_{\tilde{\mathbf{U}}_i})$ *and* $(\tilde{\mathbf{U}}_j, G_{\tilde{\mathbf{U}}_j})$ *such that* $x \in \mathbf{U}_k$ *and there are* $G_{\tilde{\mathbf{U}}_k}$*-equivariant isomorphisms*

$$\mathbf{E}_{\tilde{\mathbf{U}}_k}|_{\mathbf{U}_\alpha\cap\mathbf{U}_k} \cong \varphi^*_{\tilde{\mathbf{U}}_k\tilde{\mathbf{U}}_i}(\mathbf{E}_{\tilde{\mathbf{U}}_i}) \quad \text{and} \quad \mathbf{E}_{\tilde{\mathbf{U}}_k}|_{\mathbf{U}_j\cap\mathbf{U}_k} \cong \varphi^*_{\tilde{\mathbf{U}}_k\tilde{\mathbf{U}}_j}(\mathbf{E}_{\tilde{\mathbf{U}}_j})$$

preserving the Banach bundle structures along each stratum, where $\varphi_{\tilde{\mathbf{U}}_k\tilde{\mathbf{U}}_i}$ *is the map given in (1.1).*
(4) let $\mathbf{O}_{\mathbf{E}_{\tilde{\mathbf{U}}_i}}$ *be the zero sections of* $\mathbf{E}_{\tilde{\mathbf{U}}_i}$*, then they define a smoothly stratified suborbispace* $\mathbf{O}_\mathbf{E}$ *of* $\mathbf{E}$*. We require that it is isomorphic to* X *under the projection* $\mathbf{P}$*.*

Now let $\mathbf{E}$ be a V-fiber bundle over X. A section of $\mathbf{E}$ is a morphism $\Phi: X \to \mathbf{E}$ as stratified orbispaces so that $\mathbf{P}\circ\Phi: X \to X$ is an isomorphism. We say that Φ has compact support if $\Phi^{-1}(O_\mathbf{E})$ is compact. For simplicity, we will write $\Phi^{-1}(0)$ for $\Phi^{-1}(O_\mathbf{E})$.

We fix a V-bundle over X with a section Φ with compact support. We now describe the Fredholm structure of this bundle with section in terms of its local finite approximations.

Definition 1.6. *A local finite approximation of* $(\mathbf{E},\Phi)$ *consists of a chart* $(\tilde{\mathbf{U}}, G_{\tilde{\mathbf{U}}})$ *of* X*, a chart* $(\mathbf{E}_{\tilde{\mathbf{U}}}, G_{\tilde{\mathbf{U}}})$ *of* $\mathbf{E}$*, a representative* $\Phi_{\tilde{\mathbf{U}}}: \tilde{\mathbf{U}} \to \mathbf{E}_{\tilde{\mathbf{U}}}$ *of* Φ *and a finite rank equi-rank* $G_{\tilde{\mathbf{U}}}$*-linearized vector bundle* $\mathbf{F}$ *over* $\tilde{\mathbf{U}}$ *such that:*
(1) $\mathbf{F}$ *is a* $G_{\tilde{\mathbf{U}}}$*-equivariant smoothly stratified subspace of* $\mathbf{E}_{\tilde{\mathbf{U}}}$ *and restricting to each stratum* $\tilde{\mathbf{U}}_\alpha \subset \tilde{\mathbf{U}}$ *the inclusion* $\mathbf{F}|_{\tilde{\mathbf{U}}_\alpha} \subset \mathbf{E}_{\tilde{\mathbf{U}}}|_{\tilde{\mathbf{U}}_\alpha}$ *is a smooth vector subbundle;*
(3) $U := \Phi_{\tilde{\mathbf{U}}}^{-1}(\mathbf{F}) \subset \tilde{\mathbf{U}}$ *is an equi-dimensional smoothly stratified space of finite dimension (with strata* $U = \cup_{\alpha\in I} U_\alpha$*). Further we require* $U_0 = U \cap \mathbf{U}_0$ *is open and dense in* U *and its complement has codimension at least 2 in* U*;*
(4) $F := \mathbf{F}|_U$ *is a continuous vector bundle, and its restriction to each stratum* U_α *is a smooth vector bundle and the map* $\phi_F|_{U_\alpha}$*, where* $\phi_F = \Phi_{\tilde{\mathbf{U}}}|_U : U \to F$*, is a smooth section;*
(5) At each $w \in \Phi^{-1}(0) \cap \tilde{\mathbf{U}}_0$*, the differential*

$$d\phi_F(w) : T_w\tilde{\mathbf{U}}_0 \longrightarrow \mathbf{E}_{\tilde{\mathbf{U}}_0}|_w/\mathbf{F}|_w$$

is surjective.

Such a local finite approximation will be denoted by $(\tilde{\mathbf{U}}, G_{\tilde{\mathbf{U}}}, \mathbf{E}_{\tilde{\mathbf{U}}}, \mathbf{F})$.

An orientation of $(\tilde{\mathbf{U}}, G_{\tilde{\mathbf{U}}}, \mathbf{E}_{\tilde{\mathbf{U}}}, \mathbf{F})$ is a $G_{\tilde{\mathbf{U}}}$-invariant orientation of the real line bundle $\wedge^{\mathrm{top}}(TU)\otimes\wedge^{\mathrm{top}}(F)^{-1}$ over U_0. Note that such an orientation is given by a $G_{\tilde{\mathbf{U}}}$-invariant nonvanishing section of $\wedge^{\mathrm{top}}(TU)\otimes\wedge^{\mathrm{top}}(V)^{-1}$ over U_0.

The index of $(\tilde{\mathbf{U}}, G_{\tilde{\mathbf{U}}}, \mathbf{E}_{\tilde{\mathbf{U}}}, \mathbf{F})$ is defined to be $\operatorname{rank} F - \dim U$.

Now assume that $(\tilde{\mathbf{U}}', G_{\tilde{\mathbf{U}}'}, \mathbf{E}_{\tilde{\mathbf{U}}'}, \mathbf{F}')$ is another local finite approximation of the same index over $\mathbf{U}' \subset X$.

Definition 1.7. *We say that* $(\tilde{\mathbf{U}}', G_{\tilde{\mathbf{U}}'}, \mathbf{E}_{\tilde{\mathbf{U}}'}, \mathbf{F}')$ *is finer than* $(\tilde{\mathbf{U}}, G_{\tilde{\mathbf{U}}}, \mathbf{E}_{\tilde{\mathbf{U}}}, \mathbf{F})$ *if we have the following:*
(1) $(\tilde{\mathbf{U}}', G_{\tilde{\mathbf{U}}'})$ *is finer than* $(\tilde{\mathbf{U}}, G_{\tilde{\mathbf{U}}})$*;*

(2) Let $\varphi : \pi_{\tilde{\mathbf{U}}'}^{-1}(\mathbf{U}' \cap \mathbf{U}) \to \pi_{\tilde{\mathbf{U}}}^{-1}(\mathbf{U} \cap \mathbf{U}')$ *(which is* $\varphi_{\mathbf{U}\mathbf{U}'}$ *in (1.1)) be the covering map, then* $\varphi^*\mathbf{F} \subset \varphi^*\mathbf{E}_{\tilde{U}} \equiv \mathbf{E}_{\tilde{\mathbf{U}}'}|_{\mathbf{U}'\cap\mathbf{U}}$ *is a continuous subbundle of* $\mathbf{F}'$ *and also a smoothly stratified subspace;*
(3) Let (U, F) *and* (U', F') *be those defined in the Definition 1.6 for* $\tilde{\mathbf{U}}$ *and* $\tilde{\mathbf{U}}'$ *respectively, then for any* w *in* $U_0' \cap \varphi^{-1}(U_0)$*, the natural homomorphism*

$$T_w U'/T_{\varphi(w)}U \longrightarrow (F'/\varphi^*F)|_w$$

is an isomorphism;
(4) In case both $(\tilde{\mathbf{U}}, G_{\tilde{\mathbf{U}}}, \mathbf{E}_{\tilde{\mathbf{U}}}, \mathbf{F})$ *and* $(\tilde{\mathbf{U}}', G_{\tilde{\mathbf{U}}'}, \mathbf{E}_{\tilde{\mathbf{U}}'}, \mathbf{F}')$ *are oriented, then we also require that the orientations of* $(\tilde{\mathbf{U}}, G_{\tilde{\mathbf{U}}}, \mathbf{E}_{\tilde{\mathbf{U}}}, \mathbf{F})$ *and* $(\tilde{\mathbf{U}}', G_{\tilde{\mathbf{U}}'}, \mathbf{E}_{\tilde{\mathbf{U}}'}, \mathbf{F}')$ *are isomorphic through the isomorphism*

$$\wedge^{\mathrm{top}}(T_w U') \otimes \wedge^{\mathrm{top}}(F'|_w)^{-1} \cong \wedge^{\mathrm{top}}(T_{\varphi(w)}U) \otimes \wedge^{\mathrm{top}}(F|_{\varphi(w)})^{-1}, \qquad (1.3)$$

induced by the isomorphism in (3), where w *is any point in* $U_0' \cap \varphi^{-1}(U_0)$*.*

Note that from (5) of the Definition 1.6, $U_0' \cap \varphi^{-1}(U_0)$ is a locally closed smooth submanifold of codimension $\operatorname{rank} F' - \operatorname{rank} F$. Hence the homomorphism (3) in the Definition 1.7 is well-defined.

Now let $\mathfrak{A} = \{(\mathbf{U}_i, \mathfrak{G}_{\mathbf{U}_i}, \mathbf{E}_{\mathbf{U}_i}, \mathbf{F}_i)\}_{i\in\mathcal{K}}$ be a collection of oriented finite approximations of $(X, \mathbf{E}, \Phi)$. In the following, we will denote by $\mathbf{U}_i$ the open subsets of X such that $\Lambda_i = (\tilde{\mathbf{U}}_i, G_{\tilde{\mathbf{U}}_i})$ is a uniformization chart over $\mathbf{U}_i$. We say that $\mathfrak{A}$ covers $\Phi^{-1}(0)$ if $\Phi^{-1}(0)$ is contained in the union of $\mathbf{U}_i$ in X.

Definition 1.8. *An index* r *oriented weakly smooth structure of* $(X, \mathbf{E}, \Phi)$ *is a collection* $\mathfrak{A} = \{(\mathbf{U}_i, \mathfrak{G}_{\mathbf{U}_i}, \mathbf{E}_{\mathbf{U}_i}, \mathbf{F}_i)\}_{i\in\mathcal{K}}$ *of index* r *oriented smooth approximations such that* $\mathfrak{A}$ *covers* $\Phi^{-1}(0)$ *and that for any* $(\tilde{\mathbf{U}}_i, G_{\tilde{\mathbf{U}}_i}, \mathbf{E}_{\tilde{\mathbf{U}}_i}, \mathbf{F}_i)$ *and* $(\tilde{\mathbf{U}}_j, G_{\tilde{\mathbf{U}}_j}, \mathbf{E}_{\tilde{\mathbf{U}}_j}, \mathbf{F}_j)$ *in* $\mathfrak{A}$ *with* $x \in \mathbf{U}_i \cap \mathbf{U}_j$*, there is a local finite approximation* $(\tilde{\mathbf{U}}, G_{\tilde{\mathbf{U}}}, \mathbf{E}_{\tilde{\mathbf{U}}}, \mathbf{F}) \in \mathfrak{A}$ *such that* $x \in \mathbf{U}$ *and* $(\tilde{\mathbf{U}}, G_{\tilde{\mathbf{U}}}, \mathbf{E}_{\tilde{\mathbf{U}}}, \mathbf{F})$ *is finer than both* $(\tilde{\mathbf{U}}_i, G_{\tilde{\mathbf{U}}_i}, \mathbf{E}_{\tilde{\mathbf{U}}_i}, \mathbf{F}_i)$ *and* $(\tilde{\mathbf{U}}_j, G_{\tilde{\mathbf{U}}_j}, \mathbf{E}_{\tilde{\mathbf{U}}_j}, \mathbf{F}_j)$*.*

It follows easily from this definition that for any number of local finite approximations $\{(\tilde{\mathbf{U}}_i, G_i, \mathbf{E}_i, \mathbf{F}_i)\}_{1\le i\le l}$ in $\mathfrak{A}$ and $x \in \cap_{i=1}^{l} \mathbf{U}_\alpha$, there is a local finite approximation $(\tilde{\mathbf{U}}, G_{\tilde{\mathbf{U}}}, \mathbf{E}_{\tilde{\mathbf{U}}}, \mathbf{F})$ which contains x and is finer than all $(\tilde{\mathbf{U}}_i, G_i, \mathbf{E}_i, \mathbf{F}_i)$.

Let $\mathfrak{A}'$ be another index r oriented weakly smooth structure of $(X, \mathbf{E}, \Phi)$. We say that $\mathfrak{A}'$ is finer than $\mathfrak{A}$ if for any $(\tilde{\mathbf{U}}, G_{\tilde{\mathbf{U}}}, \mathbf{E}_{\tilde{\mathbf{U}}}, \mathbf{F}) \in \mathfrak{A}$ over $\mathbf{U} \subset X$ and $p \in \mathbf{U} \cap \Phi^{-1}(0)$, there is a $(\tilde{\mathbf{U}}', G_{\tilde{\mathbf{U}}'}, \mathbf{E}_{\tilde{\mathbf{U}}'}, \mathbf{F}') \in \mathfrak{A}'$ over $\mathbf{U}'$ such that $p \in \mathbf{U}'$ and $(\tilde{\mathbf{U}}', G_{\tilde{\mathbf{U}}'}, \mathbf{E}_{\tilde{\mathbf{U}}'}, \mathbf{F}')$ is finer than $(\tilde{\mathbf{U}}, G_{\tilde{\mathbf{U}}}, \mathbf{E}_{\tilde{\mathbf{U}}}, \mathbf{F})$. We say that two weakly smooth structures $\mathfrak{A}_1$ and $\mathfrak{A}_2$ are equivalent if $\mathfrak{A}_1$ is finer than $\mathfrak{A}_2$ and vice verse.

Definition 1.9. *A* V*-vector bundle with section* $(X, \mathbf{E}, \Phi)$ *is a weakly Fredholm V-bundle if it admits an oriented weakly smooth structure and* $\Phi^{-1}(0)$ *is compact and is contained in finitely many strata of* X*.*

Definition 1.10. *Two weakly Fredholm V-bundles* $(X, \mathbf{E}, \Phi)$ *and* $(X, \mathbf{E}', \Phi')$ *are homotopic if there is a weakly Fredholm V-bundle* $(X \times [0,1], \overline{\mathbf{E}}, \Psi)$ *such that it restricts to* $(X, \mathbf{E}, \Phi)$ *over* $X \times \{0\}$ *and to* $(X', \mathbf{E}', \Phi')$ *over* $X \times \{1\}$.

1.4 Construction of the Euler class

Let X be any stratified orbispace with the stratification $X = \cup_{\alpha \in I} X_\alpha$, let Y be a finite dimensional smooth orbifold and $f : X^{\mathrm{top}} \to Y^{\mathrm{top}}$ be a smooth map.

The following is the main result on constructing the Euler class.

Theorem 1.1. *Let* $(X, \mathbf{E}, \Phi)$ *be a WFV-bundle of relative index* r. *Let* $\mathfrak{A}$ *be the pre-weakly Fredholm structure on* $(X, \mathbf{E}, \Phi)$ *given Definition 1.8. We assume further that for any* $(\tilde{\mathbf{U}}, G_{\tilde{\mathbf{U}}}, \mathbf{E}_{\tilde{\mathbf{U}}}, \mathbf{F}_{\tilde{\mathbf{U}}})$ *in* $\mathfrak{A}$ *and* $U = \cup_{\alpha \in I} U_\alpha$ *the stratification, the representative* $f_{\tilde{\mathbf{U}},\alpha} : U_\alpha \to Y_\alpha$ *is a submersion. Then we can assign to* $(X, \mathbf{E}, \Phi)$ *an equivalence class* $e(X, \mathbf{E}, \Phi)$ *of weakly* f*-pseudocycles, called the Euler class of* $(X, \mathbf{E}, \Phi)$. *It depends only on the homotopy class of* $(X, \mathbf{E}, \Phi)$. *Further, the image of this class under* f_* *is a well-defined homology class in* $H_r(Y, \mathbb{Q})$.

The class $e(X, \mathbf{E}, \Phi)$ or its image under f_* is the virtual cycle we set to construct.

In the remainder of this section, we will outline main steps in the construction of the required Euler class by using the WFV structure of $(X, \mathbf{E}, \Phi)$. The complete account of this approach is a modification of the construction in [LT2] and will appear in [LT3].

The idea of the construction is as follows: Given a local finite approximation $(\tilde{\mathbf{U}}, G_{\tilde{\mathbf{U}}}, \mathbf{E}_{\tilde{\mathbf{U}}}, \mathbf{F})$ over $\mathbf{U} \subset X$, we can associate to it a finite dimensional model consisting of a finite dimensional stratified orbispace U, a smoothly stratified finite rank V-bundle F over U and a smooth section $\phi : U \to F$. Following the topological construction of the Euler class, we like to perturb ϕ to a new section $\tilde{\phi} : U \to F$ so that $\tilde{\phi}$ is as transversal to the zero section of F as possible. Here the difficulty with this is that F is only smooth along, not near, strata of U. Thus the notion of transversality is ill-defined near points where F is not smooth. Moreover, we have to allow $\tilde{\Phi}$ to be a multi-valued section [5] because X is only an orbispace. Here is our solution: We find a homological dimension $r-2$ set $K \subset U$ so that we can perturb ϕ to $\tilde{\phi}$ so that $\tilde{\phi}$ is transversal to the zero section of F away from K. Since the dimension of the zero locus of $\tilde{\phi}$ is r, the set K will not affect the construction of an r dimensional cycle. In the case we work with many local finite approximations, we will choose these perturbations so that their vanishing locus patch

[5] By this, we mean a section of $S^\bullet \mathbf{E}$ with certain matching properties and which can be locally represented by a finite set of unordered sections of $\mathbf{E}$, where $S^\bullet \mathbf{E}$ is the symmetric fiber product of $\mathbf{E}$. Note that $S^\bullet \mathbf{E}$ is not a vector bundle in general.

together to form a well-defined weakly f-pseudocycle in X. The Euler class of $(X, \mathbf{E}, \Phi)$ is the equivalent class it represents.

As is clear from this description, the main difficulty of this construction is to make sure that these perturbed vanishing locus patch together. We will do perturbation one chart at a time, and in the mean time make sure that the perturbation on the current chart agrees with the perturbations chosen in the preceding charts. One difficulty arises in that the intersection of charts are not open. However if the current chart is finer than all the preceding charts, then the intersection will be an open subset in the preceding charts and a locally closed subset in the current chart. Hence our first task is to pick a good set of charts. Let $\mathfrak{A} = \{(\mathbf{U}_i, \mathfrak{G}_i, \mathbf{E}_i, \mathbf{F}_i)\}_{i \in \mathcal{K}}$ be the weakly smooth structure of $(X, \mathbf{E}, \Phi)$. For simplicity, we will denote the projection $\pi_{\mathbf{U}_i} : \tilde{\mathbf{U}}_i \to \mathbf{U}_i$ by π_i, denote $\Phi_i^{-1}(\mathbf{F}_i)$ by U_i, denote $\mathbf{F}_i|_{U_i}$ by F_i and denote the restriction $\Phi_i|_{U_i} : U_i \to \tilde{\mathbf{E}}_i|_{U_i}$ by $\phi_i : U_i \to F_i$. Without loss of generality, we can assume that for any approximation $(\tilde{\mathbf{U}}_i, G_i, \mathbf{E}_i, \mathbf{F}_i) \in \mathfrak{A}$ over $\mathbf{U}_i$ and any $\mathbf{U}' \subset \mathbf{U}_i$, the restriction $(\tilde{\mathbf{U}}_i, G_i, \mathbf{E}_i, \mathbf{F}_i)|_{\mathbf{U}'}$ is also a member in $\mathfrak{A}$.

The following lemma enables us to pick a good set of charts.

Lemma 1.1. *There is a finite collection $\mathcal{L} \subset \mathcal{K}$ and a total ordering of $\mathcal{L}$ of which the following holds:*
(1) The set $\Phi^{-1}(0)$ is contained in the union $\bigcup\{\pi_i(U_i) \mid i \in \mathcal{L}\}$;
(2) For any pair $i < j \in \mathcal{L}$, the approximation $(\tilde{\mathbf{U}}_j, G_j, \mathbf{E}_j, \mathbf{F}_j)$ is finer than the approximation $(\tilde{\mathbf{U}}_i, G_i, \mathbf{E}_i, \mathbf{F}_i)$.

Proof. We now outline the proof, which is elementary.

We will use the stratified structure of $(X, \mathbf{E}, \Phi)$ as we did in [LT2]. Let X_α $(\alpha \in I_0)$ be all the strata of X that intersect $\Phi^{-1}(0)$. We give an order to I_0 so that for each $\alpha \in I_0$, $\Phi^{-1}(0) \cap \bigcup_{\beta \geq \alpha} X_\beta$ is compact. For convenience, we assume $I_0 = \{1, \cdots, k\}$.

To prove the lemma, we suffice to construct subsets $\mathcal{L}_\alpha^l \subset \mathcal{K}$, where $\alpha \in I_0$ and $0 \leq l \leq l_\alpha$, such that

$\mathbf{A}_1$: For any distinct $i, j \in \mathcal{L}_\alpha^l$, $\mathbf{U}_i \cap \mathbf{U}_j = \emptyset$;
$\mathbf{A}_2$: For each $\alpha \in I_0$ and $l \leq l_\alpha$, the set

$$\mathbf{Z}_\alpha^l = \Phi^{-1}(0) \cap X_\alpha - \bigcup_{(\beta,j) \geq (\alpha,l)} \bigcup_{k \in \mathcal{L}_\beta^j} U_k$$

is contained in the image of finitely many submanifolds of dimension less than l in X_α, where $(\beta, j) \geq (\alpha, l)$ means that either $\alpha \leq \beta$ or $\alpha = \beta$ and $l \leq j$;
$\mathbf{A}_3$: For any pair of distinct $(i, j) \in \mathcal{L}_\alpha^l \times \mathcal{L}_{\alpha'}^{l'}$ with $(\alpha, l) \leq (\alpha', l')$, the chart $(\tilde{\mathbf{U}}_i, G_i, \mathbf{E}_i, \mathbf{F}_i)$ is finer than $(\tilde{\mathbf{U}}_j, G_j, \mathbf{E}_j, \mathbf{F}_j)$.

In particular, all U_i with $i \in \bigcup_{\beta \geq \alpha} \bigcup_{0 \leq l \leq l_\beta} \mathcal{L}_\beta^l$ cover $\Phi^{-1}(0) \cap X_\alpha$.

We will construct $\mathcal{L}_\alpha^l$ inductively, starting from the largest $\alpha \in I_0$. Note that $\mathcal{L}_\alpha^l$ may be empty. We now assume that for some $\alpha \in I_0$, we have constructed $\mathcal{L}^l_{\alpha+1}, \mathcal{L}^l_{\alpha+2}, \cdots$ for all l. We now construct $\mathcal{L}^l_\alpha$. We first pick a finite $\mathcal{L}' \subset \mathcal{K}$ such that $\{\mathbf{U}_i | i \in \mathcal{L}'\}$ covers $\Phi^{-1}(0) \cap X_\alpha$. This is possible since it is compact. Let l_α be the maximum of $\{\dim(U_i \cap X_\alpha) \,|\, i \in \mathcal{L}'\}$. By assumption, $l_\alpha > 0$. Since $\tilde{\mathbf{U}}_i$ is a stratified space with strata $\pi_i^{-1}(\mathbf{U}_i \cap X_\alpha)$, we can find open subsets $\mathbf{U}'_i \subset \mathbf{U}_i$ such that

(1) let $\tilde{\mathbf{U}}'_i = \pi_i^{-1}(\mathbf{U}'_i)$ then $\overline{\tilde{\mathbf{U}}'_i} \cap \pi_i^{-1}(X_\alpha)$ is the same as the closure of $\tilde{\mathbf{U}}'_i \cap \pi_i^{-1}(X_\alpha)$ in $\pi_i^{-1}(X_\alpha)$, where $\overline{\tilde{\mathbf{U}}'_i}$ denotes the closure of $\tilde{\mathbf{U}}'_i$ in X;
(2) Let $R = \tilde{\mathbf{U}}'_i \cap U_i \cap \pi_i^{-1}(X_\alpha)$, then $\bar{R} - R$ (in $U_i \cap \pi_i^{-1}(X_\alpha)$) is a smooth submanifold in $U_i \cap \pi_i^{-1}(X_\alpha)$ of dimension less than l_α;
(3) $\{\tilde{\mathbf{U}}'_i \,|\, i \in \mathcal{L}'\}$ still covers $\Phi^{-1}(0) \cap X_\alpha$. After fixing an ordering on $\mathcal{L}'$, we define $\mathcal{L}^{l_\alpha}_\alpha$ to be the collection of charts

$$(\tilde{\mathbf{U}}_i, G_i, \mathbf{E}_i, \mathbf{F}_i)|_{\mathbf{U}'_i \setminus \bigcup_{j<i} \overline{\mathbf{U}'_j}}, \qquad i \in \mathcal{L}'. \tag{1.4}$$

Clearly, $\mathcal{L}^{l_\alpha}_\alpha$ satisfies $\mathbf{A}_1 - \mathbf{A}_3$. Now we assume that $\mathcal{L}^{l_\alpha}_\alpha, \cdots, \mathcal{L}^l_\alpha$ have been constructed. We want to construct $\mathcal{L}^{l-1}_\alpha$. For each $x \in \mathbf{Z}^l_\alpha$, we can find a finite approximation $(\tilde{\mathbf{U}}_i, G_i, \mathbf{E}_i, \mathbf{F}_i)$ such that it contains x and it is finer than all approximations in $\cup_{l' \geq l} \mathcal{L}^{l'}_\alpha$. We denote this collection by $\mathcal{L}'$ again. Since $\mathbf{Z}^l_\alpha$ is compact, we can pick a finite subcollection, still denote by $\mathcal{L}'$, such that $\{\mathbf{U}_i | i \in \mathcal{L}'\}$ covers $\mathbf{Z}^l_\alpha$. By our choice of the charts, $\mathbf{Z}^l_\alpha$ is contained in the image of finitely many submanifolds in X_α of dimension less than l. Next we choose $\mathbf{U}'_i \subset \mathbf{U}_i$ such that (1) $\pi_i^{-1}(\mathbf{U}'_i)$, $i \in \mathcal{L}'$, still cover $\mathbf{Z}^l_\alpha$; (2) let $\tilde{\mathbf{U}}'_i = \pi_i^{-1}(\mathbf{U}'_i)$ then $\overline{\tilde{\mathbf{U}}'_i} \cap \pi_i^{-1}(\mathbf{Z}^l_\alpha)$ is the same as the closure of $\tilde{\mathbf{U}}'_i \cap \pi_i^{-1}(\mathbf{Z}^l_\alpha)$ in $\pi_i^{-1}(X_\alpha)$; (3) If we let $R = \pi_i^{-1}(\mathbf{U}'_i) \cap \mathbf{Z}^l_\alpha$, then $\bar{R} - R$ is contained in a finite union of smooth submanifolds of dimensions less than $l-1$. Now fixing an ordering on $\mathcal{L}'$ as before, we define $\mathcal{L}^{l-1}_\alpha$ to be the collection of charts defined by (1.4). By continuing this procedure for $l-2, \cdots, 0$, we obtain the desired $\mathcal{L}^l_\alpha$ for $0 \leq l \leq l_\alpha$. This completes the proof of the Lemma.

We now briefly sketch the strategy of the proof of Theorem 1.1. Let $(\tilde{\mathbf{U}}_i, G_i, \mathbf{E}_i, \mathbf{F}_i)_{i \in \mathcal{L}}$ be the collection of local finite approximations given by Lemma 1.1. We denote by $\phi_i : U_i \to F_i$ the corresponding finite models. As a stratified set, $U_i = \bigcup_{\alpha \in I_0} U_{i,\alpha}$, where I_0 is the ordered index set in the proof of Lemma 1.1. Since $\Phi^{-1}(0)$ is compact, we can find open $\tilde{W}_i \subset \tilde{\mathbf{U}}_i$ with $\pi_i^{-1}(\pi_i(\tilde{W}_i)) = \tilde{W}_i$ such that $\bigcup\{\pi_i(\tilde{W}) \mid i \in \mathcal{L}\}$ covers $\Phi^{-1}(0)$ and $\tilde{W}_i \cap U_i$ is precompact in U_i.[6] By our assumption, we have that for each $\alpha \in I_0$, $f|_{U_{i,\alpha}} : U_{i,\alpha} \to Y_\alpha$ is a submersion.

As we mentioned before, we need to use multi-valued sections to perturb ϕ_i.[7] We will first define multi-valued sections associated to the collection of

[6] By this we mean that its closure is compact.

[7] We do not need to use multi-valued sections if for the given collection of local finite approximations, we can arrange φ_{ji} defined in (1.1) such that $\varphi_{kj} \circ \varphi_{ji} = \varphi_{ki}$ wherever both sides are well-defined. But the latter may not be true in general.

local finite approximations $(\tilde{\mathbf{U}}_i, G_i, \mathbf{E}_i, \mathbf{F}_i)_{\alpha\in\mathcal{L}}$. By a multi-valued section, we mean a collection of sections $\psi_i : U_i \mapsto S^{k_i}F_i$, where S^kF_i denotes the the k-th symmetric fiber product of F_i, such that (1) for any i, ψ_i is locally a sum of k_i sections of F_i near any $x \in U$ (2) for any $i > j$, we have $\psi_i^{l_i} = \psi_j^{l_j}$ [8] in $S^{k_jl_j}F_j \subset S^{k_il_i}F_i$ along $\varphi_{ji}^{-1}(U_j)$ for some $l_i, l_j \geq 1$ with $k_il_i = k_jl_j$, where φ_{ji} is defined in (1.1). Clearly, the section Φ generates a multi-valued section $\phi^{n_i} : U_i \mapsto S^{n_i}F_i$, where n_i is the order of G_i. In the following, we will abbreviate multi-valued sections as sections and identify ϕ_i with any multi-valued sections it may generate if no confusion occur.

Before we perturb the sections ϕ_i, we need to clarify by which we mean that a perturbation is generic. We begin with the general situation. Let $p: U \to V$ be a smooth map between two open smooth manifolds. Let $K \subset U$ be a pre-compact subset and $s : U \mapsto S^lE$ be a smooth multi-valued section [9] of a smooth vector bundle E over U. We will denote by $\Gamma_U(S^lE)$ the set of all smooth l-valued sections of E over U. As we said before, $\Gamma_U(S^lE)$ may not have an additive structure, so we can not simply apply the transversality theory to multi-valued sections. However, we can still add sections of E to any given multi-valued section, more precisely, if h is any section of E over U, we can get a new section $s + h^l \in \Gamma_U(S^lE)$ as follows: If $t_1, \cdots, t_l$ are any local representations of s, then $t_1 + h, \cdots, t_l + h$ represent $s + h^l$ locally. We say that $t = s + t^l \in \Gamma_U(S^lE)$ is a p-generic perturbation of s relative to K with compact support if the following holds: (1) The support of h as a section of E is pre-compact in U and (2) there is a precompact open neighborhood K^{nbfd} of $\bar{K} \subset U$ so that the graph of t is transversal to the zero section of E in K^{nbhd} and $p: K^{\mathrm{nbhd}} \cap t^{-1}(0) \to V$ is a strong immersion.[10] Here, by a strong immersion $p: A \to B$ between open manifolds, we mean that p is an immersion and further, for any $x, y \in A$ with $p(x) = p(y)$, p_*T_xA is transversal to p_*T_yA.

Lemma 1.2. *Let the situation be as before. Suppose that $f : U \to V$ is a submersion and the rank of E is bigger than* $\dim U - \dim V$. *For any pre-compact $K \subset U$ and any multi-valued section $s \in \Gamma_U(S^lE)$, we define B_s to be the set of $h \in \Gamma_U(E)$ with compact support such that $s + h^l$ is a f-generic perturbation of s relative to K. Then B_s is dense and open in the space $B = \{h \in \Gamma_U(E) \mid h$ is compact$\}$ with respect to the C^∞-topology.*

---

[8] For any vector bundle F, if a section ψ in S^kF is locally represented by k sections $t_1, \cdots, t_k$ of F, then ψ^l is locally represented by l-multiple of $t_1, \cdots, t_k$. Two sections ψ and ψ' of S^kF are equal if for any local representations $\{t_m\}$ and $\{t'_m\}$ of ψ and ψ', respectively, there is a permutation σ of $\{1, \cdots, k\}$ such that $t_{\sigma(m)} = t'_m$.

[9] By this, we mean that its local representatives are made of smooth sections of E.

[10] By this, we mean that for any local representation $t_1, \cdots, t_l$ of t over any small open subset $W \subset U$, the graph of each t_i is transversal to the zero section of E in $W \cap K^{\mathrm{nbhd}}$ and $p: K^{\mathrm{nbhd}} \cap t_i^{-1}(0) \to V$ is a strong immersion. Note that we denote by $t^{-1}(0)$ the union of the zero loci of any local representatives t_i of t.

Proof. The proof is an easy application of the ordinary transversality theorem. It is clear that $B_s \subset B$ is open. It remains to show that it is dense. First, by the standard transversality theorem, we can find a small $h \in \Gamma_U(E)$ with compact support such that the graph of $t = s + h^l$ is transversal to the zero section in a neighborhood of $\bar{K}$. It follows that $t^{-1}(0)$ is a union of submanifolds of dimension less than $\dim V$ in a neighborhood of $\bar{K}$.

Next, since f is a submersion, each $f^{-1}(x)$ is a smooth submanifold in U ($x \in V$). Since $t^{-1}(0)$ is a union of submanifolds of dimension less than $\dim V$ in a neighborhood of $\bar{K}$, by using the standard transversality arguments and perturbing h slightly if necessary, we may assume that $t^{-1}(0)$ is also transversal to fibers of f, so $f : t^{-1}(0) \to V$ is an immersion in a neighborhood of $\bar{K}$. We may further perturb h such that $f : t^{-1}(0) \to V$ is a strong immersion. The lemma is proved.

We now show how to construct the perturbations of ϕ_i one at a time. We may assume that $r \leq \dim Y$, otherwise, we can simply define $e(X, \mathbf{E}, \Phi)$ to be zero.

We begin with the first member in $\mathcal{L}$, which is 1. In this case, we only need to use ordinary sections of F. To make the notation easy to follow, we denote the finite model $\phi_1 : U_1 \to F_1$ by $\phi : U \to F$, $\tilde{W}_1 \subset \tilde{\mathbf{U}}_1$ by $\tilde{W}$ and $U_{1,\alpha}$ by U_α. We first look at the largest $\alpha \in I_0$ so that $U_\alpha \neq \emptyset$, where I_0 appeared in the proof of Lemma 1.1. Assume that it is not dense in U. Recall that $f : U_\alpha \to Y$ is a submersion for the given morphism $f : X \to Y$. We then pick a small f-generic perturbation of $\phi|_{U_\alpha}$ relative to $U_\alpha \cap \overline{\tilde{W}}$ of compact support, denoted by $\phi_\alpha : U_\alpha \to Y_\alpha$. Since U_α is not dense in U, by the assumption, it has codimension at least 2. Thus

$$\dim f(\phi_\alpha^{-1}(0) \cap \overline{\tilde{W}}) \leq r - 2.$$

Since Y is a smooth orbifold, we can find a closed neighborhood $L_\alpha \subset Y_\alpha$ of $f(\phi_\alpha^{-1}(0) \cap \overline{\tilde{W}})$ such that it has homological dimension $\leq r - 2$ and its boundary is a smooth orbifold in Y. Let $M_\alpha = f^{-1}(L_\alpha)$.

Let $\beta \in I_0$ is the next largest element so that $U_\beta \neq \emptyset$. We want to extend ϕ_α to U_β in an appropriate way. First we can extend ϕ_α to a neighborhood of $U_\alpha \subset U_\alpha \cup U_\beta$.[11] Since the vanishing locus of ϕ_α in $U_\alpha \cap \overline{\tilde{W}}$ is contained in the interior of $M_\alpha \cap U_\alpha$, we may assume that this extension of ϕ_α has its zero locus in $M_\alpha \cap (U_\alpha \cup U_\beta)$. If $\beta > 0$, then U_β is not dense in U and the rank of F is greater than $\dim U - \dim Y$. Hence, by last lemma, we can find a small perturbation of $\phi|_{U_\alpha \cup U_\beta}$ extending ϕ_α, so that it is an f-generic perturbation of $\phi|_{U_\alpha \cup U_\beta}$ relative to $(U_\alpha \cup U_\beta) \cap (\tilde{W} \backslash M_\alpha)$ with compact support. Let the new section be $\phi_\beta \in \Gamma_{U_\alpha \cup U_\beta}(F)$. Then f is a strong immersion from

[11] It follows from that F is a continuous bundle over U and any continuous function on U_α with compact support can be extended continuously to U. This extention property holds because U is a Hausdorff topological space stratified by finitely many smooth manifolds of finite dimension.

a neighborhood of $\overline{\phi_\beta^{-1}(0) \cap \tilde{W}_\alpha \backslash M_\alpha}$ into Y. Without loss of generality, we can further assume that $f(\phi_\beta^{-1}(0))$ intersects the boundary of L_α transversely in the sense of orbifold. It follows that $\overline{f(\phi_\beta^{-1}(0)) \backslash L_\alpha}$ is an immersed, closed manifold of dimension $\leq r-2$. So $f(\phi_\beta^{-1}(0)) \cup L_\alpha$ is of homological dimension $\leq r-2$. Then one can easily find a closed strong retraction neighborhood L_β of $f(\phi_\beta^{-1}(0)) \cup L_\alpha$ such that the boundary of L_β is a smooth orbifold in Y. Clearly, L_β has homological dimension $\leq r-2$ as well. We let $M_\beta = f^{-1}(L_\beta)$. Continue this procedure, we can find a subset $L_1 \subset Y$ of homological dimension $\leq r-2$ and a perturbation $\tilde{\phi}$ of ϕ, such that L_1 is the closure of an open subset in Y whose boundary is a smooth orbifold and for any $\gamma > 0$ in I_0, $f(\tilde{\phi}^{-1}(0) \cap U_\gamma \cap \overline{\tilde{W}})$ is contained in the interior of L_1. Put $M_1 = f^{-1}(L_1)$. Clearly, M_1 is invariant under the action of G_1. Let $U_1^n \subset U = U_1$ be a pre-compact neighborhood of $U \cap \overline{\tilde{W}}$. Then by using the standard transverality theorem and perturbing $\tilde{\phi}$ slightly, we may assume that
(1) the graph of $\tilde{\phi}$ is transversal to the zero section of F over a neighborhood of $\overline{U_1^n \backslash M_1}$;
(2) For any $\gamma > 0$ in I_0, $f(\tilde{\phi}^{-1}(0) \cap U_\gamma \cap \overline{U_1^n})$ is contained in the interior of L_1;
(3) The boundary of $U_0 \cap U_1^n \backslash M_1$ is transversal to $\tilde{\phi}^{-1}(0)$, where U_0 denotes the main stratum of U.

The section $\tilde{\phi}$ of F may not be G_1-invariant. However, we can get a G_1-invariant n_1-valued section $\tilde{\phi}_1$ in $\Gamma_U(S^{n_1}F)$ by putting together all sections $\sigma^*\tilde{\phi}$ ($\sigma \in G_1$), where n_l is the order of G_l ($l \geq 1$).

Suppose that we have constructed a subset $L_{i-1} \subset Y$ of homological dimension $\leq r-2$ and a G_j-invariant perturbation $\tilde{\phi}_j$ of $\phi_j^{m_j}$ in $\Gamma_{U_j}(S^{m_j}F_j)$ for each $j \leq i-1$, where m_j are some integers divisible by all m_l for $l < j$ and $\prod_{l \leq j} n_l$, such that
(1) L_{i-1} is the closure of an open subset in Y whose boundary is a smooth orbifold in Y;
(2) For each $j < i$, there is a pre-compact neighborhood $U_j^n \subset U_j$ of $U_j \cap \overline{\tilde{W}_j}$ such that $\tilde{\phi}_j$ is transversal to the zero section of F_j over a neighborhood of $\overline{U_j^n - M_{i-1}}$, where $M_{i-1} = f^{-1}(L_{i-1})$;
(3) For any $j < k < i$, $\varphi_{jk}^*(\tilde{\phi}_j^{\frac{m_k}{m_j}})$ is equal to $\tilde{\phi}_k$ over U_{jk} as m_k-valued sections of F_k, where $U_{jk} = \varphi_{jk}^{-1}(U_j)$ and φ_{jk} is defined in (1.1);
(4) For any $j < i$, the boundary of $U_{j,0} \cap U_j^n \backslash M_{i-1}$ is transversal to those submanifolds in $\tilde{\phi}_j^{-1}(0) \cap U_j^n$, where $U_{j,0}$ denotes the main stratum of U_j.

Let $\tilde{\phi}_1$ be the m_1-valued section made of those $\sigma^*(\tilde{\phi})$ for $\sigma \in G_1$. It is clear that it satisfies the above (1) - (4). So we may assume that $i \geq 2$.

Now we want to continue this procedure to U_i. Let U_{ji} be $\varphi_{ji}^{-1}(U_j)$ ($j \leq i$), where φ_{ji} are defined in (1.1). Then each $\varphi_{ji}^{-1}(U_j \cap U_j^n - M_{i-1})$ is a smooth submanifold of U_i. By our induction assumptions, for any $j < k < i$, we have $\varphi_{ji}^*(\tilde{\phi}_j^{\frac{m_i}{m_j}}) = \varphi_{ki}^*(\tilde{\phi}_k^{\frac{m_i}{m_k}})$ on $U_{ji} \cap U_{ki}$. Gluing these $\varphi_{ji}^*(\tilde{\phi}_j)$ together, we can get a section ϕ' of $S^{m_i}F_i$ over $\bigcup_{j<i} U_{ji} \subset U_i$. Note that this ϕ' is a small perturbation of $\phi_i^{m_i}$ over $\bigcup_{j<i} U_{ji} \subset U_i$. We can extend ϕ' to U_i such that it coincides with $\phi_i^{m_i}$ outside a neighborhood R of $\bigcup_{j<i} U_{ji} \subset U_i$.[12] For simplicity, we still denote this extension by ϕ'.

Next, we want to perturb ϕ' so that it satisfies the above (1) - (4) for $\tilde{\phi}_j$. As before, we assume that I_0 is the set of indices α such that corresponding stratum $U_{i,\alpha}$ is nonempty. We also arrange these $\alpha \in I_0$ such that $\bigcup_{\beta \leq \alpha} U_{i,\alpha}$ is compact for any β and $U_{i,0}$ is open and dense in U_i. First let us pick up the largest $\alpha \in I_0$. Assume that $\alpha > 0$. Since $f : U_{i,\alpha} \mapsto Y_\alpha$ is a submersion, $M_{i-1} \cap U_{i,\alpha}$ is a smooth manifold with boundary in $U_{i,\alpha}$. By last lemma, we can find a f-generic perturbation ϕ'_α of ϕ' relative to $U_i - \underline{M_{i-1}}$ such that it is unchanged near R. We may also assume that $f({\phi'_{\underline{\alpha}}}^{-1}(0) \cap \overline{\tilde{W}_i})$ is transversal to boundary of L_{i-1}. It follows that $f({\phi'_\alpha}^{-1}(0) \cap \overline{\tilde{W}_i}) \cup L_{i-1}$ is of homological dimension $\leq r - 2$. Thus we can find a closed neighborhood L_i^α of $f({\phi'_\alpha}^{-1}(0) \cap \overline{\tilde{W}_i}) \cup L_{i-1}$ such that its boundary is a smooth orbifold in Y and is of homological dimension $\leq r - 2$. Continuing this procedure as we did before for ϕ_1 on U_1, we will get L_i and $\tilde{\phi}_i$ satisfying properties (1)-(4) for L_j and $\tilde{\phi}_j$.

Keep doing this procedure until we reach the largest $m \in \mathcal{L}$. We let $L_m \subset Y$ be the corresponding closed neighborhood of homological dimension $\leq r - 2$.

Let $\Delta_i = (1/m_i)[\pi_i(\tilde{\phi}_i^{-1}(0)) \cap U_i^n]$ $(i \leq m)$. It is a rational pseudomanifold. Then by our construction, for each $i, j \in \mathcal{L}$,

$$\Delta_i|_{\mathbf{U}_i \cap \mathbf{U}_j - f^{-1}(L_m)} = \Delta_j|_{\mathbf{U}_i \cap \mathbf{U}_j - f^{-1}(L_m)}$$

as pseudomanifolds. We let Δ be the patch together of these Δ_i. Since $\{\tilde{W}_i \mid i \in \mathcal{L}\}$ covers $\Phi^{-1}(0)$, if we choose the perturbations $\tilde{\phi}_i$ sufficiently close to ϕ_i for all i, then the pair (Δ, L_m) will be a weakly f-pseudocycle. This cycle is the Euler class of $(X, \mathbf{E}, \Phi)$.

Let $(X \times [0,1], \overline{\mathbf{E}}, \Psi)$ be a homopoty between two WFV-bundles $(X, \mathbf{E}, \Phi)$ and $(X, \mathbf{E}', \Phi')$. Suppose that the above arguments have already given us two f-pseudocycles (Δ, L) and (Δ', L') representing $e(X, \mathbf{E}, \Phi)$ and $e(X, \mathbf{E}', \Phi')$, respectively. Then one can show by above arguments that there is a f-pseudocycle $(\overline{\Delta}, \overline{L})$ such that

$$\partial(\overline{\Delta}, \overline{L}) = (\Delta, L) - (\Delta', L').$$

This implies that $e(X, \mathbf{E}, \Phi)$ depends only on the homotopy class of $(X, \mathbf{E}, \Phi)$. The theorem is proved.

The following corollary is often useful and clearly true from the arguments in the above proof of Theorem 1.1

[12] We may need to raise ϕ' to certain power ${\phi'}^l$ and then get its extension.

Corollary 1.1. *Let $(X, \mathbf{E}, \Phi)$ be a WFV-bundle of relative index and $f : X \mapsto Y$ be a morphism into a smooth orbifold Y of finite dimension. Let X_α be the strata of X $(\alpha \in I)$ such that X_0 is the main stratum of X. We assume that $\Phi^{-1}(0) \cap X_0$ is a smooth submanifold in X_0 of dimension r and for any $\alpha > 0$, $\Phi^{-1}(0) \cap X_\alpha$ is contained in a smooth submanifold in X_i of dimension $\leq r - 2$. Then $f(\Phi^{-1}(0))$ is a f-pseudocycle in Y representing $e(X, \mathbf{E}, \Phi) \in H_r(Y, \mathbb{Q})$.*

Sometimes it is desirable to include the case where Y is a more general topological space. One interesting case is when $X = Y$ and $f = \mathrm{id}_X$. By identical arguments as in the proof of Theorem 1.1, one can construct the Euler class $e(X, \mathbf{E}, \Phi)$ as a homology class in $H^*(X, \mathbb{Q})$ for any WFV-bundle $(X, \mathbf{E}, \Phi)$ satisfying: It admits a pre-weakly Fredholm structure $\mathfrak{A}$ such that for any $(\tilde{\mathbf{U}}, G_{\tilde{\mathbf{U}}}, \mathbf{E}_{\tilde{\mathbf{U}}}, \mathbf{F}_{\tilde{\mathbf{U}}})$ in $\mathfrak{A}$ and $\tilde{\mathbf{U}} = \cup_{\alpha \in I} \mathbf{U}_\alpha$ the stratification, if $L \subset \overline{\mathbf{U}}_\alpha$ is a closed subset such that each $L \cap \mathbf{U}_\beta$ is either empty or a smooth submanifold, then it has a closed neighborhood M in $\tilde{\mathbf{U}}$ such that it is a deformation retract of L and for any $\beta \in I$, $\partial M \cap \mathbf{U}_\beta$ is a smooth hypersurface in $\mathbf{U}_\beta$. This condition is actually true for those WFV-bundles involved in defining the GW-invariants.

2. GW-invariants

In this section, we apply Theorem 1.1 to constructing Gromov-Witten invariants for general symplectic manifolds.

2.1 Stable maps

First we review the definition of stable maps. Let X be a smooth symplectic manifold with a given symplectic form ω, and let $A \in H_2(X, \mathbb{Z})$ and let g, $n \in \mathbb{Z}$ be fixed once and for all.

In the following, by a k-pointed prestable curve, we mean a connected complex curve with k marked points and only normal crossing singularity.

We recall the notion of stable C^2-maps [LT2, Definition 2.1].

Definition 2.1. *A k-pointed stable map is a collection $(f, \Sigma, x_1 \ldots, x_k)$, where $(\Sigma, x_1, \ldots, x_k)$ is a k-pointed connected prestable curve and $f : \Sigma \to X$ is a continuous map such that*
(1) the composite $f \circ \pi$ is C^2-smooth, where $\pi : \tilde{\Sigma} \to \Sigma$ is the normalization of Σ;
(2) Any component $R \subset \tilde{\Sigma}$ satisfying $(f \circ \pi)_*([R]) = 0 \in H_2(X, \mathbb{Z})$ must have at least $\min\{3 - 2g(R), 0\}$ the number of distinguished points on R, where the set of distinguished points on $\tilde{\Sigma}$ consists of all preimages of the marked points and the nodal points of Σ.

For convenience, we will abbreviate $(f, \Sigma, x_1, \ldots, x_k)$ to $(f, \Sigma, (x_i))$ or to (f, Σ) or simply to f, if no confusion will arise. We call Σ the domain of f, with marked points understood. Two stable maps $(f, \Sigma, (x_i))$ and $(f', \Sigma', (x_i'))$ are said to be equivalent if there is an isomorphism $\rho: \Sigma \to \Sigma'$ such that $f' \circ \rho = f$ and $x_i' = \rho(x_i)$. When $(f, \Sigma) \equiv (f', \Sigma')$, such a ρ is called an automorphism of (f, Σ). We will denote by Aut_f the group of automorphisms of (f, Σ).

We let $\mathbf{B}^X_{A,g,k}$ be the space of equivalence classes $[f, \Sigma]$ of C^2-stable maps (f, Σ) with k-marked points such that the arithmetic genus of Σ is g and $f_*([\Sigma]) = A \in H_2(X; \mathbb{Z})$. With X, A, g and k understood, we will simply write $\mathbf{B}^X_{A,g,k}$ as $\mathbf{B}$. There is a natural topology on $\mathbf{B}$, which we will define in Section 3.3. However, $\mathbf{B}$ is not a smooth Banach manifold as one hopes. It does admit a stratification with smooth strata.

Given any almost complex structure J compatible with ω, one can define a generalized bundle $\mathbf{E}$ over $\mathbf{B}$ as follows. Let (f, Σ) be any stable map and let $\tilde{f}: \tilde{\Sigma} \to X$ be the composite of f with $\pi: \tilde{\Sigma} \to \Sigma$, where $\tilde{\Sigma}$ is the normalization of Σ. We define $\Lambda^{0,1}_f$ to be the space of all C^1-smooth sections of $(0,1)$-forms of $\tilde{\Sigma}$ with values in $\tilde{f}^*TX$. Here, by a $\tilde{f}^*TX$-valued $(0,1)$-form we mean a section ν of $T\tilde{\Sigma} \otimes \tilde{f}^*TX$ over $\tilde{\Sigma}$ such that $J \cdot \nu = -\nu \cdot \mathbf{j}$, where $\mathbf{j}$ denotes the complex structure of $\tilde{\Sigma}$. Assume that (f, Σ) and (f', Σ') are two equivalent stable maps with the associated conformal map $\rho: \Sigma \to \Sigma'$, then there is a canonical isomorphism $\Lambda^{0,1}_{f'} \cong \Lambda^{0,1}_f$. It follows that the automorphism group Aut_f acts on $\Lambda^{0,1}_f$ and the quotient $\Lambda^{0,1}_f / \mathrm{Aut}_f$ is independent of the choice of the representative in the equivalence class of (f, Σ). So we can define $\Lambda^{0,1}_{[f]}$ to be $\Lambda^{0,1}_f / \mathrm{Aut}_f$. Put

$$\mathbf{E} = \bigcup_{[f] \in \mathbf{B}} \Lambda^{0,1}_{[f]}.$$

It has an obvious projection $\mathbf{P}: \mathbf{E} \to \mathbf{B}$ whose fibers are finite quotients of infinite dimensional linear spaces.

There is a natural map

$$\Phi_J : \mathbf{B} \longrightarrow \mathbf{E} \quad \text{satisfying} \quad \mathbf{P} \circ \Phi_J = \mathrm{Id}_{\mathbf{B}},$$

defined as follows: For any stable map f, we define

$$\Phi_J(f) = d\tilde{f} + J \cdot d\tilde{f} \cdot \mathbf{j} \in \Lambda^{0,1}_f.$$

Obviously, we have $\Phi_J(f) = \Phi_J(f') \cdot \rho$ if f and f' are equivalent through a conformal map ρ. Thus Φ_J descends to a map $\mathbf{B} \to \mathbf{E}$, which we still denote by Φ_J.

We will denote $\Phi_J^{-1}(0)$ by $\mathfrak{M}^{\mathfrak{X}}_{\mathfrak{A},\mathfrak{g},\mathfrak{k}}$. It is the moduli space of J-holomorphic stable maps with homology class A, having k-marked points and whose domains have arithmetic genus g. Note that a stable map $[f]$ is J-holomorphic if $\Phi_J(f) = 0$.

2.2 Stratifying the space of stable maps

In this section, we give a natural stratification of $\mathbf{B} = \mathbf{B}^X_{A,g,n}$ and study its basic properties.

Given a stable map (f, Σ), we can associate to it a dual graph Γ_f as follows: Each irreducible component Σ_α of Σ corresponds to a vertex v_α in Γ_f together with a marking (g_α, A_α), where g_α is the geometric genus of Σ_α and A_α is the homology class $f_\alpha([\Sigma_\alpha])$ in X; For each marked point x_i of Σ_α we attach a leg to v_α; For each intersection point of distinct components Σ_α and Σ_β we attach an edge joining v_α and v_β and for each self-intersection point of Σ_α we attach a loop to the vertex v_α. In the following, we will denote by $\mathrm{Ver}(\Gamma)$ the set of all vertices of Γ and $\mathrm{Ed}(\Gamma)$ the set of all edges of Γ.

Clearly, the dual graph Γ_f is independent of the representatives in $[f]$, so we can simply denote the dual graph of $[f]$ by $\Gamma_{[f]} = \Gamma_f$. Moreover, the genus g of $[f]$ is the same as the genus of $\Gamma_{[f]}$, which is defined to be the sum of g_α for all $\alpha \in \mathrm{Ver}(\Gamma)$ and the number of holes in the graph $\Gamma_{[f]}$. Also, the homology class A of $f_*([\Sigma])$ is the sum of all A_α, which is defined to be the homology class of the graph $\Gamma_{[f]}$.

Given any graph Γ with genus g and homology class A and k legs, we let $\mathbf{B}(\Gamma)$ be the space of all equivalence classes $[f]$ of stable maps in $\mathbf{B}$ with $\Gamma_{[f]} = \Gamma$. Clearly, $\mathbf{B}$ is a disjoint union of all $\mathbf{B}(\Gamma)$.

The main result of this section is the following.

Proposition 2.1. *For each dual graph Γ, the space $\mathbf{B}(\Gamma)$ is a smooth Banach orbifold and the restriction of $\mathbf{E}$ to $\mathbf{B}(\Gamma)$ is a smooth orbifold bundle. Further the restriction of Φ_J to $\mathbf{B}(\Gamma)$ is a smooth orbifold section.*

Proof. We first prove that $\mathbf{B}(\Gamma)$ is a Banach orbifold.

It suffices to construct the neighborhoods of $[f]$ in $\mathbf{B}(\Gamma)$. Let $f = (f, \Sigma, (x_i))$ and let $\tilde{\Sigma}$ be the normalization of Σ. Then

$$\tilde{\Sigma} = \bigcup \left\{ \tilde{\Sigma}_\alpha \mid \alpha \in \mathrm{Ver}(\Gamma) \right\}, \tag{2.1}$$

where $\tilde{\Sigma}_\alpha$ is the normalization of the component Σ_α corresponding to α. Recall that the distinguished points of $\tilde{\Sigma}_\alpha$ are preimages of marked points and the nodal points of Σ. Note that each edge $e \in Ed(\Gamma)$ corresponds to a node in Σ, thus has two distinguished points in $\tilde{\Sigma}$. We denote them by z_{e1} and z_{e2}.

To avoid any confusion, we denote by $\tilde{\Sigma}^{\mathrm{top}}$ the underlying real 2-dimensional manifold of $\tilde{\Sigma}$. A complex structure on $\tilde{\Sigma}^{\mathrm{top}}$ is given by an almost complex structure which is a homomorphism $\mathbf{j} : T\tilde{\Sigma}^{\mathrm{top}} \mapsto T\tilde{\Sigma}^{\mathrm{top}}$ with $\mathbf{j}^2 = -\mathrm{Id}$. Two almost complex structures $\mathbf{j}$ and $\mathbf{j}'$ give rise to the same complex structure if and only if there is a diffeomorphism ϕ of $\tilde{\Sigma}^{\mathrm{top}}$ such that $\mathbf{j}' = d\phi \cdot \mathbf{j} \cdot (d\phi)^{-1}$. Let $\mathbf{j}_t$ be a family of almost complex structures such that $\mathbf{j}_0$ is the given complex structure of $\tilde{\Sigma}$. Then the derivative $\mathbf{v}(\mathbf{j}_0) := (d/dt\, \mathbf{j}_t)_{t=0}$ is a $T\tilde{\Sigma}$-valued (0,1)-form. If $\mathbf{j}'_t$ is another family of almost complex structures such that each $\mathbf{j}'_t$ induces the same complex structure as $\mathbf{j}_t$ does, then the corresponding (0,1)-form $\mathbf{v}(\mathbf{j}'_0)$ can be written as $\mathbf{v}(\mathbf{j}_0) + \overline{\partial} u$ for some section u of $T\tilde{\Sigma}$. This shows that local complex deformations of $\tilde{\Sigma}$ are parameterized by an open subset of $H^1_{\tilde{\Sigma}}(T\tilde{\Sigma})$. Thus a local universal family $\tilde{\mathcal{U}}$ of marked curves near $(\tilde{\Sigma}, (\tilde{x}_i), (z_{e1}, z_{e2}))$, where $\tilde{x}_i \in \tilde{\Sigma}$ is the preimage of $x_i \in \Sigma$ and $e \in \mathrm{Ed}(\Gamma)$, is of the form

$$Q \times \prod_i P_i \times \prod_e P_{e1} \times P_{e2}, \tag{2.2}$$

where Q, P_i, P_{e1} and P_{e2} are small neighborhoods of 0 in $H^1_{\tilde{\Sigma}}(T\tilde{\Sigma})$ and small neighborhoods of $\tilde{x}_i$, z_{e1} and z_{e2} in $\tilde{\Sigma}$, respectively. Here, as usual, $(\tilde{x}_i)$ is the set of the k-marked points and (z_{e1}, z_{e2}) denotes $2 \cdot \#\mathrm{Ed}(\Gamma)$-marked points, after fixing an ordering of $\mathrm{Ed}(\Gamma)$.

By the Serre duality, we have $H^1_{\tilde{\Sigma}}(T\tilde{\Sigma}) = H^0_{\tilde{\Sigma}}(T^*\tilde{\Sigma}^{\otimes 2})$. The latter is the space of holomorphic quadratic differential forms.

Let $\mathcal{V}$ be a small neighborhood of 0 in the space of $\tilde{f}^*TX$-valued vector fields along $\tilde{\Sigma}$ in the C^2-topology. Here $\tilde{f} : \tilde{\Sigma} \to X$. For each $v \in \mathcal{V}$, we can associate to it a unique C^2-map $\exp_{\tilde{f}}(v)$ from $\tilde{\Sigma}$ into X, where exp is the exponential map of a fixed metric on X.

We denote by G_f the automorphism group of f. It is a finite group and acts naturally on $H^1_{\tilde{\Sigma}}(T\tilde{\Sigma})$. This is because each $\sigma \in G_f$ lifts naturally to an automorphism of $\tilde{\Sigma}$, so it induces an action on $H^1_{\tilde{\Sigma}}(T\tilde{\Sigma})$. Without loss of generality, we can assume that $\mathcal{V}$, Q and $\prod_i P_i \times \prod_e P_{e1} \times P_{e2}$ are G_f-invariant. Now we define a natural G_f-action on $\tilde{\mathcal{U}} \times \mathcal{V}$ as follows: Any point in $\tilde{\mathcal{U}} \times \mathcal{V}$ is of the form

$$(q, (x'_i), (z'_{e1}, z'_{e2}), v), \quad q \in Q, x'_i \in P_i, (z'_{e1}, z'_{e2}) \in P_{e1} \times P_{e2} \text{ and } v \in \mathcal{V},$$

given any $\sigma \in G_f$, we define

$$\sigma(q, (x'_i), (z'_{e1}, z'_{e2}), v) = (\sigma_*(q), (\sigma(x'_i)), (\sigma(z'_{e1}), \sigma(z'_{e2})), v \cdot \sigma^{-1}).$$

Next we construct a product manifold X_Γ as follows: for each edge or loop e of Γ, we associate to it two copies of X, say X_{e1}, X_{e2}. Define $X_\Gamma = \prod_e X_{e1} \times X_{e2}$. We also define $\Delta_\Gamma = \prod_e \Delta_e$, where Δ_e is the diagonal of $X_{e1} \times X_{e2}$. We can define a smooth evaluation map $\pi : \tilde{\mathcal{U}} \times \mathcal{V} \to X_\Gamma$ by

$$\pi(q, (x'_i), (z'_{e1}, z'_{e2}), v) = \prod_e \left(\exp_{\tilde{f}}(v)(z_{e1}), \exp_{\tilde{f}}(v)(z_{e2})\right).$$

One can easily show that π is transversal to Δ_Γ. So $\tilde{\mathbf{W}} = \pi^{-1}(\Delta_\Gamma)$ is smooth. The finite group G_f acts on $\tilde{\mathbf{W}}$ smoothly, thus $\tilde{\mathbf{W}}/G_f$ is a neighborhood of f in $\mathbf{B}(\Gamma)$. This proves the first part of this proposition.

We now prove the remaining part of the proposition. Note that each point in $\tilde{\mathbf{W}}$ is represented by a stable map $(f, \Sigma, (x_i))$ and $(\tilde{\Sigma}, (x_i), (z_{e1}, z_{e2})) \in \tilde{\mathcal{U}}$. Let $\tilde{\mathbf{E}}_f$ be the set of all C^1-smooth, $\tilde{f}^*TX$-valued (0,1)-forms over $\tilde{\Sigma}$. Put

$$\tilde{\mathbf{E}}_{\tilde{\mathbf{W}}} = \bigcup_{f \in \tilde{\mathbf{W}}} \tilde{\mathbf{E}}_f.$$

Clearly, $\tilde{\mathbf{E}}_{\tilde{\mathbf{W}}}$ is a smooth Banach bundle, and G_f, which acts on $\tilde{\mathbf{W}}$, lifts to a linear action on $\tilde{\mathbf{E}}_{\tilde{\mathbf{W}}}$ such that the natural projection $\tilde{\mathbf{E}}_{\tilde{\mathbf{W}}} \to \tilde{\mathbf{W}}$ is G_f-equivariant. Moreover, $\mathbf{E}|_{\mathbf{W}} = \tilde{\mathbf{E}}_{\tilde{\mathbf{W}}}/G_f$, so $\mathbf{E}$ is an orbifold bundle over $\mathbf{B}$.

Finally, the section Φ_J lifts to a section $\tilde{\Phi}_{\mathbf{W}}$ over $\tilde{\mathbf{W}}$ defined by

$$\tilde{\Phi}_{\mathbf{W}}(f) = d\tilde{f} + J \cdot d\tilde{f} \cdot \mathbf{j}_\Sigma \in \tilde{\mathbf{E}}_f.$$

Obviously, it is smooth. This proves the proposition.

2.3 Topology of the space of stable maps

Now it is time to describe the topology of $\mathbf{B}$. Let $[\mathbf{f}] \in \mathbf{B}$ be represented by a stable map $\mathbf{f} = (f, \Sigma, (x_i))$. We want to construct the neighborhoods of $[\mathbf{f}]$ in $\mathbf{B}$. Let Γ be the dual graph of $\mathbf{f}$. For each $\alpha \in \mathrm{Ver}(\Gamma)$, we define k_α to be the number of edges and legs attached to it. Here we count a loop to α as two edges. As before we still denote by Σ_α the corresponding component and $\tilde{\Sigma}_\alpha$ its normalization. We define $\mathrm{Ver}_u(\Gamma) \subset \mathrm{Ver}(\Gamma)$ to be $\{\alpha \in \mathrm{Ver}(\Gamma) \mid k_\alpha < 3 \text{ and } g_\alpha = 0\}$ and define $\mathrm{Ver}_s(\Gamma)$ the complement of $\mathrm{Ver}_u(\Gamma)$. When $\alpha \in \mathrm{Ver}_u(\Gamma)$, then $\tilde{\Sigma}_\alpha$ contains one or two distinguished points. We add two or one marked point(s) to Σ_α according to whether $\tilde{\Sigma}_\alpha$ contains one or two distinguished point. We also require that the curve Σ is smooth and the differential df is injective at these added points. Note that this is always possible since $f_*([\Sigma_\alpha]) \neq 0$, by the stability of $\mathbf{f}$. We denote by $(y_j)_{1 \leq j \leq l}$ the set of all added points to Σ. Then $(\Sigma, (x_i, y_j))$ is a Deligne-Mumford stable curve with $k + l$ marked points.

As a stable curve with $k + l$ marked points, local deformations of $(\Sigma, (x_i, y_j))$ are parameterized by admissible quadratic differentials on Σ. An admissible quadratic differential q is a meromorphic quadratic differential with at most simple poles at x_i or y_j and double poles at nodes satisfying: If w_1, w_2 are local coordinates of Σ near a node, i.e., Σ is defined by $w_1 w_2 = 0$ in $\mathbb{C}^2$ near such a node, then

$$\lim_{w_1 \to 0, w_2 = 0} w_1^2 \frac{q}{dw_1^2} = \lim_{w_2 \to 0, w_1 = 0} w_2^2 \frac{q}{dw_2^2}.$$

Neighborhoods of $(\Sigma, (x_i, y_j))$ in $\overline{\mathfrak{M}}_{g,k+l}$ can be constructed as follows: Let $\tilde{G}_f$ be the automorphism group of $(\Sigma, (x_i, y_j))$, then a neighborhood $\mathcal{U}$ of $(\Sigma, (x_i, y_j))$ in $\overline{\mathfrak{M}}_{g,k+l}$ is of the form $\tilde{\mathcal{U}}/\tilde{G}_f$, where $\tilde{\mathcal{U}}$ is a small neighborhood of the origin in the space of admissible quadratic differentials.

For each y_i, we choose a codimension two submanifold $H_j \subset X$ such that H_j intersects $f(\Sigma)$ uniquely and transversely at $f(y_j)$. We orient H_j so that it has positive intersection with $f(\Sigma)$.

We fix a compact set $K \subset \Sigma \backslash Sing(\Sigma)$ containing all marked points (x_i, y_j). We may assume that K is G_f-invariant. Let $\mathcal{CU}$ be the universal curve over $\tilde{\mathcal{U}}$. We then choose a diffeomorphism ϕ from a neighborhood of $K \times \tilde{\mathcal{U}}$ into $\mathcal{CU}$ such that ϕ preserves fibers over $\tilde{\mathcal{U}}$ and restricts to the identity map on $K \times \{(\Sigma, (x_i, y_j))\}$. We also fix a $\delta > 0$.

To each collection of $\mathcal{U}$, H_j, K, δ and ϕ given as above, we can associate to it a neighborhood $\mathbf{U} = \mathbf{U}(\mathcal{U}, H_j, K, \delta, \phi)$ as follows: Define $\tilde{\mathbf{U}} = \tilde{\mathbf{U}}(\mathcal{U}, H_j, K, \delta, \phi)$ to be the set of all tuples $(f', \Sigma', (x_i', y_j'))$ satisfying:
(1) $(\Sigma', (x_i', y_j'))$ is in $\tilde{\mathcal{U}}$;
(2) f' is a continuous map from Σ' into X with $f'(y_j) \in H_j$;
(3) f' lifts to a C^2-map $\tilde{\Sigma}' \to X$;
(4) $||f' \cdot \phi - f||_{C^2(K)} < \delta$;
(5) $d_X(f(\Sigma), f'(\Sigma')) < \delta$, where d_X denotes the distance function of a fixed Riemannian metric on X.

Note that the topology of Σ' may be different from that of Σ.

Given any tuple $(f', \Sigma', (x_i', y_j'))$ in $\tilde{\mathbf{U}}$, we call $\mathbf{f}' = (f', \Sigma', (x_i'))$ its descendant. One can show that $\mathbf{f}'$ is a stable map and gives rise to a point $[\mathbf{f}']$ in $\mathbf{B}$. Let $\mathbf{U}$ be the set of all equivalence classes of stable maps descended from tuples in $\tilde{\mathbf{U}}$. Then $\mathbf{U}$ is a neighborhood of $[\mathbf{f}]$.

Now assume that δ and $\tilde{\mathcal{U}}$ are sufficiently small and K is sufficiently large. Then there is a natural action of G_f on $\tilde{\mathbf{U}}$: let $\sigma \in G_f$ and $\tilde{\mathbf{f}}' = (f', \Sigma', (x_i', y_j')) \in \tilde{\mathbf{U}}$, then σ acts on $\tilde{\mathcal{U}}$ by sending Σ' to $\sigma(\Sigma')$ and x_i to $\sigma(x_i)$. We define

$$\sigma(\tilde{\mathbf{f}}') = (f' \cdot \sigma^{-1}, \sigma(\Sigma'), (\sigma(x_i'), y_j'')),$$

where for each j, y_j'' is the unique point in $\sigma(\Sigma')$ near y_j such that $f'(\sigma^{-1}(y_j'')) \in H_j$. The existence and uniqueness of such y_j'' are assured by the assumption that H_j are transversal to $f(\Sigma)$. Note that y_j'' may not be y_j'. Clearly, $\tilde{\mathbf{f}}'$ and $\sigma(\tilde{\mathbf{f}}')$ descend to the identical $[\mathbf{f}']$ in $\mathbf{B}$. Conversely, if $\tilde{\mathbf{f}}' = (f', \Sigma', (x_i', y_j'))$ and $\tilde{\mathbf{f}}'' = (f'', \Sigma'', (x_i'', y_j''))$ descend to the same $[\mathbf{f}']$ in $\mathbf{U}$, then there is a biholomorphic map $\tau : \Sigma' \mapsto \Sigma''$ such that $f' = f'' \circ \tau$ and $\tau(x_i') = x_i''$. When δ and $\tilde{\mathcal{U}}$ are sufficiently small, we may assume that τ induces a biholomorphic map, denoted by σ, of $(\Sigma, (x_i))$. This σ acts on $\tilde{\mathbf{U}}$ as defined above. Thus one can show that $\sigma(\mathbf{f}') = \mathbf{f}''$.[13] Hence, $\mathbf{U}$ is of the form $\tilde{\mathbf{U}}/G_f$.

This shows that $\tilde{\mathbf{U}}(\mathcal{U}, H_j, K, \delta, \phi)$ can serve as a local uniformization chart of $\mathbf{B}$ whose quotient is $\mathbf{U}(\mathcal{U}, H_j, K, \delta, \phi)$.

---

[13] First we assume that σ is the identity map. Then τ is close to the identity map on a sufficiently large open subset of Σ'. It follows that $\tau(y_j')$ is close to y_j' and consequently, $y_j'' = \tau(y_j')$, so τ is the identity map. The general case can be reduced to this special case.

The topology of $\mathbf{B}$ is generated by all such neighborhoods $\mathbf{U}(\mathcal{U}, H_j, K, \delta, \phi)$. The resulting topological space can be proven to be Hausdorff and satisfies all properties in Definition 1.3. Therefore, we have

Theorem 2.1. *Equiped with the topology described above, $\mathbf{B}$ is a smoothly stratified orbispace with strata $\mathbf{B}(\Gamma)$, where Γ runs over all dual graphs of homology class A, genus g and k legs.*

Given a local uniformization chart $\tilde{\mathbf{U}}$, we define

$$\mathbf{E}_{\tilde{\mathbf{U}}} = \bigcup_{(f', \Sigma', (x'_i, y'_j)) \in \tilde{\mathbf{U}}} \tilde{\mathbf{E}}_{(f', \Sigma', (x'_i, y'_j))},$$

where $\tilde{\mathbf{E}}_{(f', \Sigma', (x'_i, y'_j))}$ consists of all C^1-smooth, f'^*TX-valued (0,1)-forms over the normalization of Σ'. All such $\mathbf{E}_{\tilde{\mathbf{U}}}$'s form charts of $\mathbf{E}$. One can show that the conditions in Definition 1.5 are all satisfied. Therefore, we have proved

Theorem 2.2. *The above $(\mathbf{B}, \mathbf{E}, \Phi_J)$ is a V-bundle.*

2.4 Compactness of moduli spaces

In this subsection, we prove a compactness theorem for J-holomorphic stable maps. The compactness theorem of this sort first appeared in the work of Gromov [Gr], further studied and extended by Parker and Wolfson, Pansu and Ye. A proof was also given in [RT1] for holomorphic maps of any genus, following the work of Sacks and Uhlenbeck on harmonic maps in early 80's [SU]. However, a detailed proof for the general case is not available in literature. For the reader's conveniences, we will present a complete proof of the compactness theorem for moduli spaces of stable maps. Our presentation here uses ideas of [RT].

Recall that $\mathfrak{M}^{\mathfrak{X}}_{\mathfrak{A},\mathfrak{g},\mathfrak{k}}$ is the moduli space of J-holomorphic stable maps with homology class A, k-marked points and of arithmetic genus g.

Here is the main result of this subsection.

Theorem 2.3. *Assume that (X, ω) is a compact manifold with a compatible almost complex structure J. Then the moduli space $\mathfrak{M}^{\mathfrak{X}}_{\mathfrak{A},\mathfrak{g},\mathfrak{k}}$ is compact in* $\mathbf{B}$.

The rest of this subsection is devoted to its proof. The readers may skip the proof and go to the next subsection by simply taking this theorem for granted.

We always assume that J is an almost complex structure compatible with ω and g is the induced metric, i.e.,

$$g(u, v) = \omega(u, Jv), \qquad u, v \in TX.$$

We start with the monotonicity formula for holomorphic maps.

Lemma 2.1. *Let $f : D \mapsto X$ be any J-holomorphic map, where D is a disk in $\mathbb{C}$. Then there is an $\epsilon = \epsilon(X, g) > 0$ such that for any $r \leq \epsilon$ and $y \in X$, if $\partial(f(D) \cap B_r(y)) = f(D) \cap \partial B_r(y)$, where $B_r(y)$ denotes the geodesic ball of radius r and with center at y with respect to the metric g, then*

$$\frac{d}{dr}\left(e^{cr} r^{-2} \int_{f(D)\cap B_r(y)} \omega\right) \geq 0,$$

where c is a uniform constant depending only on (X, ω). In particular, if $y \in f(D)$, then

$$\int_{f(D)\cap B_r(y)} \omega \geq \pi r^2 e^{-cr}.$$

Proof. By choosing $\epsilon > 0$ sufficiently small, we may assume that $B_\epsilon(y)$ is contained in a coordinate chart. Let $\{x_i\}_{1\leq i\leq 2n}$ be coordinates such that $x_i = 0$ at y and

$$\omega = \sum_{i=1}^{n} dx_i \wedge dx_{i+n}.$$

Since J is compactible with ω, we may further assume that at y, we have

$$J_y(\frac{\partial}{\partial x_i}) = \frac{\partial}{\partial x_{n+i}}, \quad J_y(\frac{\partial}{\partial x_{n+i}}) = -\frac{\partial}{\partial x_i}, \quad i = 1, \cdots, n.$$

It follows that the distance ρ from y is different from $\sqrt{\sum_{i=1}^{2n} x_i^2}$ by a function bounded by $c\rho^2$. Note that we always denote by c a uniform constant in this section, whose actual value may vary in different places.

Write

$$\alpha = \frac{1}{2}\sum_{i=1}^{n}(x_i dx_{i+n} - x_{i+n} dx_i),$$

then

$$\omega = d\alpha \text{ and } |\alpha| \leq \frac{1}{2}\rho(1 + c\rho).$$

Let $d\theta$ be the induced volume form on $f(D) \cap \partial B_r(y)$ by the metric g. Since f is J-holomorphic, by the assumption, we obtain

$$\begin{aligned} & \int_{f(D)\cap B_r(y)} \omega \\ = \; & \int_{f(D)\cap \partial B_r(y)} \alpha \\ \leq \; & \tfrac{1}{2} r(1 + cr) \int_{f(D)\cap \partial B_r(y)} d\theta \\ = \; & \tfrac{1}{2} r(1 + cr) \tfrac{d}{dr}\left(\int_{f(D)\cap B_r(y)} \omega\right). \end{aligned}$$

This is equivalent to

$$\frac{d}{dr}\left(r^{-2} e^{2cr} \int_{f(D)\cap B_r(y)} \omega\right) \geq 0.$$

Now the lemma follows by integrating this differential inequality.

Corollary 2.1. *Let $f : R \mapsto X$ be any J-holomorphic map, where R is a connected Riemann surface with boundary ∂R. For any $\epsilon > 0$ such that $\pi e^{-c\epsilon} > 1$, where c is given in last lemma, if each connected component of $f(\partial R)$ is contained in a geodesic ball of radius ϵ in X and $\int_R |\nabla f|^2 dz \leq \epsilon^2$, then $f(R)$ is contained in a geodesic ball of radius $4m\epsilon$, where m is the number of connected components in ∂R.*

Proof. If the statement is false, then we can find at least one x in R such that $d_X(f(x), f(\partial R)) > \epsilon$. Then by last lemma,

$$\int_R |\nabla f|^2 dz \geq \int_{f^{-1}(B_\epsilon(x))} |\nabla f|^2 dz > \epsilon^2.$$

A contradiction! So the corollary is proved.

The following is due to K. Uhlenbeck and actually holds for harmonic maps from Riemann surfaces.

Theorem 2.4. *(Removable of Singularity) Let (X, ω) be a symplectic manifold with a compatible almost complex structure J. If $f : D\backslash\{0\} \mapsto X$ is a J-holomorphic map with $\int_D |\nabla f|^2(z)dz < \infty$, then f extends to a smooth map from D into X.*

Proof. We define for $x \in D$ and $r < 1 - |x|$,

$$E_r(f, x) = \int_{|z-x|<r} |\nabla f|^2(z)dz.$$

If $x = 0$, we often write $E_r(f, x)$ as $E_r(f)$.

Denoting by ρ, θ the polar coordinates of $\mathbb{C}$ centered at the origin, we can write

$$E_r(f) = \int_0^r \int_0^{2\pi} \left(\left| \frac{\partial f}{\partial \rho} \right|^2 + \frac{1}{\rho^2} \left| \frac{\partial f}{\partial \theta} \right|^2 \right) \rho d\rho d\theta,$$

so it follows that there is at least one $r' \in (r/3, r)$ such that

$$\int_0^{2\pi} \left| \frac{\partial f}{\partial \theta} \right|^2 (r', \theta)d\theta \leq E_r(f),$$

consequently,

$$\sup_{|y|=|y'|=r'} d_X(f(y), f(y')) \leq \sqrt{2\pi E_r(f)},$$

where d_V denotes the distance function of the induced metric by ω and J.

Now take $r_i = \frac{1}{3^i}$ for all sufficiently large i. By the assumption,

$$\lim_{i \to \infty} E_{r_i}(f) = 0.$$

Choose $r_i' \in (\frac{1}{3^{i+1}}, \frac{1}{3^i})$ as above, then

$$\lim_{i\to\infty} \sup_{|y|=|y'|=r'_i} d_V(f(y), f(y')) = 0.$$

Now we claim that $f(r,\theta)$ converge to a point in X. For any sufficiently small $\epsilon > 0$, by the above, there is a r'_j such that for any $i > j$, each connected component of $f(\partial(D_{r'_j}\backslash D_{r'_i}))$ is contained in a ball of radius ϵ, where $D_r = \{y \mid |y| < r\}$. Then the last corollary implies that $f(D_{r'_j}\backslash D_{r'_i})$ is contained in a ball of radius 8ϵ. Letting i go to infinity, we see that for any $x, y \in D_{r'_j}$,

$$d_V(f(x), f(y)) \le 8\epsilon.$$

It follows that $\lim_{x\to 0} f(x)$ exists, so the claim is proved and f can be extended continuously to D.

Next we want to bound the derivative ∇f. Since f is continuous, we may choose $\rho_0 < \frac{1}{2}$ so small that $f(B_{2\rho_0}(0))$ is contained in a coordinate chart U of X. For simplicity, we consider $U \subset \mathbb{R}^{2n}$. Fix any $y \in B_{\rho_0}(0)$.

In the polar coordinates (r,θ) centered at y ($r = 0$ at y), the Cauchy-Riemann equation becomes

$$\frac{\partial f}{\partial r} + \frac{1}{r} J \frac{\partial f}{\partial \theta} = 0.$$

Let c be any constant vector in $\mathbb{R}^{2n}$, then integrating by parts, we have for $\rho \le \rho_0$,

$$\begin{aligned}
0 = & \int_{B_\rho(y)} \left| \tfrac{\partial f}{\partial r} + \tfrac{1}{r} J \tfrac{\partial f}{\partial \theta} \right|^2 r dr d\theta \\
= & E_\rho(f) + 2\int_0^\rho \int_0^{2\pi} \langle \tfrac{\partial f}{\partial r}, \tfrac{1}{r} J \tfrac{\partial f}{\partial \theta} \rangle r dr d\theta \\
= & E_\rho(f) + 2\int_0^{2\pi} \langle f - \lambda, J \tfrac{\partial f}{\partial \theta} \rangle (\rho, \theta) d\theta - 2\int_0^\rho \int_0^{2\pi} \langle f - \lambda, \tfrac{\partial}{\partial r} \left(J \tfrac{\partial f}{\partial \theta} \right) \rangle dr d\theta \\
= & E_\rho(f) + 2\int_0^{2\pi} \langle f - \lambda, J \tfrac{\partial f}{\partial \theta} \rangle (\rho, \theta) d\theta \\
& -2\int_0^\rho \int_0^{2\pi} \langle f - \lambda, \nabla_{\frac{\partial}{\partial r}} J \tfrac{\partial f}{\partial \theta} \rangle dr d\theta - 2\int_0^\rho \int_0^{2\pi} \langle f - \lambda, J \tfrac{\partial^2 f}{\partial r \partial \theta} \rangle dr d\theta \\
= & E_\rho(f) + 2\int_0^{2\pi} \langle f - \lambda, J \tfrac{\partial f}{\partial \theta} \rangle d\theta + 2\int_0^\rho \int_0^{2\pi} \langle \tfrac{\partial f}{\partial \theta}, J \tfrac{\partial f}{\partial r} \rangle dr d\theta \\
& -2\int_0^\rho \int_0^{2\pi} \langle f - \lambda, \nabla_{\frac{\partial}{\partial r}} J \tfrac{\partial f}{\partial \theta} \rangle dr d\theta + 2\int_0^\rho \int_0^{2\pi} \langle f - \lambda, \nabla_{\frac{\partial}{\partial \theta}} J \tfrac{\partial f}{\partial r} \rangle dr d\theta,
\end{aligned}$$

it follows that

$$(1 - c \sup_{\partial B_\rho(y)} |f - \lambda|) E_\rho(f, y) \le -\int_0^{2\pi} \langle f - \lambda, J \frac{\partial f}{\partial \theta} \rangle d\theta,$$

where c depends only on the derivative of J.

Now choose

$$\lambda = \frac{1}{2\pi} \int_0^{2\pi} f d\theta.$$

Then by the Poincaré inequality on the unit circle, we have

$$\int_0^{2\pi} |f - \lambda|^2 d\theta \le \int_0^{2\pi} \left| \frac{\partial f}{\partial \theta} \right|^2 d\theta.$$

So by the Cauchy-Riemann equation, we have

$$(1 - c \sup_{\partial B_\rho(y)} |f - \lambda|) E_\rho(f) \leq \frac{1}{2}\rho^2 \int_0^{2\pi} |\nabla f|^2 d\theta.$$

But

$$\frac{\partial}{\partial \rho} E_\rho(f, y) = \rho \int_0^{2\pi} |\nabla f|^2 d\theta,$$

so the above is the same as

$$2(1 - c \sup_{\partial B_\rho(y)} |f - \frac{1}{2\pi} \int_0^{2\pi} f d\theta|) E_\rho(f, y) \leq \rho \frac{\partial}{\partial \rho} E_\rho(f, y).$$

Since f is continuous, we can choose ρ_0 so small that

$$\sup_{\partial B_\rho(y)} |f - \frac{1}{2\pi} \int_0^{2\pi} f d\theta| \leq \frac{1}{2c}.$$

Then

$$\frac{\partial}{\partial \rho} (\rho^{-1} E_\rho(f, y)) \geq 0.$$

It follows that for any $\rho < \rho_0$ and $y \in B_{\rho_0}(0)$,

$$E_\rho(f, y) \leq 2\rho E_1(f).$$

By the Morrey's lemma, f is 1/2-Hölder continuous. Then

$$2(1 - c\sqrt{\rho}) E_\rho(f, y) \leq \rho \frac{\partial}{\partial \rho} E_\rho(f, y).$$

It follows that $E_\rho(f, y) \leq c\rho^2$ for any $\rho \leq \rho_0$, and consequently, $|\nabla f|^2(y) \leq c$. So we have bounded the derivative of f. By the standard elliptic theory, one then deduces that f is smooth in D.

The following provides the basic gradient estimate for pseudo-holomorphic maps.

Theorem 2.5. *Let (X, ω) be a symplectic manifold with a compatible almost complex structure J. Then there are $\epsilon, c > 0$, depending only on (X, ω, J) such that for any J-holomorphic map $f : D_r \mapsto X$ with*

$$\int_{D_r} |\nabla f|^2(z) dz \leq \epsilon,$$

then

$$\sup_{D_{\frac{r}{2}}} |\nabla f|^2 \leq \frac{c}{r^2} \int_{D_r} |\nabla f|^2(z) dz,$$

where $D_r = \{|z| < r\}$ and the norm $|\cdot|$ is taken with respect to the metric induced by ω and J.

Proof. This can be proved by the same arguments as those in the proof of last theorem. So we just sketch its proof, pointing out necessary changes.

By scaling, we may assume that $r = 1$. There is a $r_0 \in (3/4, 1)$ such that

$$\int_0^{2\pi} \left| \frac{\partial f}{\partial \theta} \right|^2 (r_0, \theta) d\theta \leq 6E_1(f, 0) \leq 6\epsilon.$$

This implies

$$\sup_{0 \leq \theta, \theta' \leq 2\pi} d_X(f(r_0, \theta), f(r_0, \theta')) \leq \sqrt{6\pi\epsilon}.$$

We may assume that ϵ is sufficiently small. By Corollary 3.4, $f(D_{r_0}) \subset B_{\sqrt{6\pi\epsilon}}(p)$ for some p in X. It follows

$$d_X(f(x), f(y)) \leq 4\sqrt{6\pi\epsilon}, \quad \text{for any } \; x, y \in D_{\frac{3}{4}}.$$

We have shown in the proof of last theorem that for any $y \in D_{\frac{2}{3}}$ and $\rho < \frac{1}{12}$,

$$2(1 - c \sup_{|z-y|, |z'-y| \leq \rho} d_X(f(z), f(z'))) E_\rho(f, y) \leq \rho \frac{\partial}{\partial \rho} E_\rho(f, y).$$

Combining this with the above estimate on f, we deduce from this

$$2(1 - c\sqrt{\epsilon}) E_\rho(f, y) \leq \rho \frac{\partial}{\partial \rho} E_\rho(f, y).$$

Then by using the arguments in the proof of last theorem, we have that for any $y \in D_{\frac{1}{2}}$,

$$|\nabla f|(y) \leq c.$$

Now we prove Theorem 2.3.

First we observe: For any given metric h on Σ and $[f, \Sigma, (x_i)] \in \mathfrak{M}^{\mathfrak{X}}_{\mathfrak{A},\mathfrak{g},\mathfrak{k}}$,

$$\int_\Sigma |\nabla f|^2 dv = [\omega](A).$$

where dv denotes the volume form of h and the norm $|\cdot|$ is taken with respect to h. Note that we always denote by dv a volume form, which may vary in different places. Therefore, the energy of f depends only on its homology class A.

Next we observe: There is a positive number δ such that if $[f, \Sigma, \{x_i\}] \in \mathfrak{M}^{\mathfrak{X}}_{\mathfrak{A},\mathfrak{g},\mathfrak{k}}$ and S is an irreducible component of Σ on which f is not constant, then

$$\int_S |\nabla f|^2 dv \geq \delta > 0.$$

It follows from these and the stability that there is a uniform bound on the number of irreducible components of Σ. This bound depends only on the genus g, the homology class A and the target manifold (X, ω).

Let $[f_\alpha, \Sigma_\alpha, (x_{\alpha i})]$ be a sequence of stable maps in $\mathfrak{M}^{\mathfrak{X}}_{\mathfrak{A},\mathfrak{g},\mathfrak{k}}$. Because of the above observations, by taking a subsequence if necessary, we may assume that the topology of Σ_α is independent of α.

We will consider the following class of metrics g_α on the regular part of Σ_α. The metrics g_α have uniformly bounded geometry, namely, for each regular point p of Σ_α, there is a local conformal coordinate chart (U, z) of Σ_α containing p such that U is identified with the unit ball $D_1 = \{\, |z| < 1 \,\}$ in $\mathbb{C}$ and

$$g_\alpha|_U = e^\varphi dz d\bar{z}$$

for some $\varphi(z)$ satisfying:

$$||\varphi||_{C^k(U)} \le c_k, \quad \text{for any } k > 0,$$

where c_k are uniform constants independent of α. We also require that there are finitely many cylinder-like necks $N_{\alpha,i} \subset \Sigma_\alpha$ $(i = 1, \cdots, n_\alpha)$ satisfying:
(1) n_α are uniformly bounded independent of α;
(2) The complement $\Sigma_\alpha \backslash \bigcup_i N_{\alpha,i}$ is covered by finitely many geodesic balls $B_R(p_{\alpha,j})$ $(1 \le j \le m_\alpha$ of g_α in Σ_α, where R and m_α are uniformly bounded;
(3) Each $N_{\alpha,i}$ is diffeomorphic to a cylinder of the form $S^1 \times (a, b)$ (a and b may be $\pm\infty$) satisfying: If s, t denote the standard coordinates of $S^1 \times [0, b)$ or $S^1 \times (a, 0]$, then

$$g_\alpha|_{N_{\alpha,i}} = e^\varphi(ds^2 + dt^2),$$

where φ is a smooth function satisfying uniform bounds as stated above.

We will say that such a g_α is admissible. We will call $\{g_\alpha\}$ uniformly admissible if all g_α are admissible with uniform constants R, c_k, etc..

Admissible metrics always exist on any Σ_α. This can be seen as follows: Clearly, it suffices to construct metrics on each irreducible component of Σ_α. Let S be any connected component of the regular part of Σ_α, if the Euler number of S is nonnegative, then S is either $\mathbb{C}$ or $\mathbb{C}\backslash\{0\}$, in either of those cases, we can easily write down a metric on S. If the Euler number of S is negative, then the uniformization theorem in complex analysis gives a unique hyperbolic metric g' on S with finitely many cusps, and g_α on S is simply obtained by fatting those cusps to be cylinder-like. It is not hard to see that those admissible metrics can be chosen uniformly.

Now we fix a sequence of uniformly admissible metrics g_α on Σ_α. We will introduce a new sequence of uniformly admissible metrics $\tilde{g}_\alpha$ on Σ_α such that there is a uniform bound on the gradient of f_α. Once it is done, the theorem follows easily.

For simplicity, we write Σ for each given Σ_α and g for g_α and f for f_α. We will define $\tilde{g} = \tilde{g}_\alpha$ by induction.

If $\sup_\Sigma |df|_g \le 16$, then we simply define $\tilde{g}$ to be g. Otherwise, let $p_1 \in \Sigma$ such that

$$e = |df|_g(p_1) = \sup_\Sigma |df|_g > 16,$$

and z be the local coordinate of Σ specified above such that $z = 0$ at p_1. Write $g = e^{\varphi} dz d\bar{z}$ as above, define $\tilde{g}_1 = g$ outside the region where $|z| < 1$ and

$$\tilde{g}_1 = \frac{e}{\eta(e|z|^2)} g,$$

where $\eta : \mathbb{R} \mapsto \mathbb{R}$ is a cut-off function satisfying: $\eta(t) = 1$ for $t \leq 1$, $\eta(t) = t$ for $t \in [2, e-1]$, and $\eta(t) = e$ for $t \geq e$, moreover, we may assume that $0 \leq \eta'(t) \leq 1$. Clearly, we have $\tilde{g}_1 \geq g$. It is easy to check that $\tilde{g}$ is uniformly admissible. Moreover, we have

$$sup_{B_1(p_1, \tilde{g}_1)} |df|_{\tilde{g}_1} = 1,$$

where $B_1(p_1, \tilde{g}_1)$ denotes the geodesic ball of radius 1 and centered at p_1 with respect to the metric $\tilde{g}_1$. It follows from Theorem 3.3 that

$$\int_{B_1(p_1, \tilde{g}_1)} |df|_g^2 dv \geq \delta > 0,$$

where δ depneds only on (X, ω).

If $\sup_\Sigma |df|_{\tilde{g}_1} \leq 16$, then we take $\tilde{g} = \tilde{g}_1$, otherwise, we choose p_2 such that

$$e = |df|_{\tilde{g}_1}(p_2) = \sup_\Sigma |df|_{\tilde{g}_1} > 16,$$

then $p_2 \in \Sigma \backslash B_2(p_1, \tilde{g}_1)$. Now we can get $\tilde{g}_2$ by repeating the above construction with g replaced by $\tilde{g}_1$. Clearly, $\tilde{g}_2$ coincides with $\tilde{g}_1$ on $B_1(p_1, \tilde{g}_1)$, so

$$B_1(p_1, \tilde{g}_2) = B_1(p_1, \tilde{g}_1).$$

We also have

$$B_1(p_2, \tilde{g}_2) \cap B_1(p_1, \tilde{g}_1) = \emptyset$$

and

$$\int_{B_1(p_i, \tilde{g}_2)} |df|_g^2 dv \geq \delta > 0, \quad i = 1, 2.$$

If $\sup_\Sigma |df|_{\tilde{g}_2} \leq 16$, we simply put $\tilde{g} = \tilde{g}_2$. Otherwise, we continue the process and construct inductively $\tilde{g}_1, \cdots, \tilde{g}_L$ such that

$$\int_{B_1(p_i, \tilde{g}_L)} |df|_g^2 dv \geq \delta > 0, \quad i = 1, \cdots, L.$$

It follows that $L \leq \frac{[\omega](A)}{\delta}$, therefore, the process has to stop at some L when $\sup_\Sigma |df|_{\tilde{g}_L} \leq 16$. We then take $\tilde{g}$ to be $\tilde{g}_L$.

Now we have construct a new sequence of uniformly admissible metrics $\tilde{g}_\alpha$ such that

$$\sup_\Sigma |df_\alpha|_{\tilde{g}_\alpha} \leq 16.$$

Moreover, by scaling $\tilde{g}_\alpha$ appropriately, we may assume that $d(x_{\alpha i}, x_{\alpha i'}) \geq 1$ for $i \neq i'$. By the uniform admissibility of $\tilde{g}_\alpha$, when α is sufficiently large, we

may have m, l and R such that there are finitely many cylinder-like necks $N_{\alpha,i} \subset \Sigma_\alpha$ $(i = 1, \cdots, l)$ satisfying:
(1) The complement $\Sigma_\alpha \backslash \bigcup_i N_{\alpha,i}$ is covered by finitely many geodesic balls $B_R(p_{\alpha,j}, \tilde{g}_\alpha)$ $(1 \le j \le m)$ in Σ_α;
(2) The marked points $x_{\alpha i}$ are all contained in the union of those geodesic balls $B_R(p_{\alpha,j}, \tilde{g}_\alpha)$;
(3) Each $N_{\alpha,i}$ is diffeomorphic to a cylinder of the form $S^1 \times (a_{\alpha,i}, b_{\alpha,i})$ ($a_{\alpha,i}$ and $b_{\alpha,i}$ may be $\pm\infty$). We may further assume that for any $x \in N_{\alpha,i}$,

$$\int_{B_1(x,\tilde{g}_\alpha)} |df_\alpha|_{\tilde{g}_\alpha} dv \le \epsilon,$$

where ϵ is given in Theorem 2.5.

Now by taking a subsequence if necessary, we may assume that for each j, $(\Sigma_\alpha, p_{\alpha,j})$ converge to a Riemann surface $\Sigma^0_{\infty,j}$ as pointed metric spaces, moreover, such a $\Sigma^0_{\infty,j}$ is of the form

$$\Sigma_{\infty,j} \backslash \{q_{j1}, \cdots, q_{j\gamma_j}\},$$

where $\Sigma_{\infty,j}$ is a compact Riemann surface. More precisely, there are a natural admissible metric $g_{\infty,j}$ on $\Sigma^0_{\infty,j}$ and a point $p_{\infty,j}$ in $\Sigma^0_{\infty,j}$, such that for any fixed $r > 0$, when α is sufficiently large, there is a diffeomorphism $\phi_{\alpha,r}$ from $B_r(p_{\infty,j}, g_{\infty,j})$ onto $B_r(p_{\alpha,j}, \tilde{g}_\alpha)$ satisfying: $\phi_{\alpha,r}(p_{\infty,j}) = p_{\alpha,j}$ and the pull-backs $\phi^*_{\alpha,r}\tilde{g}_\alpha$ converge to $g_{\infty,j}$ uniformly in the C^∞-topology over $B_r(p_{\infty,j}, g_{\infty,j})$. Note that such a convergence of $\tilde{g}_\alpha$ is assured by the uniform admissibility.

Next we put together all these $\Sigma_{\infty,j}$ to form a connected curve Σ'_∞ as follows: For any two components $\Sigma_{\infty,j}$ and $\Sigma_{\infty,j'}$, we identify punctures $y_{js} \in \Sigma_{\infty,j}$ with $y_{j's'} \in \Sigma_{\infty,j'}$ (j may be equal to j') if for any α and r sufficiently large, the boundaries of $B_r(p_{\alpha,j}, \tilde{g}_\alpha)$ and $B_r(p_{\alpha,j'}, \tilde{g}_\alpha)$ specified above are contained in a cylindrical neck $N(\alpha, i)$. In this way, we get a connected curve $\Sigma_{\infty,j'}$ (not necessarily stable) since each Σ_α is connected.

Since the gradients of f_α are uniformly bounded in terms of $\tilde{g}_\alpha$, by taking a subsequence if necessary, we may assume that f_α converge to a J-holomorphic map f_∞ from $\bigcup_j \Sigma^0_{\infty,j}$ into X. By the Removable Singularity Theorem, the map f_∞ extends smoothly to a J-holomorphic map from Σ'_∞ into X. Moreover, we may assume that the marked points $x_{\alpha i}$ converge to $x_{\infty i}$ as α tends to the infinity, clearly, each $x_{\infty i}$ belongs to the regular part of Σ'_∞.

The tuple $(f_\infty, \Sigma'_\infty, \{x_{\infty i}\})$ is not necessarily a stable map, since there may be components $\Sigma_{\infty,j}$ where f_∞ restricts to a constant map and which is conformal to $\mathbb{C}P^1$ and contains fewer than three of $x_{\infty i}$ and $y_{j\beta}$ (defined above and contained in the singular set of Σ'_∞). There are three possibilities for such $\Sigma_{\infty,j}$'s. If $\Sigma_{\infty,j}$ contains no $x_{\infty i}$ but one $y_{j\beta}$, we simply drop this component; If $\Sigma_{\infty,j}$ contains no $x_{\infty i}$ but two $y_{j\beta}$ and $y_{\infty\beta'}$, then we contract this component and identify $y_{j\beta}$ and $y_{j\beta'}$ as points in other components of Σ'_∞; If $\Sigma_{\infty,j}$ contains one $x_{\infty i}$ and one $y_{j\beta}$, then we contract this component and mark the point $y_{j'\beta'}$ as $x_{\infty i}$. Carrying out this process inductively, we eventually obtain a connected curve Σ_∞ such that the induced $(f_\infty, \Sigma_\infty, \{x_\infty i\})$ is a stable map.

Clearly, this stable map has the same genus as that of Σ_α and k marked points. It remains to show that the homology class of f_∞ is the same as that of f_α. By the convengence, we have

$$\int_{\Sigma_\infty} |\nabla f_\infty|^2 dv = \lim_{r\to\infty} \lim_{\alpha\to\infty} \int_{\bigcup_j B_r(p_{\alpha,j}, \tilde{g}_\alpha)} |\nabla f_\alpha|^2 dv.$$

Since the complement of $\bigcup_j B_r(p_{\alpha,j}, \tilde{g}_\alpha)$ in Σ_α is contained in the union of cylindrical necks $N(\alpha, i)$, we suffice to show that for each i, if $N(\alpha, i) = S^1 \times (a, b)$, then

$$\lim_{r\to\infty} \lim_{\alpha\to\infty} \int_{S^1\times(a+r,b-r)} |\nabla f_\alpha|^2 dv = 0.$$

This can be seen as follows: By our choice of $\tilde{g}_\alpha$, we know that for any $p \in N(\alpha, i)$,

$$\int_{B_1(p,\tilde{g}_\alpha)} |\nabla f_\alpha|^2 dv \le \epsilon.$$

It follows from Theorem 2.5 that

$$\sup_{N(\alpha,i)} |\nabla f_\alpha|^2 \le c\epsilon,$$

where c is the uniform constant given in Theorem 2.5. Since ϵ is small, both $f_\alpha(S^1 \times \{a+r\})$ and $f_\alpha(S^1 \times \{b-r\})$ are contained in geodesic balls of radius $2\pi\sqrt{c\epsilon}$, in particular, there are two smooth maps $h_{\alpha,j} : D_1 \mapsto X$ $(j = 1, 2)$ such that

$$\sup_{D_1} |\nabla h_{\alpha,j}|_{\tilde{g}_\alpha} \le 8 \sup_{S^1\times\{a+r,b-r\}} |\nabla f_\alpha|$$

and

$$h_{\alpha,1}|_{\partial D_1} = f_\alpha|_{S^1\times\{a+r\}}, \quad h_{\alpha,2}|_{\partial D_1} = f_\alpha|_{S^1\times\{b-r\}}.$$

The maps $f_\alpha|_{N(\alpha,i)}$ and $h_{\alpha,j}$ can be easily put together to form a continuous map from S^2 into X. Since its gradient is small everywhere, this map is null homologous. It follows

$$\int_{S^1\times(a+r,b-r)} |\nabla f_\alpha|^2 dv = \int_{S^1\times(a+r,b-r)} f_\alpha^*\omega = \int_{D_1} h_{\alpha,1}^*\omega dv - \int_{D_1} h_{\alpha,2}^*\omega.$$

Therefore, we have

$$\int_{S^1\times(a+r,b-r)} |\nabla f_\alpha|^2 dv \le c \sup_{S^1\times\{a+r,b-r\}} |\nabla f_\alpha|^2.$$

This implies the required convergence.

Therefore, the stable map $[f_\infty, \Sigma_\infty, (x_{\infty i})]$ is in $\mathfrak{M}^{\mathfrak{X}}_{\mathfrak{A},\mathfrak{g},\mathfrak{k}}$. The above arguments also show that $[f_\alpha, \Sigma_\alpha, (x_{\alpha i})]$ converge to $[f_\infty, \Sigma_\infty, (x_{\infty i})]$ in the topology of **B** defined in last section. So Theorem 2.3 is proved.

2.5 Constructing GW-invariants

The main purpose of this subsection is to construct the virtual moduli cycles and GW-invarinats for general symplectic manifolds.

Let X be a smooth symplectic manifold with a given symplectic form ω of complex dimension n, and let $A \in H_2(X, \mathbb{Z})$. Let $\mathfrak{M}_{\mathfrak{g},\mathfrak{k}}$ be the empty set if $2g + k < 3$ and the moduli space of k-pointed, genus g stable curves if $2g + k \geq 3$.

Here is the main theorem of this section.

Theorem 2.6. *Let (X, ω) be a compact symplectic manifold of complex dimension n. Then for each g, k and A, there is a virtual fundamental class*

$$e_{A,g,k}(X) \in H_r(\mathfrak{M}_{\mathfrak{g},\mathfrak{k}} \times \mathfrak{X}^{\mathfrak{k}}, \mathbb{Q}),$$

where $r = 2c_1(X)(A) + 2(n-3)(1-g) + 2k$. Moreover, this $e_{A,g,k}(X)$ is a symplectic invariant.

As an application, let us define the GW-invariants now. Let $2g + k \geq 3$. We define

$$\psi^X_{A,g,k} : H^*(\mathfrak{M}_{\mathfrak{g},\mathfrak{k}}, \mathbb{Q}) \times \mathfrak{H}^*(\mathfrak{X}^{\mathfrak{k}}, \mathbb{Q}) \mapsto \mathbb{Q}, \tag{2.3}$$

to be the integrals

$$\psi^X_{A,g,k}(\beta, \alpha_1, \cdots, \alpha_k) = \int_{e_{A,g,k}(X)} \pi_1^*\beta \wedge \pi_2^*\alpha_1 \wedge \cdots \wedge \pi_{k+1}^*\alpha_k \tag{2.4}$$

where $\beta \in H^*(\mathfrak{M}_{\mathfrak{g},\mathfrak{k}}, \mathbb{Q})$, $\alpha_i \in H^*(X, \mathbb{Q})$ $(1 \leq i \leq k)$ and π_i is the projection of $\mathfrak{M}_{\mathfrak{g},\mathfrak{k}} \times \mathfrak{X}^{\mathfrak{k}}$ to its i-th component. For simplicity, we will often write $\psi^X_{A,g,k}(\beta, \alpha_1, \cdots, \alpha_k)$ as $\psi^X_{A,g,k}(\beta, (\alpha_i))$. All $\psi^X_{A,g,k}$ are symplectic invariants of (X, ω).

Let $(\mathbf{B}, \mathbf{E}, \Phi_J)$ be as before. There is a natural evaluation map

$$\mathrm{ev} : \mathbf{B} \mapsto \mathfrak{M}_{\mathfrak{g},\mathfrak{k}} \times \mathfrak{X}^{\mathfrak{k}}$$

defined by $\mathrm{ev}(f, \Sigma, (x_i)) = \Sigma_{\mathrm{red}} \times (f(x_i))$, where Σ_{red} is the empty set if $2g + k < 3$ and the stable reduction of Σ by contracting all its non-stable components if $2g + k \geq 3$.

We will apply Theorem 1.1 to constructing the virtual fundamental class $e_{A,g,k}(X)$. For this purpose, we need to show that $(\mathbf{B}, \mathbf{E}, \Phi_J)$ admits a weakly Fredholm structure with the submersion property for the evaluation map ev (as stated in Theorem 1.1).

We continue to use the notations developed so far. Let $\tilde{\mathbf{U}}$ be a chart of $\mathbf{B}$ with the corresponding group G, and let $\mathbf{E}_{\tilde{\mathbf{U}}}$ be the corresponding chart of $\mathbf{E}$ over $\tilde{\mathbf{U}}$. We know that $\tilde{\mathbf{U}}$ is of the form $\tilde{\mathbf{U}}(\mathcal{U}, H_j, K, \delta, \phi)$. Let $\tilde{\mathcal{U}}$ be the local uniformization of $\mathcal{U}$ and $\mathcal{CU}$ be the universal curve over $\tilde{\mathcal{U}}$. We may assume that $\mathcal{U}$ is sufficiently small.

Recall that a TX-valued (0,1)-form on $\mathcal{CU} \times X$ is an endomorphism

$$v : TC\mathcal{U} \mapsto TX$$

such that

$$J \cdot v = -v \cdot \mathbf{j}_{C\mathcal{U}},$$

where $\mathbf{j}_{C\mathcal{U}}$ is the complex structure on $C\mathcal{U}$. Let $\Lambda^{0,1}(C\mathcal{U}, TX)_0$ be the space of all C^∞-smooth TX-valued (0,1)-forms on $C\mathcal{U} \times X$ which vanish near $\mathrm{Sing}(C\mathcal{U})$. Here $\mathrm{Sing}(C\mathcal{U})$ denotes the set of the singularities in the fibers of $C\mathcal{U}$ over $\tilde{\mathcal{U}}$.

Given each $v \in \Lambda^{0,1}(C\mathcal{U}, TX)_0$, we can associate a section of $\mathbf{E}_{\tilde{\mathbf{U}}}$ as follows: For each $\mathbf{f} = (f, \Sigma, (x_i, y_j)) \in \tilde{\mathbf{U}}$, we define $v|_{\mathbf{f}}$ by

$$v|_{\mathbf{f}}(x) = v(x, f(x)), \qquad x \in \Sigma.$$

Clearly, $v|_{\mathbf{f}}$ is a section on the fiber of $\mathbf{E}_{\tilde{\mathbf{U}}}$ over $\mathbf{f}$. In this way, we obtain a section $\mathbf{f} \mapsto v|_{\mathbf{f}}$ over $\tilde{\mathbf{U}}$. To avoid introducing new notations, we still denote this section by v. For $\sigma \in G_f$, the pull-back $\sigma^*(v_i)$ is a section over $\tilde{\mathbf{U}}$. Let $v_1, \cdots, v_l$ be any l sections in $\Lambda^{0,1}(C\mathcal{U}, TX)_0$. Without loss of generality, we can assume that the $l \cdot |G_f|$ sections

$$\{\sigma^*(v_i) \mid 1 \le i \le l, \sigma \in G_f\}$$

of $\mathbf{E}_{\tilde{\mathbf{U}}}$ are linearly independent everywhere. We define $\mathbf{F} = \mathbf{F}(v_1, \cdots, v_l)$ to be the subbundle in $\mathbf{E}_{\tilde{\mathbf{U}}}$ generated by the above $l \cdot |G_f|$ sections. $\mathbf{F}$ is a trivial vector bundle and is a G_f-equivariant subbundle of $\mathbf{E}_{\tilde{\mathbf{U}}}$.

Lemma 2.2. *Suppose* $\mathbf{f} = (f, \Sigma, (x_i, y_j)) \in \tilde{\mathbf{U}}$ *and* $\mathrm{Sp}(v_1, \cdots, v_l)|_{\mathbf{f}}$ *are transverse to* L_f*, where* L_f *is the linearization of the Cauchy-Riemann equation at* f*. We further assume that* δ *is sufficiently small and* K *is sufficiently big in the definition of* $\tilde{\mathbf{U}}$*. Then* $(\tilde{\mathbf{U}}, G_f, \mathbf{E}_{\tilde{\mathbf{U}}}, \mathbf{F})$ *is a local finite approximation of index* r*, where* r *is the index of* L_f *which can be computed in terms of* $c_1(X)$*, the homology class of* $f(\Sigma)$*, the genus of* Σ *and the number of marked points.*[14]

This follows from the Implicit Function Theorem because $\sigma^*(v_j)$ $(1 \le j \le l, \sigma \in G_f)$ generate a subbundle in $\mathbf{E}_{\tilde{\mathbf{U}}}$ which is transverse to the cokernel of L_f for every $\mathbf{f}$ in $\tilde{\mathbf{U}}$.

Next we assign a natural orientation to the above $(\tilde{\mathbf{U}}, G_f, \mathbf{E}_{\tilde{\mathbf{U}}}, \mathbf{F})$. Let $\mathbf{U}_0$ be the main stratum of $\tilde{U}$ and $U_0 = \Phi_J^{-1}(\mathbf{F}) \cap \mathbf{U}_0$. For any $\mathbf{f} \in U_0$, we denote by $W^{1,2}(f^*TX)$ the Sobolev space of all $W^{1,2}$-sections of f^*TX and by $L^2(\Lambda_f^{0,1})$ the space of L^2-integrable (0,1)-forms with values in f^*TX (with respect to the norms induced by ω and the almost complex structure J). Notice that

$$L_f : W^{1,2}(f^*TX) \mapsto L^2(\Lambda_f^{0,1})$$

[14] In fact, one can show that $\Phi_J^{-1}(\mathbf{F})$ is a smooth manifold of dimension $r + l$.

is a Fredholm linear operator. So we have a well-defined determinant line $\det(L_f)$. It varies smoothly with $\mathbf{f}$ and gives rise to a determinant line bundle $\det(L)$ over $\mathbf{U}_0$. Let $F|_{\mathbf{f}}$ be the fiber of the bundle F at $\mathbf{f}$. Then it is a finitely dimensional subspace in $L^2(\Lambda_f^{0,1})$, so we have an orthogonal decomposition (with respect to the induced L^2-inner product induced by ω and J)

$$L^2(\Lambda_f^{0,1}) = F|_{\mathbf{f}} + F^{\perp}.$$

Let $\pi_{\mathbf{f}} : L^2(\Lambda_f^{0,1}) \mapsto F^{\perp}$ be the orthogonal projection. Then $T_{\mathbf{f}}U$ is naturally isomorphic to the kernel of $\pi_{\mathbf{f}} \circ L_f$. It follows that $\det(L)$ is naturally isomorphic to $\wedge^{\mathrm{top}}(TU) \otimes \wedge^{\mathrm{top}}(F)^{-1}$. Thus we suffice to orient $\det(L)$. By straightforward computations, we can find a canonical decomposition

$$L_f = \overline{\partial}_f + B_f$$

such that $\overline{\partial}_f$ is J-invariant and B_f is an operator of order 0. Moroever, both $\overline{\partial}_f$ and B_f vary smoothly with $\mathbf{f}$ in $\mathbf{U}_0$. Hence, L_f is homotopic to $\overline{\partial}_f$, consequently, $\det(L)$ is isomorphic to the determinant line bundle $\det(\overline{\partial})$ with fibers $\det(\overline{\partial}_f)$. On the other hand, since $\overline{\partial}_f$ is J-invariant, its determinant has a canonical orientation induced by the complex structures on $\mathrm{Ker}(\overline{\partial}_f)$ and $\mathrm{Coker}(\overline{\partial}_f)$, consequently, $\det(L)$ has a canonical orientation, so does $(\tilde{\mathbf{U}}, G_f, \mathbf{E}_{\tilde{\mathbf{U}}}, \mathbf{F})$.

Now we want to choose $v_1, \cdots, v_l$ so that the restriction of ev to each nonempty stratum $\Phi^{-1}(\mathbf{F}) \cap \mathbf{U}_\alpha$ is a submersion, where $\mathbf{U}_\alpha$ is a stratum of $\tilde{\mathbf{U}}$. The strata of $\tilde{\mathbf{U}}$ are classified by the dual graphs of genus g and homology class A and k legs. Given any $\mathcal{U}$, there are only finitely many dual graphs $\{\Gamma_\alpha\}$ such that $\mathbf{U} \cap \mathbf{B}(\Gamma) \neq \emptyset$. Let $\mathbf{U}_\alpha$ be the stratum of $\tilde{\mathbf{U}}$ corresponding to $\mathbf{U} \cap \mathbf{B}(\Gamma)$. It is obvious that the evaluation map $\mathrm{ev} : \mathbf{U}_\alpha \mapsto \mathfrak{M}_{\mathfrak{g},\mathfrak{k}} \times \mathfrak{X}^{\mathfrak{k}}$ is a submersion. It follows that there are finitely many $u_{i\alpha} \in \Lambda^{0,1}(\mathcal{CU}, TX)_0$ $(1 \leq i \leq l_\alpha)$ such that

$$\mathrm{ev} : \{\mathbf{f} \in \mathbf{U}_\alpha \,|\, \Phi_J(\mathbf{f}) \in \mathrm{Sp}(u_{i\alpha})|_{\mathbf{f}}\} \mapsto \mathfrak{M}_{\mathfrak{g},\mathfrak{k}} \times \mathfrak{X}^{\mathfrak{k}}$$

is still a submersion. Since the number of such α's is finite, we can choose $v_1, \cdots, v_l$ such that $\mathrm{Sp}(v_1, \cdots, v_l)$ contains all $\mathrm{Sp}(u_{i\alpha})$. Then for δ sufficiently small and K sufficiently large, the restriction of ev to each nonempty stratum $\Phi^{-1}(\mathbf{F}) \cap \mathbf{U}_\alpha$ is a submersion.

It is tedious, but rather straightforward, to check that those locally finite approximations defined as above provide a weakly smooth structure of $(\mathbf{B}, \mathbf{E}, \Phi_J)$. Combining this with the compactness theorem of last section, we conclude that $(\mathbf{B}, \mathbf{E}, \Phi_J)$ is actually a weakly Fredholm V-bundle.

This completes the proof of the main theorem.

2.6 Composition laws for GW-invariants

In last subsection, we have constructed GW-invariants for general symplectic manifolds. These invariants satisfy certain properties, such as the Puncture equation, the String equation and the Dilaton equation which the generating function of GW-invariants satisfy. The most useful property is the composition law for GW-invariants, which we will formulate in the following. We will drop its proof.

Assume that $2g + k \geq 4$. Given any decomposition $g = g_1 + g_2$ and $S = S_1 \cup S_2$ of $\{1, \cdots, k\}$ with $|S_i| = k_i$, where $2g_i + k_i \geq 2$, there is a canonical embedding $\mathbf{i}_S : \mathfrak{M}_{\mathfrak{g}_1, \mathfrak{k}_1+1} \times \mathfrak{M}_{\mathfrak{g}_2, \mathfrak{k}_2+1} \mapsto \mathfrak{M}_{\mathfrak{g}, \mathfrak{k}}$, which assigns stable curves $(\Sigma_i, x_1^i, \cdots, x_{k_1+1}^i)$ $(i = 1, 2)$ to their union $\Sigma_1 \cup \Sigma_2$ with $x_{k_1+1}^1$ identified to x_1^2 and remaining points renumbered by $\{1, \cdots, k\}$ according to S.

There is another natural maps $\mathbf{i}_0 : \mathfrak{M}_{\mathfrak{g}-1, \mathfrak{k}+2} \mapsto \mathfrak{M}_{\mathfrak{g}, \mathfrak{k}}$ obtained by gluing together the last two marked points.

One can define a homomorphism

$$\mathbf{i}_! : H^*(\mathfrak{M}_{\mathfrak{g}_1, \mathfrak{k}_1+1}, \mathbb{Q}) \times \mathfrak{H}^*(\mathfrak{M}_{\mathfrak{g}_2, \mathfrak{k}_2+1}, \mathbb{Q}) \mapsto \mathfrak{H}^*(\mathfrak{M}_{\mathfrak{g}, \mathfrak{k}}, \mathbb{Q})$$

as follows: For any $\beta_1 \in H^*(\mathfrak{M}_{\mathfrak{g}_1, \mathfrak{k}_1+1}, \mathbb{Q})$ and $\beta_2 \in H^*(\mathfrak{M}_{\mathfrak{g}_2, \mathfrak{k}_2+1}, \mathbb{Q})$, we represent them through the Poincare duality by rational cycles $\sum_i a_{1i} K_{1i}$ and $\sum_j a_{2j} K_{2j}$, respectively, where $a_{1i}, a_{2j} \in \mathbb{Q}$ and K_{1i} (resp. K_{2j}) are integral cycles in $\mathfrak{M}_{\mathfrak{g}_1, \mathfrak{k}_1+1}$ (resp. $\mathfrak{M}_{\mathfrak{g}_2, \mathfrak{k}_2+1}$), then $\mathbf{i}_!(\beta_1, \beta_2)$ is the Poincare dual of the homology class represented by the rational cycle $\sum_{i,j} a_{1i} a_{2j} \mathbf{i}_S(K_{1i}, K_{2i})$ in $\mathfrak{M}_{\mathfrak{g}, \mathfrak{k}}$.

Similarly, one can define a homomorphism

$$\mathbf{i}_! : H^*(\mathfrak{M}_{\mathfrak{g}-1, \mathfrak{k}+2}, \mathbb{Q}) \mapsto \mathfrak{H}^*(\mathfrak{M}_{\mathfrak{g}, \mathfrak{k}}, \mathbb{Q})$$

by using the map $\mathbf{i}_0$.

Now we state the composition law, which consists of two formulas.

Theorem 2.7. *Let (X, ω) be a compact symplectic manifold of complex dimension n. Let $\alpha_1, \cdots, \alpha_k$ be in $H_*(X, \mathbb{Q})$.*
Then for any $\beta_1 \in H^*(\mathfrak{M}_{\mathfrak{g}_1, \mathfrak{k}_1+1}, \mathbb{Q})$, $\beta_2 \in H^*(\mathfrak{M}_{\mathfrak{g}_2, \mathfrak{k}_2+1}, \mathbb{Q})$, we have

$$\psi^X_{A,g,k}(\mathbf{i}_!(\beta_1, \beta_2), (\alpha_i)) =$$

$$= \sum_{A=A_1+A_2} \sum_i \epsilon(S) \psi^X_{A_1, g_1, k_1+1}(\beta_1, (\alpha_i)_{i \in S_1}, e_i) \psi^X_{(A_2, g_2, k_2+1)}(\beta_2, e_i^*, (\alpha_j)_{j \in S_2}),$$

and for any $\beta_0 \in H^*(\mathfrak{M}_{\mathfrak{g}-1, \mathfrak{k}+2}, \mathbb{Q})$, we have

$$\psi^X_{A,g,k}(\mathbf{i}_!(\beta_0), (\alpha_i)) = \sum_i \psi^X_{(A, g-1, k+2)}(\beta_0, (\alpha_i), e_i, e_i^*),$$

where $\epsilon(S)$ is the sign of permutation $S = S_1 \cup S_2$ of $\{1, \cdots, k\}$, $\{e_i\}$ is a basis of $H^*(X, \mathbb{Q})$ and $\{e_i^*\}$ is its dual basis.

We will refer the readers to [RT2] for its proof. [15]

2.7 Rational GW-invariants for projective spaces

In this subsection, we let X be any complex projective space with standard Kähler form ω and complex structure J. All stable maps in this subsection are assumed to be of genus 0.

Lemma 2.3. *Let $(f, \Sigma, (x_i))$ be any J-holomorphic stable map of genus 0. Then $H^1(\Sigma, f^*T^{1,0}X) = 0$, where $T^{1,0}X$ denotes the holomorphic tangent bundle of X.*

Notice that $T^{1,0}X$ is a positive bundle and each component of Σ is $\mathbb{C}P^1$. This implies that $f^*T^{1,0}X$ restricts to a sum of nonnegative line bundles on any irreducible component of Σ, so lemma follows from direct computations or the standard vanishing theorem.

Since J is an integrable complex structure, one can identify $H^1(\Sigma, f^*T^{1,0}X)$ with the cokernel of the linearization L_f of the Cauchy-Riemann operator at f. Then it follows Corollary 1.1 that $\mathrm{ev}(\mathfrak{M}^{\mathfrak{X}}_{\mathfrak{A},\mathfrak{o},\mathfrak{k}})$ represents the Euler class $e_{A,0,k}(X)$ in $H_r(\mathfrak{M}_{\mathfrak{o},\mathfrak{k}} \times \mathfrak{X}^{\mathfrak{k}}, \mathbb{Q})$, where r is equal

$$2c_1(X)(A) + 2n + 2k - 6$$

and n is the complex dimension of X. In particular, we have

$$\psi^X_{A,0,k}(1,(\alpha_i)) = \int_{\mathfrak{M}^{\mathfrak{X}}_{\mathfrak{A},\mathfrak{o},\mathfrak{k}}} \mathrm{ev}^*(\pi_2^*\alpha_1 \wedge \cdots \wedge \pi_{k+1}^*\alpha_k).$$

If $n = 2$, $H_2(\mathbb{C}P^2, \mathbb{Z}) = \mathbb{Z}$. We can write $A = d[\ell]$, where ℓ is any complex line. Then we can write $\psi^X_{d,0,k}$ for $\psi^X_{A,0,k}$. If further the Poincare dual of each α_i can be represented by any point in $\mathbb{C}P^2$, then $\psi^{\mathbb{C}P^2}_{d,0,k}(1,(\alpha_i)) = 0$ whenever $k \neq 3d-1$ and otherwise $\psi^{\mathbb{C}P^2}_{d,0,k}(1,(\alpha_i))$ is the number of rational curves through $3d-1$ points in general position. This example shows that the GW-invariants are the generalization of classical enumerative invariants.

3. Some simple applications

The GW-invariants have been applied to many other branches of mathematics, such as enumerative algebraic geometry, quantum cohomology, mirror symmetry, Hamiltonian systems and symplectic topology. Because of time and space, we can not cover all these applications. Here we will give two applications briefly: (1) Construct the quantum cohomology for general symplectic manifolds; (2) Use GW-invariants to show that there are differential manifolds which admit infinitely many different symplectic structures.

[15] This theorem was proved in [RT2] for semi-positive symplectic manifolds. However, the arguments can be easily modified to give a proof in case of general symplectic manifolds.

3.1 Quantum cohomology

Let (X,ω) be a compact symplectic manifold. The quantum cohomology ring of X is the cohomology $H^*(X,\mathbb{Q}\{H_2(X)\})$ with a new ring structure defined by GW-invariants. Here $\mathbb{Q}\{H_2(X)\}$ denotes the Novikov ring. It first appeared in Novikov's study of the Morse theory for multivalued functions (cf. [No]). It can be defined as follows (cf. [HS], [MS], [RT1]): choose a basis $q_1,\cdots,q_s$ of $H_2(X,\mathbb{Z})$, we identify the monomial $q^d = q_1^{d_1}\cdots q_s^{d_s}$ with the sum $\sum_{i=1}^s d_i q_i$. This turns $H_2(X)$ into a multiplicative ring, that is $q^d\cdot q^{d'} = q^{d+d'}$. This multiplicative ring has a natural grading defined by $\deg(q^d) = 2c_1(X)(\sum d_i q_i)$. Then $\mathbb{Q}\{H_2(X)\}$ is the graded homogeneous ring generated by all formal power series $\sum_{d=(d_1,\cdots,d_s)} n_d q^d$ satisfying: $n_d\in\mathbb{Q}$, all q^d with $n_d\neq 0$ have the same degree and the number of n_d with $\omega(\sum d_i q_i)\le c$ is finite for any $c>0$. If X is a Fano manifold or a monotone symplectic manifold, then $\mathbb{Q}\{H_2(X)\}$ is just a group ring.

Now we can define a ring structure on $H^*(X,\mathbb{Q}\{H_2(X)\})$. For any α, β in $H^*(X,\mathbb{Q})$, we define the quantum multiplication $\alpha\bullet\beta$ by

$$(\alpha\bullet\beta,\gamma) = \sum_{A\in H_2(X,\mathbb{Q})} \psi^X_{A,0,3}(\alpha,\beta,\gamma)q^A \tag{3.1}$$

where $\gamma\in H^*(X,\mathbb{Q})$ and $(\cdot,\cdot)$ denotes the cup product. Equivalently, if $\{e_i\}$ is a basis of $H^*(X,\mathbb{Q})$ with dual basis $\{e_i^*\}$, then

$$\alpha\bullet\beta = \sum_A\sum_i \psi^X_{A,0,3}(\alpha,\beta,e_i)\, e_i^*\, q^A \tag{3.2}$$

Note that if $A=\sum a_i q_i$, we identify A with $(a_1,\cdots,a_s)$. In general, any α, β in $H^*(X,\mathbb{Q}\{H_2(X)\})$ can be written as

$$\alpha = \sum_d \alpha_d\, q^d,\quad \beta = \sum \beta_{d'}\, q^{d'}$$

where α_d, $\beta_{d'}$ are in $H^*(X,\mathbb{Q})$. We define

$$\alpha\bullet\beta = \sum_{d,d'} \alpha_d\bullet\beta_{d'}\, q^{d+d'} \tag{3.3}$$

Recall that the degree of $\alpha_d\, q^d$ is $\deg(\alpha)+\deg(q^d)$. It follows that the multiplication preserves the degree. However, it is not clear at all if the multiplication is associative. Given α, β, γ, δ in $H^*(X,\mathbb{Z})$, we have

$$\begin{aligned}((\alpha\bullet\beta)\bullet\gamma,\delta) &= \textstyle\sum_{A,B}\sum_i \psi^X_{A,0,3}(\alpha,\beta,e_i)\psi^X_{B,0,3}(e_i^*,\gamma,\delta)\\ (\alpha\bullet(\beta\bullet\gamma),\delta) &= \textstyle\sum_{A,B}\sum_i \psi^X_{A,0,3}(\alpha,e_i,\delta)\psi^X_{B,0,3}(\beta,\gamma,e_i^*)\end{aligned}$$

So the associativity means that for any fixed A in $H_2(X,\mathbb{Q})$ we have (up to sign)

$$\begin{array}{rl} & \sum_{A_1+A_2=A}\sum_i \psi^X_{A_1,0,3}(\alpha,\beta,e_i)\psi^X_{(A_2,0,3)}(e_i^*,\gamma,\delta) \\ = & \sum_{A_1+A_2=A}\sum_i \psi^X_{A_1,0,3}(\alpha,e_i,\delta)\psi^X_{A_2,0,3}(\beta,\gamma,e_i^*) \end{array}$$

But by the composition law of last section, we see that up to sign, both sides of (3.1) are equal to

$$\psi^X_{A,0,4}(PD(p),\alpha,\beta,\gamma,\delta),$$

where $PD(p)$ denotes the Poincare dual of the homology class of any point p in $\mathfrak{M}_{0,\mathfrak{k}}$. Therefore, we have

Theorem 3.1. *The quantum multiplication $\bullet$ is associative, consequently, there is an associative, supercommutative, graded ring structure, i.e., quantum ring structure, on $H_*(X,\mathbb{Q}\{H_2(X)\})$.*

In physics and sometimes mathematical literatures, one substitutes q by $(e^{-t\omega(q_1)},\cdots,e^{-t\omega(q_s)})$, so the quantum product becomes

$$\alpha\bullet\beta=\sum_A \psi^X_{A,0,3}(\alpha,\beta,e_i)e_i^* e^{-t\omega(A)} \tag{5.5}$$

In particular, this converges to the classical cup product as $t\to\infty$. If $c_1(X)>0$, then

$$\alpha\bullet\beta=\alpha\cup\beta+\sum_{c_1(X)(A)>0}\phi_A$$

where ϕ_A has degree $\deg(\alpha\cup\beta)-2c_1(X)(A)$.

Example 3.1. The quantum cohomology of the Grassmannian $G(r,n)$ was computed in [ST], [W12]. Let S be the tautological bundle over $G(r,n)$ of complex k-planes in $\mathbb{C}^n$. It is known that $H^*(G(r,n),\mathbb{Q})$ is given by

$$\frac{\mathbb{Q}[x_1,\cdots,x_r]}{\{s_{n-r+1},\cdots,s_n\}}$$

where s_j are Segre classes, defined inductively by

$$s_j=-x_1s_{j-1}-\cdots-x_{j-1}s_1-x_j.$$

In fact, x_i corresponds to the i-th Chern class $c_i(S)$ $(i=1,\cdots,r)$. It can be shown that

$$H^*(G(r,n),\mathbb{Q}\{H_2(G(r,n))\})=\frac{\mathbb{Q}[x_1,\cdots,x_r,q]}{\{s_{n-r+1},\cdots,s_{n-1},s_n+(-1)^rq\}}.$$

More examples of computing quantum cohomology can be found in [Ba], [GK], [Be], [CM], [KM], [CF], [Lu].

In fact, there is a family of new quantum multiplications, containing the above $\bullet$ as a special case.

Any $w\in H^*(X,\mathbb{Q})$ can be written as $\sum t_ie_i$. Clearly, $w\in H^*(X,\mathbb{Q})$ if all t_i are rational. We define the quantum multiplication $\bullet_w$ by

$$(\alpha \bullet_w \beta, \gamma) \tag{6.1}$$
$$= \sum_A \sum_{k\geq 0} \frac{\epsilon(\{a_i\})}{k!} \psi^X_{A,0,k+3}(\alpha, \beta, \gamma, e_{i_1}, \cdots, e_{i_k}) t_{i_1} \cdots t_{i_k} q^A$$

where $\alpha, \beta, \gamma \in H_*(X, \mathbb{Q})$, and $\epsilon(\{a_i\})$ is the sign of the induced permutation on odd dimensional e_i. Obviously, this multiplication reduces to $\bullet$ at $w = 0$. As we argued in the above, the associativity of $\bullet_w$ is equivalent to the so called WDVV equation. We refer the readers to [RT1] or [Ti] for more details.

Theorem 3.2. *Each quantum multiplications $\bullet_w$ is associative.*

3.2 Examples of symplectic manifolds

One application of the GW-invariants is to distinguish nondeformation equivalent symplectic manifolds. In this subsection, we will use GW-invariants to solve a special case of the following stablizing conjecture, which is due to Ruan.

Conjecture 3.1. Suppose that X and Y are two homeomorphic symplectic 4-manifolds. Then X and Y are diffeomorphic if and only if the stablized manifolds $X \times \mathbb{C}P^1$ and $Y \times \mathbb{C}P^1$ with the product symplectic structures are deformation equivalent.

It follows from a result of M. Freedman that two 4-manifolds X and Y are homeomorphic if and only if $X \times \mathbb{C}P^1$ and $Y \times \mathbb{C}P^1$ are diffeomorphic. The stablizing conjecture can be viewed as an analogy of this between the smooth and the symplectic category. The first pair of examples supporting the conjecture were constructed by Ruan in [Ru1], where X is the blow-up of $\mathbb{C}P^2$ at 8-points and Y is a Barlow surface. Furthermore, Ruan also verified the conjecture for the cases: (1) X is rational, Y is irrational; (2) X and Y are irrational but have different number of (-1) curves. In the following, we will compute certain genus one GW-invariants and prove the stablizing conjecture for simply connected elliptic surfaces $E^n_{p,q}$. This is due to Ruan and myself in [RT2].

Let's recall the construction of $E^n_{p,q}$. Let E^1 be the blow-up of $\mathbb{C}P^2$ at generic 9 points, and let E^n be the fiber connected sum of n copies of E^1. Then $E^n_{p,q}$ can be obtained from E^n by logarithmic transformations alone two smooth fibers with multiplicity p and q. Note that $E^n_{p,q}$ is simply connected if and only if p, q are coprime. Moreover, the Euler number $\chi(E^n_{p,q}) = 12n$, and hence n is a topological number.

Theorem 3.3. *Manifolds $E^n_{p,q} \times \mathbb{C}P^1$ and $E^n_{p',q'} \times \mathbb{C}P^1$ with product symplectic structures are symplectic deformation equivalent if and only if $(p,q) = (p',q')$.*

Combining with known results about the smooth classification of $E^n_{p,q}$ (cf. [FM]), we can prove

Corollary 3.1. *The stablizing conjecture holds for $E^n_{p,q}$.*

Let F_p, F_q be two multiple fibers and F be a general fiber. Let

$$A_p = [F_p], \quad A_q = [F_q].$$

Then

$$A_p = \frac{[F]}{p}, \quad A_q = \frac{[F]}{q}.$$

The primitive class is $A = [F]/pq$. Another piece of topological information is the canonical class K is Poincare dual to

$$(n-2)F + (p-1)F_p + (q-1)F_q = ((n-2)pq + (p-1)p + (q-1)q)A. \quad (3.4)$$

Then Theorem 3.3 follows from the following proposition.

Proposition 3.1. *We have*

$$\psi^{E^n_{p,q}\times \mathbb{C}P^1}_{mA,1,1}(1,\alpha) = \begin{cases} 2q\alpha(A); & m = q (mA = A_p), \\ 2p\alpha(A); & m = p (mA = A_q), \\ 0; & m \neq 0 \bmod p \text{ or } q \quad \textit{and} \quad m < pq, \end{cases} \quad (3.7)$$

where α is a 2-dimensional cohomology class.

Proof. We will outline its proof here. By the deformation theory of elliptic surfaces, we can choose a complex structure J_0 on $E^n_{p,q}$ such that all singular fibers are nodal elliptic curves. Furthermore, we can assume that the complex structures of multiple fibers are generic, i.e., whose j-invariants are neither 0 nor 1728. Let $\mathbf{j}_0$ be the standard complex structure on $\mathbb{C}P^1$. Then $X = E^n_{p,q} \times \mathbb{C}P^1$ has the product complex structure $J_0 \times \mathbf{j}_0$.

Let's describe $\mathfrak{M}^{\mathfrak{X}}_{m\mathfrak{A},1,1}$ for $m < pq$ and $m \neq 0 \bmod p$ or q. For any $f \in \mathfrak{M}^{\mathfrak{X}}_{m\mathfrak{A},1,1}$, its image $\mathrm{Im}(f)$ is a connected effective holomorphic curve. Write

$$\mathrm{Im}(f) = \sum_i a_i C_i,$$

where $a_i > 0$ and C_i are irreducibe. Since

$$mA = \sum_i a_i [C_i],$$

each C_i is of the form $C^1_i \times \{t_i\}$, where C^1_1 is a holomorphic curve in $E^n_{p,q}$ and t_i is a point in $\mathbb{C}P^1$. However, $\mathrm{Im}(f)$ is connected, so all x_i coincide and consequently, we can write

$$\mathrm{Im}(f) = (\sum a_i C^1_i) \times \{x\}, \quad (3.5)$$

where $\sum a_i C^1_i$ is a connected effective curve in $E^n_{p,q}$. By our assumption on singular fibers, each C^1_i is either a multi-section or a fiber. A multi-section has positive intersection with a general fiber. A fiber has zero intersection with a general fiber. It follows from $mA \cdot [F] = 0$ that each C_i^1 is a fiber. Since $m < pq$, each C_i^1 can only be a multiple fiber. Because of connectedness, $\mathrm{Im}(f)$ is either $F_p \times \{x\}$ or $F_q \times \{x\}$. In particular,

$$\mathfrak{M}^X_{mA,1,1} = \emptyset \text{ for } m \neq 0 \bmod p \text{ or } q \text{ and } m < pq. \tag{3.6}$$

Hence,

$$\psi^X_{mA,1,1}(1, \alpha) = 0 \text{ for } m \neq 0 \bmod p \text{ or } q \text{ and } m < pq. \tag{3.7}$$

Now assume that $m = q$, so $mA = A_p$. A straightforward computation shows

$$\mathfrak{M}^X_{A_p,1,1} = Aut(F_p) \times \mathbb{C}P^1. \tag{3.8}$$

This moduli space is not in general position since it is of dimension 4 while the expected dimension should be 2.

Let $\mathbf{f} \in \mathfrak{M}^X_{A_p,1,1}$ and f be corresponding holomorphic map, we define its normal bundle N_f to be

$$f^*TX/T(F_p \times \{x\}),$$

where $\mathrm{Im}(f) = F_p \times \{x\}$. One can easily show

$$N_f = N_{F_p}(E^n_{p,q}) \otimes T_x\mathbb{C}P^1.$$

It is clear that

$$H^1(N_{F_p}(E^n_{p,q}), \mathbb{C}) = \mathbb{C}.$$

Hence, we have

$$H^1(N_f) = T_x\mathbb{C}P^1$$

and consequently, the cokernel of L_f is $T_x\mathbb{C}P^1$, where L_f is the linearization of the Cauchy-Riemann equation at f. Putting together all these $\mathrm{Coker}(L_f)$, we obtain an obstruction bundle $\mathbf{ob} = \pi_2^* T\mathbb{C}P^1$, where π_2 denotes the projection from X onto $\mathbb{C}P^1$.

One can show by arguments in section 2 that $e_{A_p,1,1}(X)$ is simply the image of the Euler class of $\mathbf{ob}$ over $\mathfrak{M}^X_{A_p,1,1}$ under the evaluation map ev. So

$$e_{A_p,1,1}(X) = 2[F_p]$$

and consequently,

$$\psi^X_{A_p,1,1}(1, \alpha) = 2 \int_{F_p} \alpha = 2q\alpha(A). \tag{3.9}$$

The same arguments yield that

$$\Psi^X_{(A_q,1,1)}(1, \alpha) = 2p(A_0 \cdot \alpha). \tag{3.10}$$

The proposition is proved.

Now we complete the proof of Theorem 3.3. First of all, if $(p, q) = (p', q')$, $E^n_{p,q}$ and $E^n_{p',q'}$ were known to be complex deformation equivalent as Kähler surfaces regardless where we perform the logarithmic transform. It follows that $E^n_{p,q} \times \mathbb{C}P^1$ and $E^n_{p',q'} \times \mathbb{C}P^1$ with product symplectic structures are deformation equivalent.

Conversely, suppose that $E^n_{p,q} \times \mathbb{C}P^1$ and $E^n_{p',q'} \times \mathbb{C}P^1$ are symplectic deformation equivalent. Then there is a diffeomorphism

$$F : E^n_{p,q} \times \mathbb{C}P^1 \to E^n_{p',q'} \times \mathbb{C}P^1 \tag{3.11}$$

such that

$$\psi^{E^n_{p',q'} \times \mathbb{C}P^1}_{F_*(A),1,1}(1, \alpha) = \psi^{E^n_{p,q} \times \mathbb{C}P^1}_{A_q,1,1}(1, F^*(\alpha)), \tag{3.12}$$

and

$$F^* c_i(E^n_{p',q'} \times \mathbb{C}P^1) = c_i(E^n_{p,q} \times \mathbb{C}P^1) \tag{3.13}$$

and

$$F^* p_1(E^n_{p',q'} \times \mathbb{C}P^1) = p_1(E^n_{p,q} \times \mathbb{C}P^1) \tag{3.14}$$

Let $e_0 \in H^2(\mathbb{C}P^1, \mathbb{Z})$ be the positive generator. First, we claim

$$F^*(e_0) = e_0. \tag{3.15}$$

Suppose that $F^*(e_0) = ne_0 + \beta$ for some $\beta \in H^2(E^n_{p,q}, \mathbb{Z})$. Note that the first Pontrjagan class

$$p_1(E^n_{p,q} \times \mathbb{C}P^1) = p_1(E^n_{p,q}) \neq 0$$

and

$$p_1(E^n_{p',q'} \times \mathbb{C}P^1) = p_1(E^n_{p'q'}) \neq 0.$$

Let $\gamma(E^n_{p,q}) \in H^4(E^n_{p,q}, \mathbb{Z})$ be such that

$$\gamma(E^n_{p,q})[E^n_{p,q}] = 1.$$

Define $\gamma(E^n_{p',q'})$ in the same way. Then $p_1(E^n_{p,q})$ is a nonzero multiple of $\gamma(E^n_{p,q})$ and $p_1(E^n_{p',q'})$ is a non-zero multiple of $\gamma(E^n_{p',q'})$. Thus

$$F^*\gamma(E^n_{p',q'}) = \gamma(E^n_{p,q}).$$

Then,

$$\begin{aligned} 1 = &\ (\gamma(E^n_{p',q'}) \cup e_0)[E^n_{p',q'} \times \mathbb{C}P^1] \\ = &\ F^*(\gamma(E^n_{p',q'}) \cup e_0)[E^n_{p,q} \times \mathbb{C}P^1] \\ = &\ \gamma(E^n_{p,q}) \cup (ne_0 + \beta)[E^n_{p,q} \times \mathbb{C}P^1] \\ = &\ n. \end{aligned}$$

Hence $n = 1$. Furthermore, $F^*(e_0^2) = 0$. Then

$$(e_0 + \beta)^2 = 2e_0\beta + \beta^2 = 0.$$

Therefore, $2e_0\beta = 0$ and $\beta^2 = 0$, consequently, $\beta = 0$.

$$c_1(E^n_{p,q} \times \mathbb{C}P^1) = c_1(E^n_{p,q}) + 2e_0.$$

By (3.13), (3.14) and (3.15),

$$F^*(c_1(E^n_{p',q'})) = c_1(E^n_{p,q}).$$

However, F sends primitive classes to primitive classes, so $F_*(A) = A$ and

$$\psi^X_{F_*(nA),1,1}(1,\alpha) = \psi^Y_{nA,1,1}(1,\alpha).$$

where $X = E^n_{p',q'} \times \mathbb{C}P^1$ and $Y = E^n_{p',q'} \times \mathbb{C}P^1$. Suppose that $q < p$ and $q' < p'$. Then $A_p (= qA)$ and $A_q (= pA)$ are the first and the second class of $\{nA\}$ such that ψ is nonzero and so are $A_{p'}$ and $A_{q'}$. Hence

$$F_*(A_p) = A_{p'}, F_*(A_q) = A_{q'}.$$

This implies that

$$p = p', q = q'.$$

We finish the proof of Theorem 3.3.

Bibliography

[Ba] V. Batyrev, *Dual polyhedra and mirror symmetry for Calabi-Yau hypersurfaces in toric varieties,* J. of Alg. Geom., **3**(1994), no. 3, 493-535.

[Bea] A. Beauville, *Quantum cohomology of complete intersections,* Mat. Fizika, Analiz, Geometriya, **2**(1995), no. 3-4, 384-398.

[Beh] K. Behrend, *Gromov-Witten invariants in alebraic geometry,* Invent. Math. **127**(1997), no. 3, 601-617.

[BF] K. Behrend and B. Fantechi, *The intrinsic normal cone,* Invent. Math. **128**(1997), no. 1, 45-88.

[CF] I. Ciocan-Fontanine, *Quantum cohomology of Flag varieties,* IMRN (1995), no. 6, 263-277.

[CM] B. Crauder and R. Miranda, *Quantum cohomology of rational surfaces,* The moduli space of curves (Texel Island, 1994), 33-80, Birkhäuser.

[DM] P. Deligne and D. Mumford, *The irreducibility of the space of curves of given genus,* Publ. I.H.E.S. **45**(1969), 101-145.

[FM] R. Friedman and J. Morgan, *On the diffeomorphism types of certain algebraic surfaces I, II,* J. Diff. Geom., **27**(1988).

[FO] K. Fukaya and K. Ono, *Arnold conjecture and Gromov-Witten invariants,* preprint, 1996.

[GK] A. Givental and B. Kim, *Quantum cohomology of flag manifolds and Toda lattices,* preprint, 1993.

[Gr] M. Gromov, *Pseudoholomorphic curves in symplectic manifolds,* Invent. Math. **82**(1985), no. 2, 307-347.

[HS] H. Hofer and D. Salamon, *Floer homology and Novikov rings,* The Floer memorial volume, 483-524, Progress in Mathematics, Birkhäuser.

[KM] M. Kontsevich and Y. Manin, *Gromov-Witten classes, quantum cohomology, and enumerative geometry,* Comm. Math. Phys. **164**(1994), no. 3, 525-562.

[LT1] J. Li and G. Tian *Virtual moduli cycles and Gromov-Witten invariants of algebraic varieties,* J. Amer. Math. **11**(1998), no. 1, 119-174.

[LT2] J. Li and G. Tian, *Virtual moduli cycles and Gromov-Witten invariants of general symplectic manifolds*, Topics in symplectic 4-manifolds (Irvine, CA, 1996), 47–83, First Int. Press Lect. Ser., I, Internat. Press, Cambridge, MA, 199.

[LT3] J. Li and G. Tian, in preparation.

[Lu] P. Lu, *A rigorous definition of fiberwise quantum cohomology and equivariant quantum cohomology,* Comm. in Analysis and Geom., **6**(1998), no. 3, 511-588.

[MS] D. McDuff and D. Salamon, *J-holomorphic curves and quantum cohomology,* University Lec. Series, **6**, AMS.

[No] S. Novikov, *Multivalued functions and functionals - analogue of the Morse theory,* Soviet Math. Dokl. **24**(1981).

[PW] T. Parker and J. Wolfson, *Pseudo-holomorphic maps and bubble trees*, J. Geom. Anal. **3**(1993), no. 1, 63-98.

[Ru1] Y. Ruan, *Topological Sigma model and Donaldson type invariants in Gromov theory,* Duke Math. J. **83**(1996), no. 2, 461–500.

[Ru2] Y. Ruan, *Virtual neighborhoods and pseudo-holomorphic curves*, preprint, 1996.

[RT1] Y.B. Ruan and G. Tian, *A mathematical theory of quantum cohomology,* J. Diff. Geom. **42**(1995), no. 2, 259-367.

[RT2] Y.B. Ruan and G. Tian, *Higher genus symplectic invariants and sigma model coupled with gravity,* Invent. Math. **130**(1997), no. 3, 455–516.

[Si] B. Siebert, *Gromov-Witten invariants for general symplectic manifolds*, preprint, 1996, (AG/9608005).

[ST] B. Siebert and G. Tian, *On quantum cohomology rings of Fano manifolds and a formula of Vafa and Intriligator,* The Asian J. of Mathematics, **1**(1997), no. 4, 679=695.

[SU] J. Sacks and K. Uhlenbeck, *The existence of minimal immersions of 2 spheres,* Ann. of Math. **113**(1981), 1-24.

[Ti] G. Tian, *Quantum cohomology and its associativity,* Proceeding of 1st CDM conference in Cambridge, 1995.

[Wi1] E. Witten, *Two dimensional gravity and intersection theory on moduli space,* Surveys in differential geometry (Cambridge, MA, 1990), 243–310, Lehigh Univ., Bethlehem, PA, 1991.

[Wi2] E. Witten, *The Verlinde algebra and the cohomology of the Grassmannian,* Lecture Notes in Geom. and Topology, Intern. Press.

[Ye] R. Ye, *Gromov's compactness theorem for pseudo-holomorphic curves,* Trans. Amer. math. Soc., 1994.

C.I.M.E. Session on "Quantum Cohomology"

List of Participants

V. APOSTOLOV, Ecole Polytechnique, Palaiseau, France
M. BERTOLA, SISSA, Via Beirut 2-4, 34014 Trieste, Italia
R. BRUSSER, Fakultät für Mathemätik, Univ. Bielefeld, Postfach 100131, D-33501 Bielefeld, Germany
P. CARESSA, Dip. di Matematica U.Dini, Viale Morgagni 67/a, 50134 Firenze, Italia
L. DABROWSKI, SISSA, Via Beirut 2-4, 34014 Trieste, Italia
F. DELL'ACCIO, Dip. di Matematica, Univ. della Calabria, 87036 Arcavacata di Rende (CS), Italia
B. FANTECHI, Dip. di Matematica, Univ. di Trento, Via Sommarive 14, 38050 Povo, Trento, Italia
A. GHIGI, Via della Fonderia 35, 50142 Firenze, Italia
L. GOETTSCHE, Mittag-Leffler-Institut, Auravaegen 17, 18262 Djursholm, Sweden
J. HILL, Yale University, 242 Prospect Street 5, New Haven, CT 06511, USA
M. MANETTI, Scuola Normale Superiore, Piazza dei Cavalieri 7, 56126 Pisa, Italia
V. MARINO, Dip. di Matematica, Univ. della Calabria, 87036 Arcavacata di Rende (CS), Italia
L. MIGLIORINI, Dip. di Matematica Applicata, Via S. Marta 3, 50139 Firenze, Italia
V. MUNOZ, Dept. of Alg., Geom. and Top., Universidad de Malaga, 29071 Malaga, Spain
P. OLIVERIO, Dip. di Matematica, Univ. della Calabria, 87036 Arcavacata di Rende (CS), Italia
T. PACINI, Dip. di Matematica, Via Buonarroti 2, 56127 Pisa, Italia
R. PAOLETTI, Dip. di Matematica, Univ. de L'Aquila, Via Vetoio, 67010 Coppito (AQ), Italia
P. PIAZZA, Dip. di Matematica, Univ. "La Sapienza", P.le A. Moro 2, 00185 Roma, Italia
R. RE, Dip. di Matematica, Cittadella Universitaria, Viale A.Doria 6, 95125 Catania, Italia
V. SHEVCHISHIN, Fakultät f. Mathematik, Ruhr-Universität Böchum, D-44780 Bochum, Germany
X. SUN, ICTP, Mathematics Section, P.O.Box 586, 34100 Trieste, Italia
F. THOMPSON, SISSA, Via Beirut 2-4, 34014 Trieste, Italia
S. TRAPANI, Dip. di Matematica, Univ. di Tor Vergata, Via della Ricerca Scientifica, 00161 Roma, Italia
D. VISETTI, Via Avigliana 13/6, 10138 Torino, Italia
A. ZAMPA, SISSA, Via Beirut 2-4, 34014 Trieste, Italia
I. ZHARKOV, Dept. of Math., Univ. of Pennsylvania, DRL 209 S. 33 St., Philadelphia, PA 19104, USA

LIST OF C.I.M.E. SEMINARS

1954	1.	Analisi funzionale	C.I.M.E
	2.	Quadratura delle superficie e questioni connesse	"
	3.	Equazioni differenziali non lineari	"
1955	4.	Teorema di Riemann-Roch e questioni connesse	"
	5.	Teoria dei numeri	"
	6.	Topologia	"
	7.	Teorie non linearizzate in elasticità, idrodinamica, aerodinamic	"
	8.	Geometria proiettivo-differenziale	"
1956	9.	Equazioni alle derivate parziali a caratteristiche reali	"
	10.	Propagazione delle onde elettromagnetiche	"
	11.	Teoria della funzioni di più variabili complesse e delle funzioni automorfe	"
1957	12.	Geometria aritmetica e algebrica (2 vol.)	"
	13.	Integrali singolari e questioni connesse	"
	14.	Teoria della turbolenza (2 vol.)	"
1958	15.	Vedute e problemi attuali in relatività generale	"
	16.	Problemi di geometria differenziale in grande	"
	17.	Il principio di minimo e le sue applicazioni alle equazioni funzionali	"
1959	18.	Induzione e statistica	"
	19.	Teoria algebrica dei meccanismi automatici (2 vol.)	"
	20.	Gruppi, anelli di Lie e teoria della coomologia	"
1960	21.	Sistemi dinamici e teoremi ergodici	"
	22.	Forme differenziali e loro integrali	"
1961	23.	Geometria del calcolo delle variazioni (2 vol.)	"
	24.	Teoria delle distribuzioni	"
	25.	Onde superficiali	"
1962	26.	Topologia differenziale	"
	27.	Autovalori e autosoluzioni	"
	28.	Magnetofluidodinamica	"
1963	29.	Equazioni differenziali astratte	"
	30.	Funzioni e varietà complesse	"
	31.	Proprietà di media e teoremi di confronto in Fisica Matematica	"
1964	32.	Relatività generale	"
	33.	Dinamica dei gas rarefatti	"
	34.	Alcune questioni di analisi numerica	"
	35.	Equazioni differenziali non lineari	"
1965	36.	Non-linear continuum theories	"
	37.	Some aspects of ring theory	"
	38.	Mathematical optimization in economics	"

1966	39. Calculus of variations	Ed. Cremonese, Firenze
	40. Economia matematica	"
	41. Classi caratteristiche e questioni connesse	"
	42. Some aspects of diffusion theory	"
1967	43. Modern questions of celestial mechanics	"
	44. Numerical analysis of partial differential equations	"
	45. Geometry of homogeneous bounded domains	"
1968	46. Controllability and observability	"
	47. Pseudo-differential operators	"
	48. Aspects of mathematical logic	"
1969	49. Potential theory	"
	50. Non-linear continuum theories in mechanics and physics and their applications	"
	51. Questions of algebraic varieties	"
1970	52. Relativistic fluid dynamics	"
	53. Theory of group representations and Fourier analysis	"
	54. Functional equations and inequalities	"
	55. Problems in non-linear analysis	"
1971	56. Stereodynamics	"
	57. Constructive aspects of functional analysis (2 vol.)	»
	58. Categories and commutative algebra	»
1972	59. Non-linear mechanics	"
	60. Finite geometric structures and their applications	"
	61. Geometric measure theory and minimal surfaces	"
1973	62. Complex analysis	"
	63. New variational techniques in mathematical physics	"
	64. Spectral analysis	"
1974	65. Stability problems	"
	66. Singularities of analytic spaces	"
	67. Eigenvalues of non linear problems	"
1975	68. Theoretical computer sciences	"
	69. Model theory and applications	"
	70. Differential operators and manifolds	"
1976	71. Statistical Mechanics	Ed. Liguori, Napoli
	72. Hyperbolicity	"
	73. Differential topology	"
1977	74. Materials with memory	"
	75. Pseudodifferential operators with applications	"
	76. Algebraic surfaces	"
1978	77. Stochastic differential equations	"
	78. Dynamical systems	Ed. Liguori, Napoli & Birkhäuser
1979	79. Recursion theory and computational complexity	"
	80. Mathematics of biology	"

Year	No.	Title	LNM	Publisher
1980	81.	Wave propagation		Ed. Liguori, Napoli & Birkhäuser
	82.	Harmonic analysis and group representations		"
	83.	Matroid theory and its applications		»
1981	84.	Kinetic Theories and the Boltzmann Equation	(LNM 1048)	Springer-Verlag
	85.	Algebraic Threefolds	(LNM 947)	"
	86.	Nonlinear Filtering and Stochastic Control	(LNM 972)	"
1982	87.	Invariant Theory (LNM 996)		"
	88.	Thermodynamics and Constitutive Equations	(LN Physics 228)	"
	89.	Fluid Dynamics	(LNM 1047)	"
1983	90.	Complete Intersections	(LNM 1092)	"
	91.	Bifurcation Theory and Applications	(LNM 1057)	"
	92.	Numerical Methods in Fluid Dynamics	(LNM 1127)	"
1984	93.	Harmonic Mappings and Minimal Immersions	(LNM 1161)	"
	94.	Schrödinger Operators	(LNM 1159)	"
	95.	Buildings and the Geometry of Diagrams	(LNM 1181)	"
1985	96.	Probability and Analysis	(LNM 1206)	"
	97.	Some Problems in Nonlinear Diffusion	(LNM 1224)	"
	98.	Theory of Moduli	(LNM 1337)	"
1986	99.	Inverse Problems	(LNM 1225)	"
	100.	Mathematical Economics	(LNM 1330)	"
	101.	Combinatorial Optimization	(LNM 1403)	"
1987	102.	Relativistic Fluid Dynamics	(LNM 1385)	»
	103.	Topics in Calculus of Variations	(LNM 1365)	»
1988	104.	Logic and Computer Science	(LNM 1429)	"
	105.	Global Geometry and Mathematical Physics	(LNM 1451)	"
1989	106.	Methods of nonconvex analysis	(LNM 1446)	"
	107.	Microlocal Analysis and Applications	(LNM 1495)	"
1990	108.	Geometric Topology: Recent Developments	(LNM 1504)	"
	109.	H_∞ Control Theory	(LNM 1496)	"
	110.	Mathematical Modelling of Industrial Processes	(LNM 1521)	"
1991	111.	Topological Methods for Ordinary Differential Equations	(LNM 1537)	"
	112.	Arithmetic Algebraic Geometry	(LNM 1553)	"
	113.	Transition to Chaos in Classical and Quantum Mechanics	(LNM 1589)	"
1992	114.	Dirichlet Forms	(LNM 1563)	»
	115.	D-Modules, Representation Theory, and Quantum Groups	(LNM 1565)	"
	116.	Nonequilibrium Problems in Many-Particle Systems	(LNM 1551)	"
1993	117.	Integrable Systems and Quantum Groups	(LNM 1620)	"
	118.	Algebraic Cycles and Hodge Theory	(LNM 1594)	"
	119.	Phase Transitions and Hysteresis	(LNM 1584)	"

1994	120. Recent Mathematical Methods in Nonlinear Wave Propagation	(LNM 1640)	Springer-Verlag
	121. Dynamical Systems	(LNM 1609)	"
	122. Transcendental Methods in Algebraic Geometry	(LNM 1646)	"
1995	123. Probabilistic Models for Nonlinear PDE's	(LNM 1627)	"
	124. Viscosity Solutions and Applications	(LNM 1660)	"
	125. Vector Bundles on Curves. New Directions	(LNM 1649)	"
1996	126. Integral Geometry, Radon Transforms and Complex Analysis	(LNM 1684)	"
	127. Calculus of Variations and Geometric Evolution Problems	(LNM 1713)	"
	128. Financial Mathematics	(LNM 1656)	"
1997	129. Mathematics Inspired by Biology	(LNM 1714)	"
	130. Advanced Numerical Approximation of Nonlinear Hyperbolic Equations	(LNM 1697)	"
	131. Arithmetic Theory of Elliptic Curves	(LNM 1716)	"
	132. Quantum Cohomology	(LNM 1776)	"
1998	133. Optimal Shape Design	(LNM 1740)	"
	134. Dynamical Systems and Small Divisors	to appear	"
	135. Mathematical Problems in Semiconductor Physics	to appear	"
	136. Stochastic PDE's and Kolmogorov Equations in Infinite Dimension	(LNM 1715)	" "
	137. Filtration in Porous Media and Industrial Applications	(LNM 1734)	"
1999	138. Computational Mathematics driven by Industrial Applications	(LNM 1739)	"
	139. Iwahori-Hecke Algebras and Representation Theory	to appear	"
	140. Theory and Applications of Hamiltonian Dynamics	to appear	"
	141. Global Theory of Minimal Surfaces in Flat Spaces	(LNM 1775)	"
	142. Direct and Inverse Methods in Solving Nonlinear Evolution Equations	to appear	"
2000	143. Dynamical Systems	to appear	"
	144. Diophantine Approximation	to appear	"
	145. Mathematical Aspects of Evolving Interfaces	to appear	"
	146. Matheamtical Methods for Protein Structure	to appear	"
	147. Noncummutative Geometry	to appear	
2001	148. Topological Fluid Mechanics	to appear	
	149. Spatial Stochastic Processes	to appear	
	150. Optimal Transportation and Applications	to appear	
	151. Multiscale Problems and Methods in Numerical Simulations	to appear	

Fondazione C.I.M.E.
Centro Internazionale Matematico Estivo
International Mathematical Summer Center
http://www.math.unifi.it/~cime
cime@math.unifi.it

2002 COURSES LIST

Real Methods in Complex and CR Geometry

June 30 - July 6 - Martina Franca (Taranto)

Course Directors:

Prof. Dmitri Zaitsev (Università di Padova), zaitsev@math.unipd.it
Prof. Giuseppe Zampieri (Università di Padova), zampieri@math.unipd.it

Analytic Number Theory

July, 10-19 - Cetraro (Cosenza)

Course Directors:

Prof. C. Viola (Università di Pisa), viola@dm.unipi.it
Prof. A. Perelli (Università di Genova), perelli@dima.unige.it

Imaging

September, 15 - 21 - Martina Franca (Taranto)

Course Directors:

Prof. George Papanicolaou (Stanford University), papanico@georgep.Stanford.edu,
Prof. Giorgio Talenti (Università di Firenze), talenti@math.unifi.it

Printing: Strauss GmbH, Mörlenbach
Binding: Schäffer, Grünstadt

4. Lecture Notes are printed by photo-offset from the master-copy delivered in camera-ready form by the authors. Springer-Verlag provides technical instructions for the preparation of manuscripts. Macro packages in TEX, LATEX2e, LATEX2.09 are available from Springer's web-pages at

http://www.springer.de/math/authors/b-tex.html.

Careful preparation of the manuscripts will help keep production time short and ensure satisfactory appearance of the finished book.

The actual production of a Lecture Notes volume takes approximately 12 weeks.

5. Authors receive a total of 50 free copies of their volume, but no royalties. They are entitled to a discount of 33.3% on the price of Springer books purchase for their personal use, if ordering directly from Springer-Verlag.

Commitment to publish is made by letter of intent rather than by signing a formal contract. Springer-Verlag secures the copyright for each volume. Authors are free to reuse material contained in their LNM volumes in later publications: A brief written (or e-mail) request for formal permission is sufficient.

Addresses:

Professor J.-M. Morel
CMLA, Ecole Normale Supérieure de Cachan
61 Avenue du Président Wilson
94235 Cachan Cedex France
E-mail: Jean-Michel.Morel@cmla.ens-cachan.fr

Professor B. Teissier
Université Paris 7
UFR de Mathématiques
Equipe Géométrie et Dynamique
Case 7012
2 place Jussieu
75251 Paris Cedex 05
E-mail: Teissier@ens.fr

Professor F. Takens, Mathematisch Instituut,
Rijksuniversiteit Groningen, Postbus 800,
9700 AV Groningen, The Netherlands
E-mail: F.Takens@math.rug.nl

Springer-Verlag, Mathematics Editorial, Tiergartenstr. 17
D-69121 Heidelberg, Germany
Tel.: *49 (6221) 487-701
Fax: *49 (6221) 487-355
E-mail: lnm@Springer.de